Noise, Vibrations and Harshness of Electric and Hybrid Vehicles

Noise, Vibrations and Harshness of Electric and Hybrid Vehicles

LIJUN ZHANG, DEJIAN MENG, AND GANG CHEN

INTERNATIONAL®

Warrendale, Pennsylvania, USA

400 Commonwealth Drive
Warrendale, PA 15096-0001 USA
E-mail: CustomerService@sae.org
Phone: 877-606-7323 (inside USA and Canada)
724-776-4970 (outside USA)
FAX 724-776-0790

Chief Growth Officer
Frank Menchaca

Publisher
Sherry Dickinson Nigam

Director of Content Management
Kelli Zilko

**Production and
Manufacturing Associate**
Erin Mendicino

Library of Congress Catalog Number 2020936332
http://dx.doi.org/10.4271/9780768099669

ISBN-Print 978-0-7680-9964-5
ISBN-PDF 978-0-7680-9966-9

To purchase bulk quantities, please contact: SAE Customer Service

E-mail: CustomerService@sae.org
Phone: 877-606-7323 (inside USA and Canada)
724-776-4970 (outside USA)
Fax: 724-776-0790

Visit the SAE International Bookstore at books.sae.org

contents

CHAPTER 8

Sound Quality of BEV/HEV/FCEV   307

Introduction

1.1 Introduction

The noise, vibration, and harshness (NVH), also known as noise and vibration (N&V), is a critical feature for customers to assess the performance and quality of vehicles. NVH characteristics are higher among factors that customers use to judge the vehicle's quality. NVH is one of the strongest brand differentiators. NVH can cause discomfort, fatigue, and even injury to vehicle passengers. Therefore, reducing NVH levels to a tolerable range is one of the most important goals in vehicle design and development.

NVH refinement is the comprehensive study and modification of the noise and vibration characteristics of vehicles. NVH refinement has been conducted as an engineering process for internal combustion engine (ICE) vehicles where objective analysis and measurements have been together with subjective evaluation. The psychoacoustics have been used for the correlation and the efforts in enhancing sound quality by adding or subtracting particular harmonics in addition to making the vehicle quieter. Interior NVH deals with noise and vibration experienced by the occupants of the cabin, whereas exterior NVH is largely concerned with the noise radiated by the vehicle and includes drive-by noise testing.

The vibrations and noise sources in vehicles include the systems such as doors, emission systems, engines, fans, heating, ventilation, and air conditioning (HVAC) systems, instrument panels, seats, suspensions, tires, transmissions, windows, and seats. Acoustic disturbances can also be generated by many other subsystems or components, such as engine and transmission mounts, headliners, mufflers, oil and water pumps, prop shafts, rear axle differentials, steering yokes, and tailpipes.

In terms of physical nature, the sources of noise in a vehicle can be classified as aerodynamic, mechanical, and electrical ones. The noise of wind and cooling fans of HVAC are aerodynamic. The noise of engine, driveline, tire contact patch and road surface, brakes are mechanical. The electromagnetically-excited acoustic noise and vibration coming from electrical actuators, alternator, or traction motor in electrical cars are electrical.

Many problems are generated as either vibration or noise, transmitted via a variety of paths, and then radiated acoustically into the cabin. These are classified as "structure-borne" noise. Others are generated acoustically and propagated by airborne paths as "air-borne" noise. Structure-borne noise is attenuated by isolation, whereas airborne noise is reduced by absorption or through the use of barrier materials. Vibrations are sensed at the steering wheel, the seat, armrests, or the floor and pedals. Some problems are sensed visually - such as the vibration of the rear-view mirror or header rail on open-topped cars.

Techniques used to help identify NVH include part substitution, modal analysis, rig squeak and rattle tests (complete vehicle or component/system tests), lead cladding, acoustic intensity, transfer path analysis, and partial coherence. Most NVH work is done in the frequency domain, using fast Fourier transforms to convert the time domain signals into the frequency domain. Wavelet analysis, order analysis, statistical energy analysis, and subjective evaluation of signals modified in real time are also used. Integration and iteration of computer aided engineering (CAE) and test processes are critical to develop and deliver a smooth, quiet vehicle with excellent attributes.

In terms of the NVH propagation in a vehicle, NVH energy from a source travels through the mounts, into the structure, and through the cabin as vibration perceived by the body of driver. But energy from the same source can take a similar path through the structure to become acoustic noise perceived by the driver when it is amplified by the cabin. Optimizing these factors is therefore of significance for the overall experience and perception of the vehicle. There are three basic means of improving and refine NVH:

1. Reducing the source strength to make an NVH source smaller by using optimized design to reduce excitation force or improving the balance, or using a muffler;

2. Blocking the noise/vibration transfer path, with barriers for noise or isolators for vibration;

3. Using absorption/counter measures of vibration/noise such as foam noise absorbers, tuned vibration dampers, and active control systems.

Deciding which of these or what combination to use for a particular problem is one of the challenges facing the NVH engineer. For example, specific methods for improving NVH include the use of tuned mass dampers, subframes, balancing, modifying the stiffness or mass of structures, retuning exhausts and intakes, modifying the characteristics of elastomeric isolators, adding sound deadening or absorbing materials, or using active noise control. In some circumstances, substantial changes in vehicle architecture may be the only way to solve specific problems cost effectively.

Electric powertrains become more common as there is an industry wide push for better fuel economy. The increasing trend towards electrified vehicles, including battery electric vehicles (BEV), hybrids electric vehicles (HEV), and fuel cell electric vehicles (FCEV), have created new unique challenges for NVH development and refinement beyond the above-mentioned due to the electric powertrain applications.

BEV only use an electric motor and battery with electricity stored in a battery pack to power an electric motor and drive the wheels. When depleted, the batteries are recharged using grid electricity, either from a wall socket or a dedicated charging station. HEV combine both a gasoline engine with an electric motor. While these vehicles have an electric motor and battery, they can't be plugged in and recharged. Instead their batteries are charged from capturing energy when braking, using regenerative braking that converts kinetic energy into electricity. Hybrid electric vehicles that can be recharged from a wall outlet or a charging station are Plug-in hybrids. The vehicle typically runs on electric power until the battery is depleted, and then the car automatically switches over to use the ICE. FCEV power an electric motor and battery by converting stored hydrogen to electricity using a fuel cell, which can be refueled at a filling station similar to a gasoline vehicle.

BEV/HEV/FCEVs are usually very complicated with many mechanical and electrical system integrated together. For example, a typical HEV consists of the following systems.

Electric traction motor uses power from the traction battery pack, this motor drives the vehicle's wheels. Some vehicles use motor generators that perform both the drive and regeneration functions.

ICE has fuel injected into either the intake manifold or the combustion chamber, where it is combined with air, and the air/fuel mix is ignited by the spark from a spark plug.

Traction battery pack stores electricity for use by the electric traction motor.

Auxiliary battery provides electricity to start the car before the traction battery is engaged and also powers vehicle accessories. Exhaust system channels the exhaust gases from the engine out through the tailpipe.

Fuel filler is a filler or "nozzle" used to add fuel to the tank. Fuel tank stores gasoline on board the vehicle until it is needed by the engine. Charge port allows the vehicle to connect to an external power supply in order to charge the traction battery pack. DC/DC converter converts higher-voltage DC power from the traction battery pack to the lower-voltage DC power needed to run vehicle accessories and recharge the auxiliary battery. Electric generator generates electricity from the rotating wheels while braking, transferring that energy back to the traction battery pack. Some vehicles use motor generators that perform both the drive and regeneration functions. Onboard charger takes the incoming AC electricity supplied via the charge port and converts it to DC power for charging the traction battery. It monitors battery characteristics such as voltage, current, temperature, and state of charge while charging the pack. Power electronics controller manages the flow of electrical energy delivered by the traction battery, controlling the speed of the electric traction motor and the torque it produces. Thermal cooling system maintains a proper operating temperature range of the engine, electric motor, power electronics, and other components. The transmission gearbox transfers mechanical power from the engine and/or electric traction motor to drive the wheels.

Obviously, the technical, system and architectural complexities in BEV/HEV/FCEV result in many new NVH challenges compared with conventional ICE vehicles, due to

© Siemens 2017. Reprinted with permission.

FIGURE 1.1   NVH engineering issues in the development of EV/HEVs.

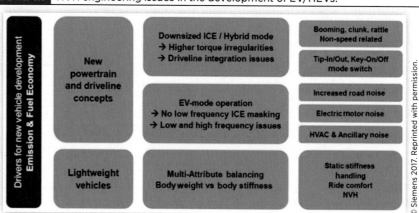

the introduction of battery, electronics, alternative powertrains, new materials/systems, and new vehicle performance [1–8]. It is noted that the downsized powertrain and more application of lightweight materials such as plastics and aluminums structures in BEV/HEV/FCEV pose new NVH challenges. As an example, **Figure 1.1** shows certain new NVH engineering issues in the development of BEV/HEVs.

There are many new NVH issues associated with BEV/HEV/FCEVs. BEV has low overall noise levels as it has no ICE noise. However, this also reduces the masking of wind and tire noise, and reduces masking of noise from ancillaries, allowing wind and tire noise, and noise from ancillaries to be more prominent. BEV/HEV has new NVH issues due to motor operations and controls, which exhibit tonal, high frequency noise of complex harmonic nature related to the number of poles on the electric motor. Moreover, there exit broadband high-frequency noise, which is not linked to operating condition. HEV has hybrid mode switching noise as transient phenomena, less dependency of the interior noise on the load. BEV/HEV/FCEV have new demands for describing sound quality and setting targets, needing to quantifying loudness, sharpness, and tonality, etc., assessing the customer values of annoyance, quality, etc. Particularly, due to the quietness, electric vehicle warning sounds must be designed and implemented in vehicles to alert pedestrians to the presence of BEV/HEV/FCEVs travelling at low speeds.

This book sets out to introduce the basic concepts, principles, and applications of the NVH development and refinement of BEV/HEV/FCEV.

1.2 Challenges and Significances of Studies on the BEV/HEV/FCEV NVH

Worldwide there are tendency towards the more electrified vehicle BEV/HEV/FCEV. The electrified cars bring new challenges of NVH such as high frequency electric motor generator noise, control electronics high frequency switching noise, power-split system gear whine and frequent engine start/stop noise and vibration. The technology used in electrified vehicle

FIGURE 1.2 Vehicle interior noise (at driver's ears) for conventional (ICE) and electric vehicles (EV).

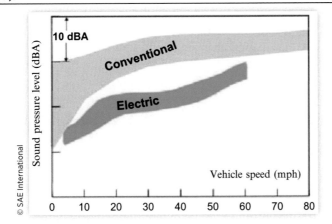

concepts is significantly different from conventional vehicle technology with consequences also for the noise and vibration behavior, which is dramatically different from conventional ICE vehicles [9-50]. NVH refinement is an important aspect of powertrain development and the vehicle integration process, being often critical to satisfy customer expectations. Attention should be paid to the NVH performance and sound quality of the electrified vehicles as the masking effect of engine is reduced substantially.

The extensive use of electric drives and actuators brings different sound than in the ICE vehicles as the conventional combustion engines are replaced or complicated by adding electrical motors and new types of components. These components often generate noise of higher frequency. At the same time, the masking effect from the engine noise is totally or partly lost for electrified vehicles, meaning that noise from these new components will be more dominant.

As a typical example to illustrate the interior noise of electric vehicle, at low and medium vehicle speeds, the electric vehicle is quieter by 10 dBA as shown in **Figure 1.2**. At higher speeds, the shares of wind and tire noise are increasing, causing the difference to be reduced.

The forces generated by electric motors are generally lower than the combustion and reciprocating mass forces of ICE and have significantly higher frequency.

In addition to the motor itself, other parts of the electronic system create noise, in particular, the power electronics unit. For power electronic unit systems, the radiated noise spectrum is very high in frequency. The normal levels of wind noise and tire/road noise could be unacceptable due to the absence of ICE noise, especially at low to medium speeds. As such, achieving a new balance between electric motor system noise, wind noise, and tire/road noise is one of fundamental works of electrified vehicles NVH.

Compared to a conventional ICE powertrain, BEV/HEV/FCEV powertrain features additional components such as electric motors, electronic control units, and a high voltage battery, resulting in these components. Each electrified vehicle configuration brings unique NVH challenges that result from a variety of sources. For example, the main HEV NVH problems with a negative effect on comfort include: low-frequency vibrations of the powertrain during start/stop of the ICE at load change; modified moments of inertia and eigen-frequencies in the powertrain; magnetic noise of the motor/generator during electric driving and regenerative braking; aerodynamic noises of the battery cooling system; and switching noise of the power control unit.

In HEV, the driving condition is often decoupled from the operation state of the ICE, which leads to unusual and unexpected acoustical behavior. Very often there will be no clear connection between vehicle speed and engine speed because the last one is both influenced by the load demand and the state of the battery. As such, start/stop event happens frequently and influence the noise and vibration behavior.

The electric motor torque ripple can excite the propulsion system resonance when the frequency of the torque ripple is close to the propulsion mount resonance frequencies. This results in NVH issues and unacceptable performance for the electric-drive propulsion system.

Depending upon the design of the motor, the electromagnetic (EM) pulses and corresponding torque pulses from the motor can be very strong. The magnetic radial force excites vibrations and noise directly from the motor housing and can also be transmitted structurally to the support structure through the motor mounts.

The noise from electric machines such as motors and generators manifests in the form of whine noise, i.e., tonal noise, which can be annoying to the customer.

Noise and vibration sources of motor stems from the magnetic field. The attraction due to magnetic field is the main cause of vibration in rotating electrical machines. It occurs in the direction of the flux lines and is due to the sinusoidal flux in the air gap. It changes as the motor rotates and is characteristically harmonic. Even though the radial force vectors are undesirable, they are unavoidable. The radial forces result in the mechanical vibration and resonance of the motor systems.

Ripple torque, cogging torque as well as magnetic radial forces are the main electromagnetic sources of noise and vibration in most of the electrical machines.

The ripple torque is produced from the harmonic content of the current and voltage waveforms in the electric machine. In the case of multi-phase motors, torque ripple is a crucial factor in the performance and operation because it determines the magnitude of the machine's vibration and acoustic noise.

The complicated operational conditions of HEVs have created unique challenges for NVH development and refinement. Traditionally, characterization of in-vehicle powertrain noise and vibration has been assessed through standard operating conditions such as fixed gear engine speed sweeps at varied loads. Given the multiple modes of operation, which typically exist for HEVs, characterization and source-path analysis of these vehicles can be more complicated than conventional ICE vehicles. In-vehicle NVH assessment of an HEV powertrain requires testing under multiple operating conditions for identification and characterization of the various issues, which may be experienced by the driver. Generally, it is necessary to assess issues related to ICE operation and electric motor operation (running simultaneously with and independent of the ICE), under both motoring and regeneration conditions. Additionally, mode transitions, including ICE start/stop must be assessed.

Decades of experience in designing brand-specific sound, based on NVH refinement of ICE vehicles, cannot be simply transferred to HEVs. Although electric vehicles are significantly quieter, their interior noise is marked by high-frequency tonal noise components, which can be subjectively perceived as annoying. Disturbing noise shares from other components (e.g. oil pump, HVAC system, battery fan, alternator, transmission systems) are no longer masked by ICE noise and give rise to complex sound signatures. With lightweight design stretched to its limits to maximize driving range and performance while moderating battery cost, achieving acceptable NVH performance becomes a greater challenge. **Figure 1.3** shows a summary of certain NVH challenges in HEVs.

Figure 1.4 shows a summary of frequency range of varied sources [5].

FIGURE 1.3 Summary of HEV-NVH challenges [1].

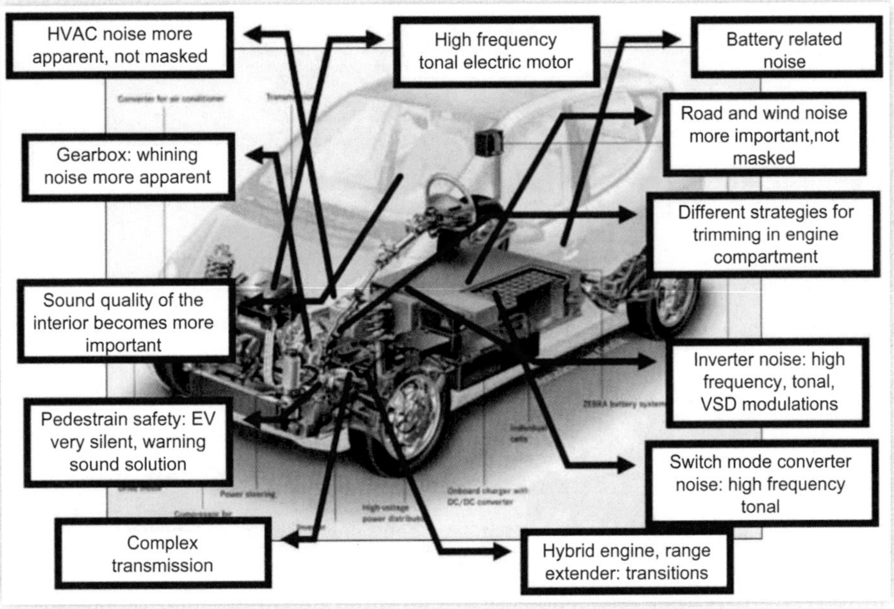

HVAC noise more apparent, not masked
High frequency tonal electric motor
Battery related noise
Road and wind noise more important,not masked
Gearbox: whining noise more apparent
Different strategies for trimming in engine compartment
Sound quality of the interior becomes more important
Inverter noise: high frequency, tonal, VSD modulations
Pedestrain safety: EV very silent, warning sound solution
Switch mode converter noise: high frequency tonal
Complex transmission
Hybrid engine, range extender: transitions

FIGURE 1.4 Summary of frequency range of varied sources [5].

(km/h)
ICE vehicle speed
100
60
Wind Noise
Road Noise – Tire Noise
Engine – Structure B.
Engine – Air B.
200
400
1000
2000
8000
(Hz)
(km/h)
EV vehicle speed
100
40
Wind Noise
Road Noise – Tire Noise
Inverter
Ancill. Noise
EM/TM
200
400
1000
2000
8000
(Hz)
Frequency range of interest

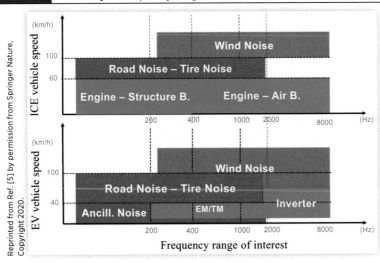

The BEV/HEV/FECV NVH development and refinement are significant for vehicle design and development. The first critical need is interior NVH, which is determined by market needs and customer expectations as NVH directly addresses two out of five human senses (hearing and feeling). It is important to maintain low NVH levels in order to ensure the competitiveness of the developed vehicles. The second critical need is external sound, which is mostly determined by the social concern about environmental and transportation issues. Vehicle external sound emission including passing-by noise and warning sound are attributes of the vehicle that cannot be perceived and assessed by the vehicle driver alone and is governed by governmental regulations.

1.3 Organization of the Book

The book has been set out with a two fold aim in view. The first aim is to give a general introduction to the principles of BEV/HEV/FCEV NVH, by offering a physical picture of the fundamental theory. The second aim is to give a series of examples of the applications of the theoretical NVH approaches in BEV/HEV/FCEV. The authors are expected to provide contemporary coverage of the primary concepts and techniques in the treatment of BEV/HEV/FCEV NVH.

This book consists of nine chapters. The basic principles have been introduced in the first four chapters. Chapter 1 introduces the entire book.

Chapter 2 briefly introduces BEV/HEV/FCEV architectures.

Chapter 3 provides a comprehensive introduction to the fundamental of automotive NVH. This portion covers from vehicle NVH target and refinement, ICE NVH, driveline NVH, powertrain isolation, road/tire-induced NVH, to wind-induced NVH.

Chapter 4 presents electric motor vibrations and noise. This chapter briefs the motors for BEV/HEV/FCEV. It introduces the motor-induced vibrations and noise, electromagnetic forces and torque pulsation, and motor system NVH.

Chapter 5 presents NVH of BEV, covering varied NVH issues of BEV: metrics to assess BEV NVH; NVH development and targeting, motor NVH/whine; gearbox NVH and driveline NVH; power electronics NVH; inverter high frequency side band/tonal noise; NVH of specific components/accessories; sound characteristics of BEV, with particular focus on the exterior warning sound considerations for BEV.

Chapter 6 presents the NVH development and refinement of HEV. This chapter includes ICE start/stop vibration and noise in HEV; HEV driveline NVH; transmission NVH and gear whine; motor and power electronics NVH; ancillary noise.

Chapter 7 introduces the unique features of FCEV NVH. It includes NVH development of FCEV; NVH of blowers/compressor/pump in FCEV; noise and vibrations of positive displacement (PD) type pumps/compressors; sound characteristics and NVH refinement of FCEV.

Chapter 8 gives an introduction of sound quality of EV/HEV/FCEV. After a brief overview of sound quality techniques of vehicles, it presents basic psychoacoustic parameters and sound quality matric, loudness, roughness, sharpness, fluctuation strength, etc. Then it presents sound quality index for high frequency tonal noise, prominence ratio, tone-to-noise ratio. Finally, it presents the sound quality development of BEV/HEV/FCEV

Last chapter presents a case study as an example of the application of NVH development on HEV, including NVH development of HEV powertrain, system modeling, and NVH study.

Complete references given in each chapter in the book provide comprehensive perspectives on the progress in NVH development and refinement of BEV/HEV/FCEV, as well as coverage of various applications. For didactical reasons, the text is not interrupted by the inclusion of references. However, at the end of each chapter, the relevant literatures published are comprehensively cited.

References

1. EARPA Position Paper, "Noise, Vibration and Harshness Research Needs, Priorities & Challenges for Road Transport inHorizon2020," 2016, https://www.earpa.eu/earpa/22/position_papers_per_task_force.html.

2. Tousignant, T., Govindswamy, K., Stickler, M., and Lee, M., "Vehicle NVH Evaluations and NVH Target Cascading Considerations for Hybrid Electric Vehicles," SAE Technical Paper 2015-01-2362, 2015, https://doi.org/10.4271/2015-01-2362.

3. Genuit, K., "Future Acoustics of Electric-Vehicle," SAE Technical Paper 2012-36-0612, 2012, https://doi.org/10.4271/2012-36-0612.

4. Journal of Society of Automotive Engineers of Japan, "Vibration, Noise and Ride Quality," https://www.jsae.or.jp/en/publications/yearbook_e/2013/docu/18_vibration_noise_ride_quality.pdf, accessed Aug. 2018.

5. Vibro-acoustic engineering challenges in (hybrid and) electric vehicles, Unrestricted @ Siemens AG 2017

6. Zhang, L., Chen, W., Meng, D., Gu, P. et al., "Vibration Analysis of Series-parallel Hybrid Powertrain System under Typical Working Condition and Modes," SAE Technical Paper 2018-01-1291, 2018, doi:https://doi.org/10.4271/2018-01-1291.

7. Jin, X. and Zhang, L., *Analysis of Vehicle Vibration* (Shanghai: Tongji University Press, 2002), isbn:978-7-5608-2406-2.

8. Zhang, L., "Elementary Investigation into Road Simulation Experiment of Powertrain and Components of Fuel Cell Passenger Car," SAE Technical Paper 2008-01-1585, 2008, https://doi.org/10.4271/2008-01-1585.

9. Chen, G., "ICE/EV/HEV NVH: Challenges of New Parts and Subsystems," presented at *AITech*, Michigan, Dec. 2015.

10. Chen, G., "Vehicle Dynamics: Chapter 9- Electrified Vehicles," MU-ME515, 2016.

11. Chen, G., "Brake NVH from ICEV to EV/HEV," presented at *Akabono*, Michigan, Jun. 2014.

12. Chen G, "Accessory NVH: From ICE Vehicles to Electrified Vehicles," presented at *Dayco*, Michigan, Jul. 2013.

13. Chen, G. and Sheng, G., "Vehicle Noise, Vibrations and Sound Quality," (SAE International R-400, 2012), 978-0768034844.

14. Lazăr, F., Lucache, D., Simion, A., Borza, P. et al., "Regarding the NVH Behaviour of the more Electric Vehicles Study Case of a Small PM Motor," *AnaleleUniversitățiiEftimieMurguReșița. Fascicula de Inginerie* 21, no. 2: 401–412, 2014.

15. Chau, K.T., Chan, C.C., and Chunhua, L., "Overview of Permanent-Magnet Brushless Drives for Electric and Hybrid Electric Vehicles," *IEEE Transactions of Industrial Electronics* 55, no. 6: 2246–2257, 2008.

16. Ozturk, C., "How to Silence Motor Noise," Machine Design: S46–S48, Feb. 25, 1999.

17. Wolschendorf, J., Rzemien, K., and Gian, D.J., "Development of Electric and Range-Extended Electric Vehicles through Collaboration Partnerships," SAE Technical Paper 2010-01-2344, 2010, https://doi.org/10.4271/2010-01-2344.

18. Goetchius G., "Leading the Charge - The Future of Electric Vehicle Noise Control," *Sound & Vibration* 45: 5–8, Apr. 2011.

19. Eisele, G. et al., "NVH of Hybrid Vehicles," *SIA Conf. on NVH of Hybrid and Electric Vehicles*, Saint-Ouen, Paris, France, Feb. 2010.

20. Govindswamy, K., Wellmann, T., and Eisele, G., "Aspects of NVH Integration in Hybrid Vehicles," SAE Technical Paper 2009-01-2085, 2009, https://doi.org/10.4271/2009-01-2085.

21. D'Anna, T., Govindswamy, K., Wolter, F., and Janssen, P., "Aspects of Shift Quality With Emphasis on Powertrain Integration and Vehicle Sensitivity," SAE Technical Paper 2005-01-2303, 2005, https://doi.org/10.4271/2005-01-2303.

22. Javadi, H., Lefevre, Y., Clenet, S., and Mazenc, M.L., "Electro-Magneto Mechanical Characterizations of the Vibration of Magnetic Origin of Electrical Machines," *IEEE Transactions on Magnetics* 31, no. 3: 1892–1895, 1995.

23. Studer, C., Keyhani, A., Sebastian, T., and Murthy, S.K., "Study of Cogging Torque in Permanent Magnet Machines," *IEEE IAS Annual Meeting*, New Orleans, LA, Oct. 4–9, 1997.

24. Park, J.B. et al., "Integrated Torque Ripple Analysis Method for Multi-Phase Motors," *IEEE International Electric Machines & Drives Conference (IEMDC)*, Chicago, IL, 2013.

25. Cui, G., "Study of Cogging Torque in Shaft Permanent Magnet Synchronous Generator," *15th International Conference on Electrical Machines and Systems (ICEMS)*, Sapporo, Japan, 2012.

26. Zhu, L., Jiang, S.Z., Zhu, Z.Q., and Chan, C.C., "Analytical Methods for Minimizing Cogging Torque in Permanent-Magnet Machines," *IEEE Transactions on Magnetics* 45, no. 4: 2023–2031, 2009, 10.1109/TMAG.2008.2011363.

27. Lee, G.C., Kang, S.M., and Jung, T.U., "Permanent Magnet Structure Design of Outer Rotor Radial Flux Permanent Magnet Generator for Reduction Cogging Torque with Design of Experiment," *International Conference on Electrical Machines and Systems*, Korea, Oct. 26-29, 2013.

28. Islam, R. and Husain, I., "Analytical Model for Predicting Noise and Vibration in Permanent-Magnet Synchronous Motors," *IEEE Transactions on Industry Applications* 46, no. 6: 2346–2354, 2010.

29. Yang, Z., Krishnamurthy, M., and Brown, I.P., "Electromagnetic and Vibrational Characteristic of IPM over Full Torque-Speed Range," *IEEE International Electric Machines & Drives Conference (IEMDC)*, Chicago, IL, 2013.

30. Han, S.H., Jahns, T., and Zhu, Z.Q., "Design Tradeoffs Between Stator Core Loss and Torque Ripple in IPM Machines," *IEEE Transactions on Industry Applications* 46, no. 1: 187–195, 2010.

31. Zhang, J., "Torque Ripple Compensation of Interior Permanent Magnet Synchronous Machine Propulsion System," *Colloquia at School of Electrical Engineering and Computer Science*, Oregon State University, Sep. 2012.

32. Hong, J.P., Ha, K.H., and Lee, J., "Stator Pole and Yoke Design for Vibration Reduction of Switched Reluctance Motor," *IEEE Transactions on Magnetics* 38, no. 2: 929–932, 2002.

33. Inderka, R., Menne, M., and De Doncker, R., "Control of Switched Reluctance Drives for Electric Vehicle Applications," *IEEE Transactions on Industrial Electronics* 90, no. 1: 48–53, 2002.

34. Pollock, C. and Wu, C.Y., "Acoustic Noise Cancellation Techniques for Switched Reluctance Drives," *IEEE Transactions on Industry Applications* 33, no. 2: 477–484, 1997.

35. Lecointe, J.P., "Etude et réduction active du bruit d'originemagnétique des machines a réluctance variable a double saillance," PhD thesis, Universite′ d'Artois, Arras Cedex, 2003.

36. Mininger, X. et al., "Vibration Damping with Piezoelectric Actuators for Electrical Motors," *COMPEL: The International Journal for Computation and Mathematics in Electrical and Electronic Engineering* 26, no. 1: 98–113, 2007.

37. Gillijns, S., Anthonis, J., Wyckaert, K., and Neves, W., "Balancing NVH and Energy Management of Eco-Efficient Powertrains through a Model Based Systems Approach, Integrating Functional Performance and Controls Optimization," *VDI Berichte* 2187: 605–620, 2013, Dusseldorf.

38. Herman Van der Auweraer Karl Janssens, "A Source-Transfer-Receiver Approach to NVH Engineering of Hybrid/Electric Vehicles," SAE Technical Paper 2012-36-0646, 2012, https://doi.org/10.4271/2012-36-0646.

39. Crewe, A., Distler, H., and Heinz, T., "Simulator Sound Objects - A Proposal for an Open standard on Sound Components for Driving Simulators," SAE Technical Paper 2003-01-1440, 2003, https://doi.org/10.4271/2003-01-1440.

40. Allman-Ward, M., Venor, J., Williams, R., Cockrill, M., Distler, H., Crewe, A., and Heinz, T., "The Interactive NVH Simulator as a Practical Engineering Tool," SAE Technical Paper 2003-01-1505, May 2003, https://doi.org/10.4271/2003-01-1505.

41. Allman-Ward, M., Williams, R., Dunne, G., and Jennings, P., "The Evaluation of Vehicle Sound Quality Using an NVH Simulator", *Proceedings of 33rd International Congress on Noise Control Engineering (Internoise 2004)*, Prague, Czech, 2004.

42. Otto, N., Amman, S., Eaton, C., and Lake, S., "Guidelines for Jury Evaluations of Automotive Sounds," SAE Technical Paper 1999-01-1822, May 1999, https://doi.org/10.4271/1999-01-1822.

43. Baker, S., Jennings, P., Dunne, G., and Williams, R., "Improving the Effectiveness of Paired Comparison Tests for Automotive Sound Quality," presented at the *11th International Congress on Sound and Vibration*, Petersburg, Russia, Jul. 5–8, 2004.

44. Fry, J., Jennings, P., Williams, R., and Dunne, G., "Understanding How Customers Make Their Decisions on Product Sound Quality," *Proceedings of the 33rd International Congress and Exposition on Noise Control Engineering*, Prague, Czech, 2004.

45. Allman-Ward, M., Balaam, M.P., and Williams, R., "Source Decomposition for Vehicle Sound Simulation," presented at *2001 CETIM Conference*, Senlis, France, 2005.

46. Otto, N., Simpson, R., and Wiederhold, J., "Electric Vehicle Sound Quality," SAE Technical Paper 1999-01-1694, 1999, doi:https://doi.org/10.4271/1999-01-1694.

47. Shin, E., Ahlswede, M., Muenzberg, C., Suh, I. "Noise and Vibration Phenomena of On-Line Electric Vehicle*," SAE Technical Paper 2011-01-1726, 2011, doi:https://doi.org/10.4271/2011-01-1726.

48. Govindswamy, K. and Eisele, G., "Sound Character of Electric Vehicles," SAE Technical Paper 2011-01-1728, 2011, doi:https://doi.org/10.4271/2011-01-1728.

49. Van der Auweraer, H. and Janssens, K., "A Source-Transfer-Receiver Approach to NVH Engineering of Hybrid/Electric Vehicles," SAE Technical Paper 2012-36-0646, 2012, doi:https://doi.org/10.4271/2012-36-0646.

50. Shiozaki, H., Iwanaga, Y., Ito, H., and Takahashi, Y., "Interior Noise Evaluation of Electric Vehicle: Noise Source Contribution Analysis," SAE Technical Paper 2011-39-7229, 2011, https://doi.org/10.4271/2011-39-7229.

2

BEV/HEV/FCEV Architectures

2.1 Introduction of BEV/HEV/FCEV

The electrified vehicles or electric vehicles (EVs) are vehicles that can run solely on electric motor propulsion, or they can run by using an internal combustion engine (ICE) together with motors [1–5]. Having only batteries as an energy source makes the basic kind of battery EVs (BEVs), but there are other kinds that can employ other modes of energy source. These can be called hybrid EVs (HEV).

EVs can be categorized as following: Battery electric vehicle (BEV); hybrid electric vehicle (HEV)/plug-in hybrid electric vehicle (PHEV); and fuel cell electric vehicle (FCEV)

Battery-powered electric vehicle (BEV): EVs with only batteries to provide power to the drive train are known as BEVs. BEVs have to rely solely on the energy stored in their battery packs; therefore the range of such vehicles depends directly upon the battery capacity. Typically, they can cover 100–250 km on one charge, whereas the top-tier models can go a lot further, from 300 to 500 km. These ranges depend upon the driving condition and style, vehicle configurations, road conditions, climate, battery type, and age. Once depleted, charging the battery pack takes quite a lot of time as compared to refueling a conventional ICE vehicle.

Hybrid electric vehicle (HEV): HEVs employ both an ICE and an electrical power train to power the vehicle. The combination of these two can come in different forms. An HEV uses the electric propulsion system whenever the power demand is low. It is a great advantage in low speed conditions like urban areas, and it also reduces the fuel consumption as the engine stays totally off during idling periods. This feature also reduces the emission.

When higher speed is needed, the HEV switches to the ICE. The two drivetrains can also work together to improve the performance. Hybrid power systems are used extensively to reduce or to completely remove turbolag in turbo charged car. It enhances performance also by filling the gaps between gear shifts and providing speed boosts whenever required. The ICE can charge up the batteries, HEVs can also retrieve energy by means of regenerative braking.

Plug-in hybrid electric vehicle (PHEV): The concept of PHEV came to extend all-electric range of the HEVs. It uses both an ICE and an electrical power train, like HEV, but the difference between them is that PHEV uses electric propulsion as the main driving force. So, these vehicles require a bigger battery capacity than HEVs. PHEVs start in 'all electric' mode, runs on electricity, and whenever the batteries are low in charge, it calls in the ICE to provide a boost or to charge up the battery pack. The ICE is used here to extend the range. PHEVs can charge their batteries directly from the grid, whereas HEVs cannot. They also have the facility to utilize regenerative braking. PHEVs' ability to run solely on electricity for most of the time makes its carbon footprint smaller than the HEVs. They consume less fuel as well and thus reduce the associated cost.

Fuel cell electric vehicle (FCEV): FCEVs got the name as the heart of such vehicles is fuel cells that use chemical reaction to produce electricity. Hydrogen is the fuel of choice for FCEVs to carry out this reaction, so they are often called "hydrogen fuel cell vehicles." FCEVs carry the hydrogen in special high-pressure tanks. Electricity generated from the fuel cells goes to an electric motor that drives the wheels. Commercially available FCEVs like Toyota Mirai or Honda Clarity use batteries for this purpose. FCEVs only produce water as a byproduct of its power generating process that is ejected out of the car through the tailpipes. An advantage of such vehicles is that they can produce their own electricity, which emits no carbon, enabling it to reduce its carbon footprint further than any other EV. Another major advantage of these is that refilling these vehicles takes the same amount of time required to fill a conventional vehicle at a gas pump. It makes adoption of these vehicles more likely in the near future [6–25].

EVs are vehicles wholly or partially driven by electricity. Specifically, they are battery-powered electric vehicles (BEVs), fuel cell electric vehicles (FCEVs), and hybrid electric vehicles (HEVs). A basic classification of EVs is shown in **Figure 2.1**. In a globalized automotive market, the major vehicle manufacturers have launched their own commercial EV products, such as Toyota Prius, Toyota Mirai, Honda Insight, GM Volt, Nissan Leaf, and so on. In addition, nontraditional vehicle companies, such as Tesla and Google, have also entered the EV market and have launched a series of distinctive EV products.

Typically, customers' perception of vehicle quality closely parallels the noise, vibration, and harshness (NVH) characteristics of the vehicle. Drivers of the conventional ICE powered vehicles have come to expect a high level of refinement in vehicles and will continue to demand similar levels of refinement in all BEVs/HEVs/FCEVs. NVH refinement is an important aspect of the powertrain development and vehicle integration process. In particular, BEVs/HEVs/FCEVs present unique NVH challenges in the vehicle integration process.

FIGURE 2.1 EV classification: ICEV (internal combustion engine vehicle); HEV (hybrid electric vehicle); BEV (battery-powered electric vehicle); FCEV(fuel cell electric vehicle).

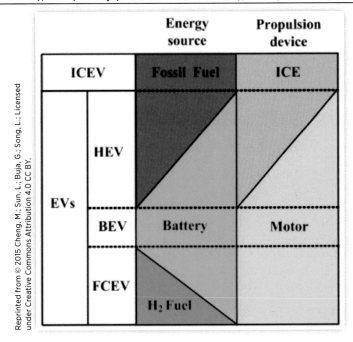

2.2 BEV/HEV/FCEV Powertrain Architectures and Operation Module

In the following, the architecture and configuration of different ICEs, BEVs, and HEVs vehicles are presented as shown in **Figure 2.2**, which include: (a) ICE vehicle; (b) battery-powered electric vehicle; (c) series hybrid vehicle; (d) parallel hybrid vehicle; (e) series-parallel hybrid vehicle [1–4]:

1) *The conventional ICE vehicles* as shown in Figure 2.2a have a long driving range and a short refueling time but face challenges related to pollution and oil consumption.

2) *BEVs:* BEVs as shown in Figure 2.2b relies on the electric machine/motor (EM1) power train. Because the vehicle is powered only by batteries or other electrical energy sources, zero emission can be achieved. However, the high initial cost of BEVs, as well as its short driving range and long refueling time has limited its use.

3) *HEVs:* Depending upon the way the two power trains of ICE and motor are integrated, there are generally three basic HEV architectures: a) series hybrid; b) parallel hybrid; and c) series-parallel hybrid.

   a. *Series HEVs:* In series HEVs as shown in Figure 2.2c, all the traction power is converted from the electricity, and the sum of energy from the two power

FIGURE 2.2   The architecture and configuration of different EV vehicles [1–4].

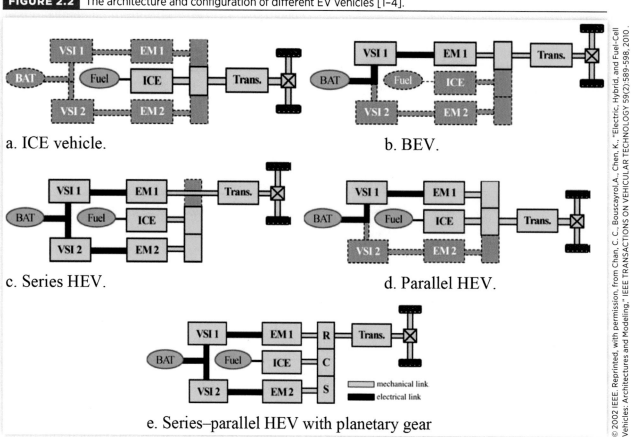

a. ICE vehicle.

b. BEV.

c. Series HEV.

d. Parallel HEV.

e. Series–parallel HEV with planetary gear

© 2002 IEEE. Reprinted, with permission, from Chan, C. C., Bouscayrol,A., Chen, K., "Electric, Hybrid, and Fuel-Cell Vehicles: Architectures and Modeling," IEEE TRANSACTIONS ON VEHICULAR TECHNOLOGY 59(2):589–598, 2010 .

sources is made in an electric node. The ICE has no mechanical connection with the traction load that means it never directly powers the vehicle. The connection between the ICE and the electric machine/motor (EM2) can be a simple gear. In series HEVs, the ICE mechanical output is first converted into electricity by the EM2. This converted electricity can either charge the battery or directly go to propel the wheels via EM1 and the transmission thus bypassing the battery. Due to the decoupling of the ICE and the driving wheels, series HEVs have the definite advantage of being flexible in terms of the location of the ICE generator set. For the same reason, the ICE can operate in its very narrow optimal region, independent of the vehicle speed. Controlling series HEVs is simple due to the existence of a single torque source (EM1) for the transmission.

Because of the inherently high performance of the characteristic torque speed of the electric drive, series HEVs do not need a multi gear transmission and a clutch. However, such a cascade structure leads to relatively low efficiency ratings, and thus all three motors are required. All these motors need to be sized for the maximum level of sustained power. Although, for short trips, the ICE can relatively easily be downsized, sizing the EMs and the battery is still a challenge, which makes the series HEVs expensive.

b. *Parallel HEVs:* A parallel HEV is shown in Figure 2.2d. In a parallel powertrain, the energy node is located at the mechanical coupling that may be considered as one common shaft, or two shafts connected by gears, a pulley-belt unit, etc.

The traction power can be supplied by ICE alone, by EM1 alone, or by both acting together. EM1 can be used to charge the battery through regeneration when braking or to store power from the ICE whenever its output is greater than the power required to drive the wheels. More efficient than the series HEV, this parallel HEV architecture requires only two motors: the ICE and the EM1. In addition, smaller motors can be used to obtain the same dynamic performance. However, because of the mechanical coupling between the ICE and transmission, the ICE cannot always operate in its optimal region, and thus clutches are often necessary.

c. *Series-Parallel HEVs:* The series-parallel hybrid architecture is also called the power-split architecture as shown in Figure 2.2e with a planetary gear system. It possesses a maximal number of subsystems in HEVs, thatallows series and parallel operations, or a combination of the two. A planetary gear set as shown in **Figure 2.3** can be used in a series-parallel HEV. As shown in Figure 2.1e, electric machine/motor 1 (EM1) and the transmission shaft (Trans.) are connected to the planetary ring gear set (R), whereas the ICE is connected to the carrier (C), and EM2 is connected to the sun gear (S).

A series-parallel HEV can operate as either a series HEV or a parallel HEV in terms of energy flow. The energy node can be located in the electric coupling components or in the mechanical coupling components. Because of the planetary gear, the ICE speed is a weighted average of the speeds of EM1 and EM2. The EM1 speed is proportional to the vehicle speed. For any given vehicle speed (or any given EM1 speed), the EM2 speed can be chosen to adjust the ICE speed. The ICE can thus operate in an optimal region by controlling the EM2 speed. Although series-parallel HEVs have the features of both the series and parallelHEVs, they still require three motors and a

FIGURE 2.3 Planetary gear setused in HEVs [23].

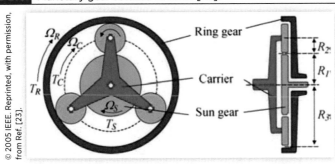

planetary gear set, which makes the powertrain some what complicated and costly. In addition, controlling this architecture is quite complex. Instead of a planetary gear set, a second type of series-parallel HEV uses a combination of two concentric machines EM1 and EM2 as a power-split device. To reduce the system weight and size, the two machines can be merged, creating a single machine with double rotor. In both systems (planetary gear set and concentric machines), the speed ratio between the ICE shaft and transmission shaft is continuous and variable. Both kinds of systems can replace the constantly variable transmission (CVT). For this reason, they are also called electric CVT (E-CVT or EVT). Two kinds of E-CVT have been developed: double-rotor induction machines and double rotor permanent-magnet machines.

4) *FCEVs:* From a structural viewpoint, an FCEV can be considered as a type of BEV because an FCEV can also be equipped with batteries. FCEVs can be considered as a type of series hybrid vehicle, in which the fuel cell acts as an electrical generator that uses hydrogen. The onboard fuel cell produces electricity that is either used to provide power to the machine EM1 or is stored in the battery or the super-capacitor bank for future use.

It is noted that the previous HEV architectures provide different levels of functionality. These levels can be further classified by the power ratio between the ICE and EMs [1–4].

1) *Micro Hybrid:* Micro hybrid vehicles use a limited-power EM as a starter alternator, and the ICE insures the propulsion of the vehicle. The EM helps the ICE to achieve better operations at startup. Because of the fast dynamics of EMs, micro hybrid HEVs employ a stop-and-go function, it means that the ICE can be stopped when the vehicle is at a standstill (e.g., at a trafficlight). Fuel economy improvements are estimated to be in the range of 2%–10% for urban drive cycles.

2) *Mild Hybrid:* In addition to the stop-and-go function, mild hybrid vehicles have a boost function that means that they use the EM to boost the ICE during acceleration or braking by applying a supplementary torque. The battery can also be recharged through regenerative braking. However, the EMalone cannot propel the vehicle. Fuel economy improvements are estimated to be in the range of 10%–20%.

3) *Full Hybrid:* Full hybrid vehicles have a fully electric traction system that means that the electric motor can insure the vehicle's propulsion. When such a vehicle uses this fully electric system, it becomes a "zero-emission vehicle" (ZEV). The ZEV mode can be used, for example, in urban centers. However, the propulsion of the vehicle can also be insured by the ICE or by the ICE and the EM together. Fuel economy improvements are estimated to be in the range of 20%–50%.

4) *Plug-in Hybrid:* Plug-in hybrid electric vehicles (P-HEVs) can externally charge the battery by plugging into the electrical grid. In some cases, the plug in vehicle may simply be a BEV with a limited-power ICE. In other cases, the driving range

FIGURE 2.4 Schematic of Toyota Hybrid System.

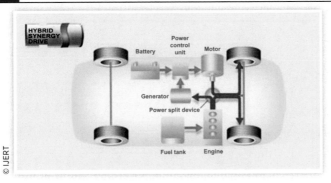

can be extended by charging the batteries from the ICE to extend the EV autonomy. This type of P-HEV is also called "range-extend EV." P-HEVs are promising in terms of fuel economy. For example, the fuel economy of P-HEVs can be improved by 100% if the ICE is not used to charge the battery (e.g., in urban drive cycles). Increasing the battery size allows EV operation for small trips and thus results in an important reduction in fuel consumption and greenhouse gas emissions. Many studies are being conducted about P-HEVs. Although the impact of P-HEVs on electrical grid loads needs to be examined, the initial studies have shown that P-HEVs could reduce peak demand without requiring more power plants.

A plug-in hybrid electric vehicle is similar to the hybrid electric vehicles (HEVs) on the market today, but it has a larger battery that is charged both by the vehicle's gasoline engine and from plugging into a standard 110 V/230 V electrical outlet for a few hours each day. Plug-in hybrid electric vehicles (PHEV) offer the potential to charge the battery from the power grid and hence have increased electric vehicle (EV) range.

As an example, Toyota Prius series, Toyota hybrid system (THS) is shown in **Figure 2.4**. The Toyota hybrid system consists of ICE along with two motor/generators, a power control unit, and a battery. When the car is started, it runs solely on the electric motor. Later, when the car achieves a higher speed the ICE kicks in and the car runs both on the motor and the ICE. Moreover, the engine also operates a generator with the help of a power split device, which in turn, drives the electric motor. This power splitting is controlled by the power control unit that manages the power for the maximum efficiency. During braking, the motor acts as a generator and the energy recovered is stored in the battery. The battery doesn't need any external charging. If the battery is drained, the car runs on the ICE in "stand mode", which charges the battery.

2.3 Characteristics of BEV/HEV/FCEV

The characteristics of BEV/HEV/FCEV are summarized in **Table 2.1** [3, 24].

TABLE 2.1 The characteristics of BEV/HEV/FCEV [3, 24].

Characteristics	Propulsion system	Energy storage	Energy source infrastructure	Advantages	Drawbacks	Important issues
ICE vehicles	• ICE based drives	• Fuel tank	• Petroleum products with refueling station	• Matured technology • Fully commercialized • Integration of components • Better performance • Simple operation • Reliable • Durable	• Less efficient • Poor fuel economy • Harmful emission • Comparatively bulky	• Fuel economy • Harmful emission • Highly dependency on petroleum products
EVs	• EPS based drive	• Battery • Ultra capacitor • Flywheel	• Electrical energy with charging facilities	• Energy efficient • Zero emission • Independency from petroleum products • Quite • Smooth operation • Commercialized	• Limited driving range • Higher recharging time • Poor dynamic response	• Size and weight of battery pack • Vehicle performance • Infrastructure for charging station
Hybrid EVs	• EPS and ICE based drive	• Fuel tank • Battery • Ultra capacitor • Flywheel	• Electrical power • Petroleum products with refueling station	• Higher fuel economy • Very low emission • Long electric driving range • Reliable • Commercialized • Durability	• Costly • Bulky • Complex system • Complexity in control algorithm • Increased component count	• Power management of multi input source • Size and weight of battery pack and ICE • Integration of components
Plug in HEVs	• EPS and ICE based drive	• Fuel tank • Battery • Ultra capacitor • Fly wheel	• Electrical power with charging station • Petroleum products with refueling station	• Lower emission • Higher fuel efficient • Extended electric driving range • V2G or G2V capability • Partially commercialized • Quite and smooth operation	• Higher complexity • Impact on grid • Higher initial cost • Sophisticated electronic circuitry • Battery technology	• Size and weight of battery pack and ICE • Charging station infrastructure • Power flow control and management • Impact on grid
FCVs	• EPS based drive	• Fuel cell stack • Battery • Ultra capacitor	• Hydrogen cylinder or hydrogen enriched fuel • Hydrogen refiner and refueling station	• Ultra low emission • Highly efficient • Independency from petroleum products • Competent driving range • Reliable • Durable • Under development	• High cost • Slow transient response • Not commercialized yet • Sophisticated electronic controllers	• Cost of fuel cell • Cycle life and reliability • Infrastructure for hydrogen conditioning, storage, and refilling system

2.4 **BEV Powertrain System and Other Subsystems**

Figure 2.5 shows a functional block diagram of BEV/HEV/FCEV including possible types of electronic controller, control hardware, software algorithms, energy storage systems, power converter devices/ topologies, and electrical motors and their computer-aided design (CAD) methodologies. Nowadays induction motors (IM) and permanent magnet (PM) motors are favored for which the comprehensive CAD or finite element method (FEM)-based engineering and experimental verification and validations are usually conducted. In the power converter technology, pulse-width modulated/insulated gate bipolar transistor (PWM/IGBT) inverters are the most popular along with bidirectional dc/dc boost converter. With regards to control technology, micro-processor or digital signal processor (DSP)-based vector control and direct torque control technology are very common.

Induction machines (IM), switched reluctance machines (SRM), and permanent magnet synchronous machines (PMSM) have been used for EVs. The typical torque/power vs. speed characteristic of motor in BEV/HEV/FCEV is shown in **Figure 2.6**.

EVs can be considered as a combination of different subsystems. Each of these systems interact with each other to make the EV work and there are multiple technologies that can be employed to operate the subsystems. In **Figure 2.7**, key parts of these subsystems and their contribution to the total system are illustrated. Some of these parts must work extensively with some of the others, whereas some have to interact very less. Whatever the case may be, it is the combined work of all these systems that make an EV operate. There are quite a few configurations and options to build an EV. EVs can be solely driven with stored electrical power, some can generate this energy from an ICE, and there are also some vehicles that employ both the ICE and the electrical motors together.

Figure 2.8 shows a view of the technology roadmap [25].

FIGURE 2.5 Architectural overview of electric propulsion system [4].

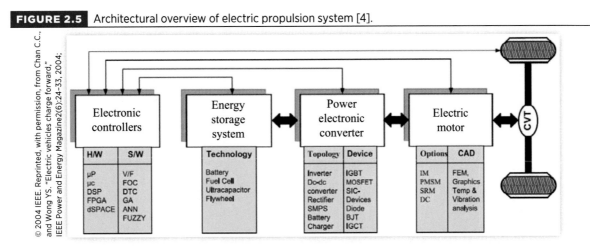

FIGURE 2.6 Torque-speed and power-speed characteristics for a typical traction electric motor.

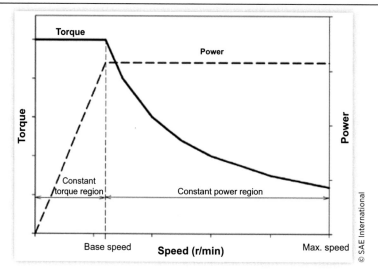

© SAE International

FIGURE 2.7 Major EV subsystems and their interactions [1, 2, 23].

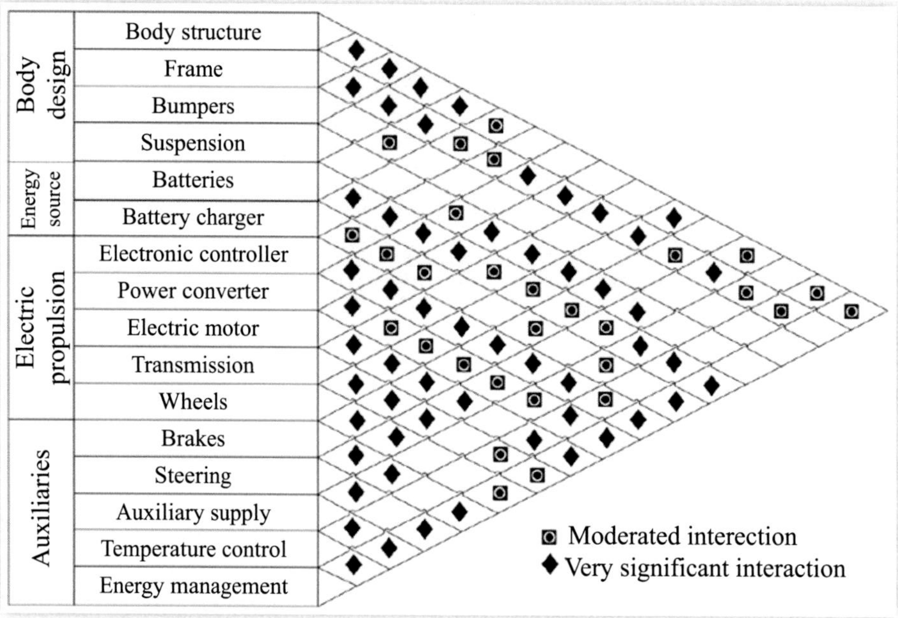

© 2002 IEEE. Reprinted, with permission, from Chan, C. C., "The state of the art of electric and hybrid vehicles," Proceedings of the IEEE 90(2):247-275, 2002.

FIGURE 2.8   View of the technology roadmap [25].

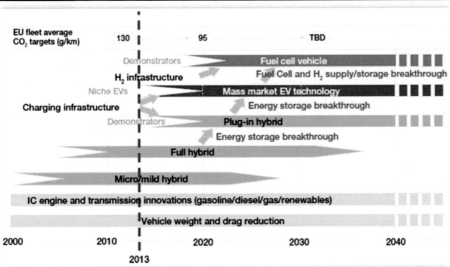

References

1. Chan, C.C., "The State of the Art of Electric and Hybrid Vehicles," *Proceedings of the IEEE* 90, no. 2: 247–275, 2002.

2. Chan, C.C., Bouscayrol, A., and Chen, K., "Electric, Hybrid, and Fuel-Cell Vehicles: Architectures and Modeling," *IEEE Transactions on Vehicular Technology* 59, no. 2: 589–598, 2010.

3. Chan, C.C., "The State of the Art of Electric Hybrid, and Fuel Cell Vehicles," *Proceedings of the IEEE* 95, no. 4: 704–718, 2007.

4. Chan, C.C. and Wong, Y.S., "Electric Vehicles Charge Forward," *IEEE Power and Energy Magazine* 2, no. 6: 24–33, 2004.

5. Un-Noor, F., Padmanaban, S., Mihet-Popa, L., Mollah, M.N. et al., "A Comprehensive Study of Key Electric Vehicle (EV) Components, Technologies, Challenges, Impacts, and Future Direction of Development," *Energies* 10, no. 8: 1217, 2017. 10.20944/preprints201705.0090.v1.

6. Tousignant, T., Govindswamy, K., Eisele, G., Steffens, C. et al., "Optimization of Electric Vehicle Exterior Noise for Pedestrian Safety and Sound Quality," SAE Technical Paper 2017-01-1889, 2017, doi:https://doi.org/10.4271/2017-01-1889.

7. Larminie, J. and Lowry, J., *Electric Vehicle Technology Explained* (Chichester, UK: John Wiley & Sons, 2003), ISBN:978-0470851630.

8. Yong, J.Y., "A Review on the State-of-the-Art Technologies of Electric Vehicle, Its Impacts and Prospects," *Renewable and Sustainable Energy Reviews* 49: 365–385, 2015.

9. Arfa, G.E. and Thiringer, T., "Performance Analysis of Current BEVs Based on a Comprehensive Review of Specifications," *IEEE Transactions on Transportation Electrification* 2, no. 3: 270–289, 2016.

10. SAE, "Electric Vehicle and Plug-in Hybrid Electric Vehicle Conductive Charge Coupler," SAE Standard J1772, Jan. 2010.

11. Murat, Y. and Krein, P.T., "Review of Battery Charger Topologies, Charging Power Levels, and Infrastructure for Plug-in Electric and Hybrid Vehicles," *IEEE Transactions on Power Electronics* 28, no. 5: 2151–2169, 2013.

12. Bayindir, K. C., Gözüküçük, M.A., and Teke, A., "A Comprehensive Overview of Hybrid Electric Vehicle: Powertrain Configurations, Powertrain Control Techniques and Electronic Control Units," *Energy Conversion and Management* 52, no. 2: 1305–1313, 2011.

13. Marchesoni M. and Vacca, C., "New DC-DC Converter for Energy Storage System Interfacing in Fuel Cell Hybrid Electric Vehicles," *IEEE Transactions on Power Electronics* 22, no. 1: 301–308, Jan. 2002.

14. Schaltz, E.A., Khaligh, T., and Resmussen, P.O., "Influence of Battery/Ultracapacitor Energy-Storage Sizing on Battery Lifetime in a Fuel Cell Hybrid Electric Vehicle," *IEEE Transactions on Vehicle Technology* 58, no. 8: 3882–3891, 2009.

15. Kramer, B. Chakraborty, S., and Kroposki, B., "A Review of Plug-in Vehicles and Vehicle-to-Grid Capability," *Proc. 34th IEEE Ind. Electron. Annu. Conf.*, Orlando, FL, Nov. 2008.

16. Williamson, S.S., "Electric Drive Train Efficiency Analysis Based on Varied Energy Storage System Usage for Plug-in Hybrid Electric Vehicle Applications," *Proc. IEEE Power Electron. Spec. Conf.*, Orlando, FL, Jun. 2007.

17. Wirasingha, N., Schofield, B., and Emadi, A. "Plug-in Hybrid Electric Vehicle Developments in the US: Trends, barri S. G. ers, and Economic Feasibility," *Proc. IEEE Vehicle Power Propulsion Conf.*, Harbin, China, Sep. 2008.

18. Gao, Y. and Ehsani, M. "Design and Control Methodology of Plug-in Hybrid Electric Vehicles," *Proc. IEEE Vehicle Power Propulsion Conf.*, Harbin, China, Sep. 2008.

19. U.S. Dept of Energy, *Fuel Cell Handbook* (Morgantown: University Press of the Pacific, 2005).

20. Kumar, L. and Jain, S., "Electric Propulsion System for Electric Vehicular Technology: A Review," *Renewable and Sustainable Energy Reviews* 29: 924–940, 2014.

21. Chau, K.T. and Chan, C.C., "Emerging Energy-Efficient Technologies for Hybrid Electric Vehicles," *Proceedings of the IEEE* 95, no. 4: 821–835, 2007.

22. Tousignant, T., Govindswamy, K., Stickler, M., and Lee, M., "Vehicle NVH Evaluations and NVH Target Cascading Considerations for Hybrid Electric Vehicles," SAE Technical Paper 2015-01-2362, 2015, doi:https://doi.org/10.4271/2015-01-2362.

23. Kessels, J., "Energy Management for Automotive Power Net," Ph.D. dissertation, Technische Universiteit Eindhoven, Eindhoven, The Netherlands, Feb. 2007.

24. Emadi, A.,Rajashekara, K., Wiliamson, S., and Lukic, S., "Topological Overview of Hybrid Electric and Fuel Cell Vehicular Power System Architectures and Configurations," *IEEE Transactions on Vehicular Technology* 54, no. 3: 763–770, 2005.

25. Department for Business, Enterprise and Regulatory Reform, "An Independent Report on the Future of the Automotive Industry in the UK," Department for Business, Enterprise and Regulatory Reform, 2009.

CHAPTER 3

Automotive NVH Fundamentals

3.1 Vehicle NVH Target and Refinement

Noise, vibration, and harshness (NVH) is a very important factor in vehicle customer satisfaction. Good NVH performance is a necessity for vehicles development. To have competitive NVH goals at the vehicle level we may need to produce systems/subsystems and components that have best-in-class NVH characteristics. Cascading vehicle level NVH targets into system levels and component levels quantitatively are the first step to achieving this goal, schematically as shown in **Figure 3.1** [1].

Figure 3.2 shows two examples of NVH target cascading and benchmark.

Especially for vehicle overall NVH targets, benchmarking information is useful to define a proper NVH performance that is competitive in the selected vehicle segment and in line with all other vehicle targets like performance, emission, weight, or costs.

Conventionally, NVH targets from vehicle level to component level have been set up by using benchmarks based on competitors, the best of practice, or available technology. This is true for the overall vehicle performance as well as component targets. **Table 3.1** shows an example of vehicle NVH targets [2, 3].

FIGURE 3.1 A generic schematic of the NVH process [1].

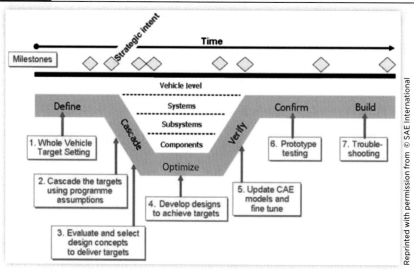

Reprinted with permission from © SAE International

FIGURE 3.2 Two examples of NVH target cascading and benchmark.

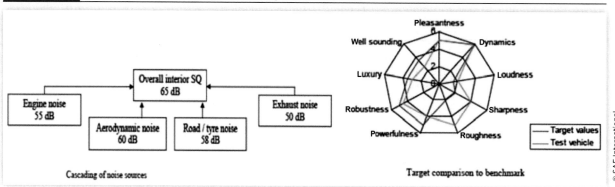

© SAE International

Cascading of noise sources

Target comparison to benchmark

TABLE 3.1 Vehicle NVH targets.

Vehicle NVH problem	Targets
Pass by noise of a car	<74 dBA (IS 3028) or 72 dBA (ECE R51:03)
Interior overall (OA) noise at high speed of road driving	<X* dBA
Seat vibration at vehicle idle	<0.09 m/s² (ISO 2631–1) for a long exposure
Articulation index of in-cab noise at speeds above 100kmph	>55%
Tyre/road rolling noise of a car	<72 dBA (ECE R–117)
Engine noise level	<X* dBA at rated speed/load of a dynamometer
Vibration isolation efficiency of engine mounts	>90% at idling
Sound transmission loss across the firewall	>25 dB across all frequency bands
The first natural frequency of a car BIW global mode	>X* Hz
Intake orifice noise	<X* dBA at various engine speeds
Tailpipe noise rise	<5 dB per 1000 rpm of engine rpm

© SAE International

At the start of a vehicle development, targets must be set at a whole vehicle level, not only to ensure functional performance but also to reinforce brand identity. For NVH, these whole vehicle targets must address the three main attributes: powertrain noise, road noise, and wind noise.

Both road and wind noise can be considered as error states, with the aim being to achieve a spectral balance and loudness which is appropriate for the vehicle program but within powertrain noise there lies a unique opportunity to enhance the character of the vehicle to reinforce the brand identity.

It is noted that there are many benchmarking procedures used, for example, three frequently used procedures are constant speed drives, part throttle (PT) drives, and wide-open throttle (WOT) drives. Constant speed drives provide a platform to benchmark vehicles at specific speeds, for example, the known vehicle speeds that induce maximum wind and tyre noise contributions. PT drives assess gradual vehicle acceleration and can be used to simulate daily driving conditions. WOT drives provoke the maximum response of the vehicle drivetrain as a result of the full-throttle acceleration, and thus it is a preferred method for vehicle benchmarking.

To set targets for systems, an extensive benchmarking attribute needs to be assessed. Early involvement of NVH during the concept definition is necessary to identify and highlight critical operating conditions. During this phase, simple simulation models can be used to get a principle NVH performance estimation for various operating conditions.

3.2 ICE NVH

The powertrain consists of an engine, intake and exhaust subsystems, transmission, and drivetrain systems. The powertrain is one of the major sources of a vehicle sound and vibrations. In general, powertrain vibration and sound sources consist of the following ones: (1) engine, including combustion-related sound and vibrations, reciprocating unbalance, rotating unbalance, crankshaft torsional oscillations, etc. [4, 5]; (2) valve train system, including valves, cam system, timing belt or chain; (3) accessories, including their unbalance and resonance; (4) intake system and exhaust system vibrations; (5) driveshaft first-order resonance; (6) universal joint second order bending vibrations and torsional vibrations; (7) axle vibrations due to gear tooth conjugation error, transmission error, pinion par eccentricity, slip-stick between pinion and ring gear, etc.

3.2.1 Engine Vibration

Engine vibration can be classified as internal vibration and external vibration. The internal vibration are referred to as the vibration of internal components of engine induced by the inertia force of motion parts and the variable pressure of combustion. These vibration are usually needed to be suppressed to avoid the malfunctions of the engine and the damage/fracture and noise of parts. The frequently encountered vibration are the torsional and bending vibration of the crankshaft and the vibration of valves-cam shaft system [6, 7]. The severe torsional and bending vibration of crankshaft could lead to the fracture of the shaft and/or the damage of bearings. Most of the internal vibrations result in noise and are unlikely to cause dangerous stress of parts. The external vibrations are referred to as the vibration of the whole engine system as a black which is usually integrated with transmission case. The external vibration are due to unbalanced moment, inertial moment, or variable output torsional torque.

Engine vibration are mainly due to the variable gas pressure in the cylinder, and the inertial force from the motion of the crank mechanism. Engine vibrations have detrimental effects on the internal parts, which allow them to have malfunctions thus have lifetime reduction. Engine vibration could also be transmitted to its supportive base-like frames, or its accessories, and therefore lead to the vibration of other systems.

Internal vibration mainly lie in the torsional vibration of crankshaft systems. The crankshaft has mass and elasticity; therefore, it constitutes a torsional vibration system. Under the excitation of the periodically modulated torsional torque, crankshaft system is capable of making torsional vibration. In operation, the crankshaft rotates and has an average torque applied on it. A static torsional deformation is associated with the crankshaft system. Torsional vibration of the crankshaft system is a periodically varied torsional deformation superposed on the static deformation of the crankshaft. Torsional vibration always exists for an operational engine crankshaft. But it is not readily perceptible as the whole engine vibration. The strength of the torsional vibrations can be estimated through measurement. If the frequency of harmonic excitation torque coincides with the natural frequency of crankshaft, the resonance occurs. If the resonance situation is severe, the disaster consequence such as crankshaft failure could occur.

The external vibration is referred to as the black vibration of the whole engine system together with the transmission case, in which the bending resonance has been critical. In the engine crank system, the reciprocal motion of the piston system leads to reciprocal inertial force; the rotating crank generates centrifugal force. In multiple cylinder engines, the cranks are arranged uniform to attain uniform firing of each cylinder. The first and second inertial forces are usually balanced (except for second inertial forces in 4 cylinder/4 stroke engine). The balance of reciprocal inertia moment and centrifugal moment depend on the configuration of cranks. On the other hand, the output torque of the engine and the turnover moment applied to the whole engine are periodic due to the periodic variation of gas pressure and reciprocal inertial force. These forces or torques result in vibrations of the whole engine black, which is usually treated as six-degree-of-freedom, including the up-down, front-rear, left-right, pitch, yaw, and roll. Thus, the isolation of whole engine black is necessary which will be discussed in a later section. The engine is usually mounted on the supportive base such as frame or subframe of the vehicle which is connected with the body.

In a real application, engine undergoes varies impact excitations and periodic excitations. The impact excitations are due to gap effect in motion parts such as bearings and piston-cylinders. The impact-induced response only lasts a short period and decays quickly due to damping dissipation. The main excitations leading to steady vibrations come from varies of force or torques, including the following:

1. Periodic tangential and radial forces acting on crankshaft emerge from the gas pressure in the engine cylinder, the inertial and gravity forces from the piston, connecting rod, and crankshaft. These forces are the primary resources of engine vibrations.

2. Excitation forces and torques due to the rotation parts, such as the centrifugal force or torque due to static or dynamic unbalance of rotating parts. The engines of most road vehicles use four-stroke process. An engine finishes a full working process by four steps (strokes): air intake, air compression, combustion (explosion or firing), and exhaust.

Among the four steps, the combustion process produces force when compressed gas is combined and fired. Consider an ideal model of a single-cylinder engine, a firing pulse appears in every two cycles, or there is "half" firing pulse for each cycle. The firing order, or simply called order, is defined as the firing number in each cycle. Thus, the basic firing order for a single-cylinder engine is the half order. For an engine with two cylinders, there is a half firing order for each cylinder in one cycle, i.e., there exists a whole firing order in one cycle for the two-cylinders. Hence, the basic firing order for a two-cylinder engine is 1st order. Similarly, there are two firing pluses, three firing pulses and four firing pulses for 4-cylinder engine, 6-cylinder engine, and 8-cylinder engine in one cycle, respectively. Thus, their corresponding firing orders are 2nd order, 3rd order, and 4th order, respectively.

The major excitation force is due to gas pressure in the cylinder.

In an operational engine, gas pressure within cylinder experiences an abrupt change in each working circle. **Figure 3.3** is a diagram of the variation of gas pressure with a crank angle. Its circulation period is 4π.

In practice, the history of gas pressure can be obtained by measurement. The coefficient of harmonics and phase angles in the above formulations can be calculated by Fourier analysis. **Figure 3.4** shows the variation of gas pressure concerning the crank angle and the first seven order harmonic components.

FIGURE 3.3   Variation of gas pressure of single-cylinder concerning the crank angle.

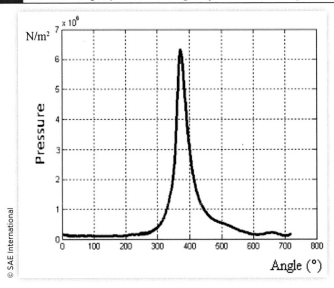

© SAE International

FIGURE 3.4 Variation of gas pressure with the crank angle and the first seven harmonic components.

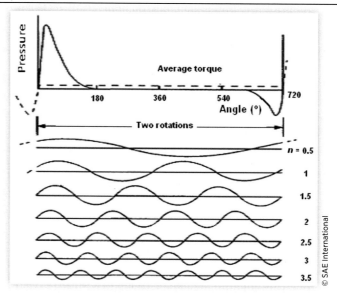

The firing frequency f_{firing} for each engine speed can be calculated by the following formula,

$$f_{firing} = \frac{\text{Engine speed}}{60} * \text{Firing order} = \frac{\text{Engine speed}}{60} * \frac{\text{Number of cylinder}}{2} \quad (3.1)$$

Gas pressure acts on piston and is transmitted to crank through the connecting rod. The force on the connecting rod can be decomposed as two portions: one is the tangent force perpendicular to the direction of crank radius; the other is the radial force along the crank radius.

When the engine operates at a steady speed, the gas pressure P_g can be considered as periodic. Accordingly, the tangent and radial forces F_t, F_r can be assumed as periodic forces, whose period is identical to that of pressure. Applying Fourier series on F_t, F_r, we have,

$$F_t = F_{t0} + \sum_{\overline{v}=1}^{\infty} \left(F_{t\overline{v}} \sin\left(\overline{v}\Omega t + \varphi_{\overline{v}}\right) \right) \quad (3.2a)$$

$$F_r = F_{r0} + \sum_{\overline{v}=1}^{\infty} \left(F_{r\overline{v}} \sin\left(\overline{v}\Omega t + \varphi_{\overline{v}}\right) \right) \quad (3.2b)$$

where

F_{t0}, F_{r0} are respectively the constant portions of F_t, F_r, or their average values

$\overline{v}$ is the mathematical harmonic order, $\overline{v} = 1, 2, 3, \ldots$

$F_{t\overline{v}}$, $F_{r\overline{v}}$ are respectively the coefficients of $\overline{v}$ order harmonic forces of tangent and radial forces F_t, F_r

$\varphi_{\overline{v}}$ the initial phase angle of $\overline{v}$ order harmonic of tangent and radial forces F_t, F_r

$\Omega = 2\pi/\tau$ is the fundamental angular frequency of the periodic variation of tangent and radial forces F_t, F_r, which τ is the period of the period variation

The excitation force is contributed by the inertia of moving parts.

The inertial forces of moving parts such as piston, connecting rod, crank includes reciprocating inertial force and centrifugal inertia force, as well as the inertial force from swaying motion of the connecting rod. The precise analysis of connecting rod is very complicated, we can approximate it by using an equilibrium equivalence approach in which the motion of connecting rod is approximated by the motion of two equivalent masses. The total mass of the connecting rod m_c is approximated by two equivalent concentrated mass m_{c1}, m_{c2}, with m_{c1} located at the smaller end of connecting rod performing the reciprocating motion, whereas m_{c2} at the larger end of a connecting rod having rotational motion.

The excitation force is also contributed by reciprocating inertia force. The tangent force from reciprocating inertia consists of five primary harmonic components with the angular frequency being one to five times of crank angular velocity, whereas the radial force consists of a constant component and four harmonic components with the angular frequency being one to four times of crank angular velocity [8].

Conventionally, torsional vibrations were considered as the main root cause of fracture of the crankshaft, damage of gear and the camshaft. With the advance of weight reduction over the last two decades, the torsional vibration has become one of the major root causes of noise instead of failure. The treatment of torsional vibration involves the following steps of modeling, analysis and the countermeasures: (1) build up lumped mass model or finite element model of crankshaft system; (2) calculate system natural frequency and mode shape; (3) estimate the magnitude of torsional vibration and stress; (4) employ vibration countermeasures such as dynamic absorbing [9, 10].

In conventional torsional vibration analysis of crank, the lumped parameter method has been applied. Finite element method has now been used extensively. The approach of generating a lumped parameter is briefed in the following:

1. The motion parts of each cylinder (including unit crank, connecting rod, and piston assembly) are idealized to be a uniform rigid component located at the cross point of crankshaft axis and cylinder centerline. For an engine with multiple array cylinders, the motion parts of each cylinder on the same array are combined to be as a rigid component, the stiffness of connecting spring between the rigid components of each cylinder is assumed to be equal to the stiffness of unit crank.

2. Transmission gear, flywheel, and clutch are idealized to be uniform rigid component located at the gravity center of the flywheel.

3. Accessory and timing systems are usually not taken into account. The belt can be considered as a de-coupled component without effect on the torsional vibrations of the crankshaft.

4. Both driven and driving parts of rubber damper (dynamic absorber) are simplified as rigid components with a spring in between. The stiffness of the spring equals to the stiffness of the rubber ring in between driving and driven part of damper.

5. Silicon oil damper is simplified as a uniform rigid component, with a rotational moment of inertia equaling to the moment of inertia of case plus half of the moment of inertia ring.

6. Constraints of bearing to the torsional vibrations are ignored.

7. In free vibration calculation, the damping effect can be ignored. In forced vibration analysis, the linear viscous damping model is usually applied.

Typical damping models [11–16] include linear viscous damping and hysteresis damping model. The equivalent linear viscous model usually offers some mathematical advantages. In the analysis of torsional vibration of the engine, the system damping is usually treated in terms of components, such as engine, torsional damper, etc.

Torsional vibration damping can be expressed as one of the following forms: damping coefficient, (equivalent linear viscous damping coefficient), damping torque, damping work, or amplification factor.

Torsional vibrations of the automotive engine crankshaft and driveline system can be reduced by several approaches. When the driven line system encounters resonance, the dynamic load could exceed two times of the maximum rotational torque of the engine. For light vehicle, the resonance leads to strong noise. For heavy vehicle, the resonance reduces the life span of the driveline system.

The approaches to suppress resonance of torsional vibration include tuning natural frequency, employing damper, and reducing input vibration energy of the system.

Frequency tuning is to regulate natural frequencies of the system to allow critical speeds out of operational speed range.

There are two kinds of dampers: the viscous damper that mainly relies on viscous damping to reduce vibration amplitude and dynamic absorber type dampers that uses a subsystem to reduce the primary resonance.

Crankshaft vibration is a primary source of engine radiated noise. It is important to analyze crankshaft vibration early in the design process to prevent durability and vibration and noise concerns. Most of the efforts in the past focused on crankshaft torsion vibration and bending and analysis and some attentions have also been paid to axial vibrations. For many encountered cases, the crankshaft has the fundamental natural frequency of torsional mode about 200 Hz, bending mode about 100 Hz, and an axial resonance around 500 Hz.

The bending vibrations of engine crankshaft and driveshaft have significant effects on vehicle vibrations and noise. In low-frequency range it affects the ride comfortable performance; whereas in the range of 50–250 Hz, the elastic vibration could result in body and structure resonance and sound resonance.

The excitation force leading to bending vibrations of crankshaft and driveshaft includes the inertia force due to reciprocating mass of engine; the inertia force due to unbalance of the driving shaft (proportional to rotating speed); and the exciting force of universal joint due to its angle being proportional to the square of rotating speed. For convenience, the crankshaft and driving shaft can be idealized to be an elastic beam attached with discrete lumped inertia. The bending vibration modes and natural frequencies can be obtained from this model.

The occurrence of resonance of bending vibrations depends on the engine and vehicle structures. For instance, the firing frequency such as the 3rd order harmonics of 4 wheel drive vehicle is likely to excite the natural modes of bending vibration of driveline, it is likely to result in body's resonance with associated boom noise. In V6 engine, there is secondary order unbalanced force due to reciprocating inertia and 3/2 order unbalanced torque of valve system. The external damper installed on the torsional tube near to support brace can help suppress bending vibrations.

The bending vibration of the crankshaft can also be suppressed by using torsional-bending mode damper. The reduction of bending vibration has a significant effect on automotive response [17–22].

Finally, we brief the engine vibrations on the frame. To reduce the influence of engine vibrations on the frame/subframe, the engine must be installed on a frame by using elastic isolations such as elastic mounts. The elastic mounts of the engine generally employ rubber cushion with one end fixed on the engine and the other end fixed on the frame. The rubber mount has flexibility in three directions. As the distance between mounts is larger, the torsional elasticity of mount can be ignored. The rubber mount can be simplified as a spring with three elastic axes in space. The three axes are referred to as the elastic principal axis of the mount. In free vibration analysis, the damping can be ignored, and rubber mount is idealized as a linear elastic component. Therefore, the engine mount system can be simplified as a system with six-degrees-of-freedom [23–35]. The frame is assumed as a rigid body. From vibration isolation point of view, the natural frequency of the elastic system of power plant and mounts are in the range of 6-20 Hz. In this frequency range engine vibration only has rigid modes. Therefore engine assembly can be simplified as a rigid body in space, whose location can be defined by three coordinates of gravity center x, y, z, and three angular coordinates of $\theta_x, \theta_y, \theta_z$. Therefore, the engine assembly has six-degree-of-freedom.

3.2.2 Engine Noise

The engine is one of the major noise sources of a vehicle. This section we present three kinds of major engine noise, including combustion noise, mechanical noise, and fluid dynamic noise [36–43]. The typical noise resources of the engine includes valve train, timing chain (or belt) noise (radiate from its cover), intake noise, exhaust noise, combustion noise, as well as piston slap noise, bearing noise, structural noise of valve cover, oil pan (sump), engine block, injection pump, the noise from accessories such as oil pump, belt/pulley, and fan system.

Combustion noise intensity is proportional to the square of the pressure rise rate. The sound pressure level of noise is proportional to the logarithm of heating generation or release rate in the cylinder. Combustion noise is also dependent on ignition delay, speed and torque load. Mechanical noise mainly comes from the piston slap, the friction, and impact response of valve train, the meshing of gear and tooth belt, belt slippage, bearing operations, timing system and accessory systems, oil pump systems. Mechanical noise is proportional to engine speed. The resonance of engine block structure also radiates noise. For the noise radiated from engine surface, it is mainly from the radiation of engine block surface and bottom oil pan. The head of the cylinder and cylinder cover also radiates noise. Aerodynamic noise includes the fan noise, intake, and exhaust noise [44–46]. Fan noise is determined by speed, blade dimensions and configurations, and several blades. The intake noise and exhaust noise are due to the pressure pulse, flow friction, and turbulence. The effect of tailpipe and surface radiation of silencer vibration are also primary sources of exhaust noise.

The wide variations in engine design make it difficult to give a general ranking of engine noise in terms of sound pressure level.

For different operational conditions, the contributions of the different sources are quite different and are highly dependent on the engine type.

The specific frequency ranges of primary noise sources of a typical engine are illustrated in **Table 3.1**. The frequency ranges of the primary noise sources of the automotive engine are dependent not only on engine and system structure but also on operating speed and load, therefore the estimation and identification of specific frequency need testing and analysis. By comparing the fundamental frequency and harmonics of these identified individual source noise with the spectrum of total noise, the contribution of individual noise source to total noise can be determined.

The testing of automotive engine noise is usually conducted in the semi-anechoic chamber according to SAE-J1074. The room constant should be larger than four times of measurement surface. In measurement, to avoid the disturbance of vibration and exhaust noise, the engine is installed together with dynamometer on a base-isolated from the rest of floor; the exhaust tailpipe is connected with a pipe toward the outside of the test room. The background noise of the test room should be at least 3 dBA lower than the noise to be measured. The testing of total engine noise needs to consider overall testing conditions. The steady test conditions should cover real operational conditions. In ramp-up testing from idle, the speed increase rate should be smaller than 15 rpm/s.

Engine combustion noise originates from the combustion in the cylinder. When fuel is injected into the engine cylinder chamber where high-pressure air exists, gasoline spark ignites the mixed gas, then part of the ignitable gas starts to burn. The pressure and temperature increase rapidly. Then the combustion propagates from firing part to other districts, which is associated with the continuous increase of temperature and pressure in the cylinder, and the combustible gas could have circular flow motion. The pressure wave in-cylinder impacts the wall of the combustion chamber, which results in the structural resonance of the chamber. The cylinder parts usually have high stiffness and their natural frequencies are very high. The frequencies of radiated noise are accordingly in the high-frequency range. The pressure within the cylinder also exhibits periodic variation, which results in low-frequency vibrations of the cylinder.

The combustion of mixed gas results in gas pressure change, which results in the structural vibration of the engine. The vibration radiates to air through engine surface and is perceived as the combustion noise.

In practice, it is difficult to distinguish combustion noise from mechanical noise. For convenience, we assume that the combustion noise is the noise due to combustion, originated from the pressure vibration within cylinder and piston, transmitted to cylinder cover, and piston-connecting rod-crankshaft-engine block to surroundings. We assume that the mechanical noise is the noise from mechanical interaction, impact and friction in piston-cylinder impact, timing gear or belt, valve train, injection mechanism, accessories, and belt. Generally, in direct injection diesel engine, combustion noise is higher than mechanical noise; in non-direct injection diesel engine, mechanical noise is higher than combustion noise; in low-speed operation, combustion noise is always higher than mechanical noise. A gasoline engine has less severe combustion; both of its combustion noise and mechanical noise are lower than that of a diesel engine.

The generation of combustion noise has been attributed to the rapid change of cylinder pressure in the combustion process. The effect of combustion consists of the dynamic load due to rapid pressure change and the high-frequency gas vibration and impulsive wave. The strength of noise from gas dynamic load depends on the rate of pressure rise and the timing of the maximum pressure rise rate. If the pressure remains constant, the noise cannot be generated. The variation of cylinder pressure is characterized by the rate of pressure rise, dP/dt.

Combustion noise is generated mainly in the phase of rapid combustion. When cylinder pressure increases severely, cylinder parts experience a sudden dynamic load with a certain strength, the effect of which equals to a slapping excitation. The engine is an intricate mechanical vibration system, different parts have different natural frequencies and most fall in the high-frequency range. Therefore, the combustion noise radiated to air through the transmission of engine parts is in the middle- and high-frequency range, which happens to be the most sensitive range of human hearing capability.

Slow combustion phase also affects engine high-frequency vibrations and noise. Late combustion phase has a small effect on combustion noise.

Normally combustion noise of gasoline engine accounts for a small part of its total noise; however, when combustion-related knock occurs, cylinder pressure has a rapid increase and leads to high frequency knocking noise.

The spectrum plot of cylinder pressure can be derived from the graph of cylinder pressure versus time. Cylinder pressure can be measured by using a pressure sensor mounted on the cylinder head and allowing the sensor to relate to the inside of the combustion chamber.

A minor change in pressure curve has no significant effect on engine power but it has a significant effect on noise. Engine's power is determined by the averaged pressure curve from multiple cycles, whereas combustion noise is dependent on the actual curve reflecting transient pressure variation in each cycle. The cylinder pressure curve usually is a variation of cylinder gas pressure concerning time. To understand the frequency signature of cylinder pressure, the spectrum of cylinder pressure has been used.

The spectrum of cylinder pressure of some engines exhibits a line spectrum at low-frequency range, which has several peaks at specific frequencies. These specific frequencies are the firing frequency and harmonics. In mediate and high-frequency ranges, the spectrum is continuous due to the rapid elevation of cylinder pressure in an impulsive way.

The shape of the spectrum curve of cylinder pressure at low-frequency range is not affected by engine speed, except for that the curve shafts towards high frequency when rotating speed increases. This is because that when rotating speed varies, the pressure curve shape remains unchanged concerning crank rotating angle. The high-frequency vibration of gas is mainly dependent on the size of the combustion chamber and the propagation speed of the impulsive wave. The frequencies corresponding to the pressure peaks in the third region of the curve are almost independent of engine speed.

From the spectrum of cylinder pressure, we learn that cylinder pressure is essentially the sum of a series of harmonics with different frequencies and amplitudes. Based on superposition principle, the quantity of cylinder equals to the sum of the individual effect of respective harmonics, therefore, the excitations of combustion gas to the cylinder can be considered as the sum of the individual excitation of this series of harmonics.

The excitation of harmonics can be transmitted from inside of the cylinder to engine surface through three major paths, which results in the surface vibrations and radiates noise. The first path goes through the piston, connecting rod, crankshaft, main bearing, through which the vibration transmits to the surface of the engine block. The second path goes through the cylinder head to cover. The third path is the transmission from the sidewall of the cylinder to the outside of the cylinder and block. Many experiments demonstrate that most of the vibration energy from combustion is transmitted from the larger ends of the connecting rod and main bearing to the structure of the engine, and result in the surface vibration of the engine and radiate noise.

The magnitude of combustion noise is not only dependent on the spectrum of cylinder pressure, but also dependent on the structural response and damping property of engine.

This is because that noise is due to vibrations, and the vibrations depend on the properties of excitation and the structural response of the vibration system. The difference of pressure level between cylinder inside and engine outside is characterized by a decay, which is an attenuation quantity reflecting the inherent characteristics of engineering structures. The decay is a constant value for a specific engine. The typical structural attenuation property of the engine is independent of the property of excitations and the spectrum of cylinder gas pressure. The engine's operating parameters such as speed, load, and the adjustment of the fuel supply system have no substantial effect on this property.

For different engines, structural attenuation property exhibits a larger difference at the same frequency.

The structural attenuation curve could be divided into three regions:

(i) Below 1000 Hz, the attenuation is quite high. This is because most of the structural parts of the engine has relatively larger stiffness, and their natural frequencies are at middle and high-frequency range. Therefore, vibration response in low-frequency range is relatively small due to larger structural decay, despite that the excitation of pressure is larger.

(ii) In the middle-frequency range from 1000 to 4000 Hz, the structural attenuation is small. This is because that most of parts' natural frequencies fall in this frequency range, which gives rise to low attenuation property.

(iii) Above 4000 Hz, structural attenuation is very high. This high-frequency range is above the natural frequencies of most parts; therefore, the structural attenuation is very high.

The engine's structure is an attenuator to combustion noise. The attenuation is larger at both low and high-frequency ranges. The sound pressure level of engine is high in the range 800–4000 Hz, which is corresponding to the range of low structural decay of engine. In the low-frequency range (below 800 Hz), despite that cylinder pressure level is high, the noise radiated by the engine is low due to high structural attenuation of the engine. In the high-frequency range (above 4000 Hz), the structural attenuation is high, and the cylinder pressure is small, the noise sound pressure level is low. In this range, cylinder pressure level decreases with the increase of frequency, and structural attenuation increases with the increase of frequency. Therefore, noise decreases rapidly with the increase of frequency. In the middle frequency range (800–4000 Hz), cylinder pressure is not as high as that in the low-frequency range but the structural decay has a minimum in this frequency range, therefore the structural response is strong, and the sound level pressure attains peak in this range. From the above observations, we learn that combustion noise can be suppressed by reducing cylinder pressure through combustion optimization and by increasing structural decay of engine structure.

The factors influencing combustion noise are briefed as follows. The rate of pressure rise is a fundamental factor controlling combustion noise, which mainly depends on the retarded spark timing and the quantities of mixture formation of combustible gas formed during the delay. A shorter retarded spark timing means that if the initial point of fuel injection is same, and if the start point of combustion is relatively earlier, the injected fuel quantity before the combustion is relatively smaller; therefore the amount of combustible gas formed before firing is less, then the pressure increase after the firing is slow; whereas the longer the period of ignition delay, the more the quantities of the combustible gas formed before firing. The fuel could combustion simultaneously in the second phase of the combustion process, which results in a higher rate of pressure increase and higher maximum combustion pressure, which accordingly leads to higher combustion noise. Therefore, in the design of the combustion system, the retarded combustion usually needs to be reduced as much as possible in the point of view of noise control. For specific engine structure, many factors are affecting retarded combustion. In normal operational conditions, compression temperature and pressure are the major factors influencing retarded combustion. The advance angle of fuel injection and the features of combustion also have significant influence. The influence of the structure of the combustion chamber and operation parameters on combustion noise is due to their influence on retarded spark timing through compression temperature and pressure.

a. The structure and layout of the combustion chamber and whole combustion system design have obvious influence on the rate of pressure increase, maximum combustion pressure, and spectrum of cylinder pressure.

b. *Temperature and pressure*: When compression temperature and pressure are increased, the physical and chemical preparation process of fuel spark will be improved; the retarded spark also reduced. The final temperature of compression mainly depends on compression ratio, and on cooling water temperature, piston temperature, cylinder head temperature, and intake temperature. The increase of compression ratio allows the gas temperature in retarded combustion to increase, and the final temperature and pressure at the end of compression to increase. This accordingly reduces retarded combustion, the rate of pressure increase, and the combustible fuel quantity accumulated in the period of retarded combustion, which also reduces the maximum value of thermal release rate and combustion noise. But the increase in compression ratio results in the increase of cylinder pressure which leads to the increase of piston impact noise, therefore it will not lead to a significant reduction of total engine noise. The compression-increase results in higher intake temperature, accordingly, reducing the combustion noise of direct injection diesel.

The higher the intake temperature and the later the fuel injection, the higher the gas temperature, and the shorter the retarded spark timing.

The higher the load and the higher the temperature of cooling liquid, the higher the temperature of the cylinder, and the shorter the retarded spark timing.

c. *Fuel injection parameters*: The parameters of the fuel system including the advance angle of fuel injection, injection pressure, number of injection nozzles, and the fuel supply law all influence the combustion process. If the other conditions remain the same, the increase of injection pressure results in the increase of injection rate and the increase of fuel quantity in combustion delay. The high-pressure injection improves the mixture of fuel and air and increases the rate of combustible fuel generation. This leads to the increase of combustible fuel accumulated in the period of ignition delay, therefore, increases combustion noise. Under the condition that the other parameters are unchanged for injection system, the reduction of injection fuel area results in an increase in the resistance of fuel injection hole and reduces the rate of fuel injection, accordingly reduces the quantity of fuel injection in combustion delay and reduces the noise of direct injection diesel.

d. *Engine speed*: If the other conditions remain unchanged, the increase of speed reduces the fuel injection time, increase fuel injection speed and the fuel quantity injected in the period of combustion delay. The increase of speed also increases the maximum cylinder pressure, the maximum of the rate of pressure increase, and the combustion noise. But usually, the effect of engine speed on combustion noise is not the most significant.

e. *Load*: For indirect injection diesel and gasoline engine, as their pressure increase is relatively smooth, when the load varies, the maximum combustion pressure change is relatively small and remains at a small value, and the impact of the cylinder from the piston is small. The sound pressure level of noise under full load could be smaller than that under no-load case by a couple of dB.

With the increase of load, the thermal release quantity will increase; the combustion pressure peak and the rate of pressure rise will increase. This results in a higher noise level. On the other hand, with the increase of load, the temperature of the combustion chamber will increase; the gap between cylinder and piston will be reduced, which could suppress noise. Overall, the load has a small effect on engine noise.

The basic approaches to reduce combustion noise are summarized as follows: in principle, the combustion noise can be reduced in the following two aspects. The first one is from its root cause, which includes reducing the spectrum of cylinder pressure, particularly the magnitude in the middle or high-frequency range; to reduce the period of ignition delay or reduce the quantity of mixed gas in combustion delay. The second approach is from noise transmitting path: to increase the attenuation of engine structure, particularly in the middle and high-frequency range. The approach includes increasing the stiffness of engine block and cylinder, and employing vibration isolation and sound insulation; reducing the gaps between parts such as piston and cylinder, cranks and connecting rod; increasing the thickness of oil film; using a cylinder with smaller diameter; using more number of cylinders or using design with a larger ratio of the stroke to cylinder diameter, to remain the output power to be less varied; changing materials of plate or shell parts (e.g., oil pan) by adding damping treatment. In general, the combustion noise control needs trade-off of thermal efficiency and emissions.

Several approaches used to reduce engine combustion noise include:

i. *Piston with thermal insulation*: The application of piston with thermal insulation can increase the temperature of the cylinder wall, reduce the period of ignition delay and reduce combustion noise of direct injection diesel.

ii. *Injection delay*: Generally, the earlier the injection time, the larger the combustion noise. If the injection time is postponed, the combustion noise can be reduced. This is because the compression temperature and pressure in cylinder vary with crank angles. The injection time affects the firing delay (combustion delay) through the compression temperature and pressure. If the injection time is set earlier, then the temperature and pressure are lower when the fuel enters the chamber. Then the period of firing delay is increased, which leads to an increase in combustion noise. However, if the injection time is set too late when the fuel enters chamber both the temperature and pressure become lower, and accordingly the firing delay is increased, which leads to the increase of combustion noise. There exists an optimal time for injection delay.

iii. *Preinjection*: Preinjection is to separate the injection process to be two stages, which allows fuel to be injected twice instead of once within one circle. In the first stage, a small portion of the fuel is injected, to precede the prereaction of firing before major injection, to reduce the quantity of combustible fuel accumulated in combustion delay time. This is one of the effective approaches to reduce the noise of direct injection diesel.

iv. *Improve the structure, layout, and parameters of combustion chamber*: The formation of air mixture and combustion is influenced by the structure and layout of the combustion chamber, which not only affects the performance of diesel but also affects the firing delay, the rate of pressure rise thus the combustion noise. For the same condition, the sphere combustion chamber and biased cylinder chamber of direct injection diesel engine yield comparatively lower combustion noise. Diesel engine with separation chamber generally has lower noise. The optimization of the parameters of the combustion chamber can reduce combustion noise. Compared with the lower crankcase, the cylinder and cylinder head are usually very stiff to resist the combustion pressures to prevent from movement. It has been found that some special structure design can attain better stiffness performance including the lower crankcase, flat panels on the upper crankcase, and optimal sub-structures for the oil pan and valve cover, etc. The extra ribs applied to reinforce crankcase walls could affect noise reduction. It has been estimated that the total engine noise can be reduced by 3 dB by using the treated covers (sump, valve cover, etc.). The crankcase walls and the main-bearing caps has been recommended to be integrated to form a ladder-type structure.

v. *Optimization of fuel pump*: The injection rate has a significant effect on combustion noise. The certain experiment illustrated that doubling the injection rate increases combustion noise by 6 dB. Therefore, the combustion noise can be reduced by decreasing the injection rate of the fuel pump. But this approach may worsen the high-speed performance and increase idle noise.

vi. *Employ turbocharge technique*: Turbocharge can increase the air density of the air entering cylinder, increase the temperature, and pressure of air in the cylinder at the end of compression, thus improve the firing condition for the mixed gas and reduce the firing delay. The higher the turbocharging pressure, the shorter the firing delay period and the lower the pressure elevation rate, thus the lower the combustion noise. Some experiment demonstrated that turbocharge allows combustion noise to be reduced by 2–3 dB.

vii. *Increase compression ratio*: Increase compression ratio can increase the gas temperature and pressure at the end of compression, shorten the period of firing delay, and reduce pressure rise rate, thus reduce combustion noise. On the other hand, increase compression ratio could increase cylinder pressure and increase piston slap noise.

viii. *Increase the quality of fuel*: Some ingredient in the fuel may influence the physical and chemical process of the gas mixture before firing, thus leads to different firing delay. Therefore, some high-quality fuel gives rise to short firing delay, lower pressure rise rate and thus lower combustion noise.

ix. *Electronic control*: Diesel engine with electronic control injection can optimize injection in terms of speed, load, air temperature, turbocharge pressure, and fuel temperature, thus effectively reduce combustion noise. The common rail injection system has been applied widely. The application of common rail injection can help to reduce the injection rate in the first injection period. The component of the high-frequency range has improved after the application of common rail injection, thus reduce combustion noise. The attributes of the common rail injection system are that the injection pressure is independent on engine load and speed, there are multiple timing and injection volume, the variable profile of injection rate, flexible design, fewer constraints of cylinder number, and improved start-up properties.

Mechanical noise of the engine is referred to as the vibration or impact-induced noise of motion components of the engine under the effect of cylinder gas pressure and inertia forces.

Mechanical noise of engine consists of piston slap noise (it has also been referred to as indirect combustion noise), gear noise, valve train and timing system noise, accessory noise, bearing noise, block structure noise, etc. Mechanical noise is usually the main noise resource of the engine in high-speed operations. A typical engine has several hundred motion pairs. In operation, the impact, friction, wear, unbalance in rotation result in vibration, and noise. The resonance due to the coincidence of natural frequency and excitation frequency leads to severe noise.

In the reciprocal motion process of engine crank/piston, when it passes upper or lower dead ends, the transversal force changes direction. This allows the contact zone between the piston and cylinder to switch from one side to the other, which induces impact and cylinder vibrations. Each motion pair has a certain gap, which results in impact when it undergoes oscillatory motion; for intake or exhaust valve in alternative close and open motions, when valve seats, it creates impact and noise. The frequency of vibration depends on the numbers of valve operation per seconds. In general, the mechanical noise of the engine increases rapidly with the increase operational speed. With the application of high speed and light engine and the implementation of more strict noise regulation, the major difficulty to reduce engine noise lies in the reduction of mechanical noise.

When the piston reaches its upper or lower dead ends, the position of the rod will make a change, and the direction of the lateral force applied on the piston will switch from one side to the other. The change of the direction of lateral force results in the change of piston's motion direction from one side to the other side, which leads to the piston slap to cylinder inner wall. This secondary motion is due to the collective effect of the force and moment applied on the piston and the gaps between piston and cylinder, which causes the transversal motion and rotation motion of the piston. It results in the periodic impact of the piston to the cylinder. When the impact is significant, it causes perceived noise called piston slap noise. The major factors influencing slap noise include the offset of the pinhole of the piston, the design of piston skirt, the gap between piston and cylinder under non-operation condition, and the piston stiffness. The minor changes in the geometric dimension or operation condition can lead to significant differences in the piston slap noise.

1. Impact of the piston to cylinder could be the largest source of mechanical noise. The strength of the slap mainly depends on the maximum pressure in the cylinder and the gap between piston and cylinder, thus it is dependent on both combustion design and detailed structural design of piston.

2. *Cylinder*: In cold start and idle operation situations, this slapping noise is very salient if the gap between piston and cylinder is relatively large.

   When gas pressure, inertial force and friction on piston experience periodic change, piston undergoes a periodic lateral force on the plane perpendicular to crank axis. Accordingly, it causes the piston's motion from one side to the other side. When the engine operates at high speed, this lateral motion has high speed thus form severe impact to the cylinder wall. This periodic impact approaches its strongest state when compression stroke ends and when the work stroke starts.

   Moreover, due to the effects of the sway of piston around its pin, the friction between piston and cylinder, the deformation of the piston and the cylinder vibrations, the piston slap to cylinder occurs not only at the upper and lower points but also at other points in the stroke with smaller impact force.

3. *Factors influencing piston slap noise*: There are many factors influencing piston slap noise, which includes piston gap, offset of the piston pin hole, number of piston rings, the thickness of cylinder, the diameter of the cylinder, lubrication conditions, and engine speed.

    The reduction of the clearance or gap between piston and cylinder can reduce slap strength thus the cylinder and block vibrations, and accordingly reduce piston slap noise.

    There exist two kinds of offsets: the offset of the pinhole to the piston axis and the offset of the piston axis to the crankshaft center. When piston pin hole offsets toward the direction of main thrust force, the moment for piston switching from one contact side to the other contact side in the proximity of upper-end point is altered from the moment of severe pressure increase of cylinder, then vibrations and noise can be reduced.

    Besides depending on impact energy that is related to piston force and gap magnitude, piston slap noise is also dependent on the amplification factor between piston and cylinder, including the number and tension of piston ring, the amount and temperature of the lubricant, and cylinder thickness.

    The proper increase of the length of the piston skirt can reduce the swaying magnitude of the piston, increasing the contact area between the piston and cylinder wall during impact, and reduce slap noise.

4. The countermeasure for the abatement of piston slaps noise includes:

    (i) *Reduce the gap between piston and cylinder*: employ piston with skirt having eclipse taper profile; tweak piston design; have skirt embedded steel cylinder; assembled piston; improve piston materials.

    (ii) *Reduce the number of piston rings*: the friction between the piston ring and cylinder wall could induce cylinder vibrations thus increase noise.

    (iii) *Offset piston pin hole towards preliminary push force area*: generally, the piston pin hole is set in the center of the piston. If the piston pin hole offsets toward the preliminary thrust force area properly, the piston slap noise can be reduced.

    (iv) Employ quality lubricant.

    (v) *Optimize the cylinder stiffness*: friction, vibration and noise can be reduced by optimizing the cylinder stiffness.

    (vi) *Increase damping*: deploy a layer of special coating material on the piston skirt.

In some engine, when piston reverses direction in operation, the piston could experience stick-slip oscillations. The noise named as diesel sounding knock has been reported to be associated with the stick-slip motion of piston caused by the abrupt change of friction coefficient from static to dynamic state when the engine is in idling speed. The excitation acting on the piston, in turn, excites the crankshaft at its resonant frequency through the connecting rods. Some of the identified noise frequency is about 1000 Hz and is thought to be fundamental to the crankshaft. The diesel sounding knock usually occurs for a new vehicle and disappears for mileage up to thousands of miles. The reduced sound pressure level with mileage is a result of a friction force change between the piston/rings and cylinder, due to the surface run-in from original manufacture condition. To remedy this, piston parameters can be optimized, and some friction modifiers can be used to the engine lubricant and their effects were verified. Some modifiers showed significant noise reduction up to several dB.

Crankshaft main bearings and connecting rod bearings have motion clearance space. Under the external excitation of combustion and inertia forces, they are likely to generate vibration and noise at some operation conditions. The radial motion of the journal and the vibration of the crankshaft system are likely to result in impact due to the existence of clearance.

In the process of load transmission from combustion pressure to engine block, bearing clearances had a critical role. The existence of clearances allows the system to have impacts between components. The impact of clearance complicates the vibration of the system which is dominated by vibration at multiples of half order of crankshaft rotation frequency and allows it to have intricate nonlinear vibrations. The tangible results are the impact and self-excited vibration associated with rough "rumbling" noise.

There are some measures proposed to optimize the main bearing clearance. In some cases, the peak noise level could be reduced through the efforts, but the signature of rumbling noise level remains the same. It had been shown that bearing clearance has no significant effect on impact-induced or parameter-excited half order noise.

In some special situation, reducing clearance leads to an increase of noise, which could be due to oil film effects on bearing vibration. Oil film with thinner thickness gives rise to rigid cushion for the journal, which is likely to cause comparatively larger vibration. This disadvantageous effect may offset the advantageous effect of smaller clearance; the collective effect is that the vibration is increased. Moreover, the existence of the oil film has the likelihood to generate parameter-excited vibrations. For most the cases, the reduction of bearing clearances can be used for noise reduction design, and there is an optimum value for bearing clearance.

One of the noises due to bearing effect is rumble noise or roughness, which is a modulated sound with carrier frequency about 300 Hz and modulation frequency of about 100 Hz (low engine order, 1/2, 1 order) as well as peak modulation up to 50%. This kind of noise could be due to the engine torsional and bending resonance induced by the bearing clearances.

The methods to reduce bearing noise include the optimization of the clearances, the application of optimal crankshaft damper with vibration reduction performance of both torsional and bending modes, and the application of flexible or dual flywheel design.

Besides, the crankshaft vibration, other kinds of sources could also result in the rumbling noise perceived in the compartment. These include gas column resonance of intake system, powertrain system torsional and bending resonance, non-uniformity of combustion, and structural transmission with carrier frequency about 300–800 Hz (system natural frequency) and modulation frequency of 20–200 Hz. Usually, the tuning of one part is not solvable to rumble noise. The optimization of the following system could be effective to reduce rumble noise: engine and transmission mount, driveshaft, exhaust hangers, intake and manifold system, exhaust pipes, and mufflers.

In general, bearing itself does not generate large noise. However, bearing has a significant effect on the supporting stiffness of the whole system and system natural frequency. The bearing vibration results in the resonance of the shaft system. Typically, sliding bearing has less noise than roller bearings.

For sliding bearing, the increase in bearing gap can significantly influence oil film pressure and the path of the shaft axis thus increases vibration and noise. For roller bearing, when loaded radically, roller and case generate elastic deformation. In rotating operation, the shaft has periodic oscillatory motion which allows roller and case/fixture to generate impact. The structural and manufacturing accuracy has a significant effect on bearing noise. In the installation, the inaccuracy and compliance of the bearing bracket can make the additional deformation of bearing parts [47–49]. The poor accuracy of bearing causes the deformation of bearing inner ring. If the natural frequency of the shaft system is close to the natural frequencies of bearing, resonance is likely to occur. Moreover, contamination, damage, debris, wear, and corrosion affect bearing vibration and noise.

To suppress bearing noise, manufacturing accuracy and ring stiffness should be enhanced. Pretension can be used to reduce the gap. The tight control of assembly error and tolerance, use of quality lubricant cream, and improved seal can reduce bearing noise. Other approaches to reduce noise include to enhance manufacturing accuracy, to ensure the smoothness and non-defaults on all sliding/rolling interfaces, to control dimensional accuracy of all parts, to enhance the accuracy and stiffness of bearing bracket, and improve application condition to reduce debris and corrosion. As an example, the following is a case of boom noise due to four-stroke engine bearing. The four-stroke engine in full acceleration process was perceived having an intermittent boom noise. It is due to the instability caused by the interaction between coupled torsional-transversal vibrations of the crankshaft and the bearing having a relatively larger radial gap. The recorded signal is processed to obtain a time-frequency spectrum of the booming noise, from which the magnitude of the dominant frequency component of the booming noise is illustrated with magnitude modulation. The magnitude modulation is concerned with the 0.5 and 1.5 order of shaft speed, whereas the noise frequency is independent of speed. It is believed this booming noise is concerned with the parameter-excited vibration of bearing.

The typical cam system or valve train consists of gas valves, pushrods, rocker arms (swaying arm), lifters, and camshaft which are synchronized to the crankshaft by a timing chain or timing belt. Overhead camshaft engines use fingers or bucket tappets, upon which the cam lobes contact, whereas cam-in-block engines, use rocker arms. Rocker arms are actuated by a pushrod, and pivot on a shaft or individual ball studs to actuate the valves. Camshaft system could exhibit bending and torsional vibrations. The vibrations of valve train system are influenced by many factors, including valve seating velocities, acceleration and its change rate, clearances at tappets and valve guides, the influence of timing chain or belt, spring surge frequencies, mass of components, and friction effects, as well as system natural frequency. Particularly, in opening operation, valve train vibration leads to the deviation of valve motion relative to the designed profile. In closure operation, excessive vibration may cause impact and noise. Noise propensity is governed by valve acceleration and its change rate, "jerk," in the near vicinity of valve opening and closing events. The key points of low noise cam system design are to avoid the loss of contact of two engages components and to achieve quiet valve liftoff and seating process. Cam and follower noise excitation are only one source of valve train noise. Another important source is the impact upon valve seating against the head.

The acceleration at the cam-follower interface is the source of the noise. The components determine the noise spectrum and the path of energy transmission. The transfer path is along with cam, camshaft, and its supporting bearings. The valve train is likely to generate vibration and noise due to multiple components of small stiffness. In the valve train, noise is mainly due to the friction vibration between cam and bar, impact of the arm with bar, non-uniformity of the valve, and the seating of the valve.

When the engine operates at low speed, the inertial force of the cam system is not significant, and the system can be treated as a rigid system in terms of the simple kinetics of cam-follower component. The noise source is mainly due to friction and impact of valve opening and closing. The larger noise occurs at the instantaneous moment when the top of cam pushes tappet upward, and the moment when the valve opens and closes. The valve open noise is due to the impact of the valve system and the noise frequency is related to the natural frequencies of the valve system. The valve closure noise is due to the impact when the valve seats. The associated noise has a higher frequency than that of open noise. Its frequency is dependent on the design of the valve system and the head of the cylinder. The sound pressure level of valve noise is proportional to the speed of the valve.

When the engine operates at high speed, the inertial force of the cam system is significant which causes the whole system vibrations. Cam system behaves as an elastic system. The elastic deformation of each component renders the motion of the valve to have large deformation from the standard profile. It causes the motion of the valve to be delayed from tappet. The detachment of the transmission chain or belt and the abnormal closing of the valve results in abnormal motion, the bounce of the valve.

The high-speed operation exaggerates the abnormal motion, which results in strong noise. The severe situation can even malfunction engine's operation.

The noise of the cam system at high speed is mainly due to the abnormal motion of the valve. This abnormal motion due to deformation can result in significant deviation of real motion profile from the designed profile. During the high-speed phase in the acceleration process of a camshaft, the valve chain or belt can store some elastic deformation energy, and the actual elevation is smaller than the designed one. When cam acceleration changes from positive region to a negative region, the stored energy gets released and renders the valve open at a relatively high speed. This causes the jump of the valve and leads to the actual elevation to be higher than designed elevation. This leads to the valve beyond the control of the cam system and the discontinuity of chain or belt. This generates impact and noise at individual non-offset points.

The factors influencing cam system noise include lubricant, the gap of the tappet of the valve, engine speed, cam profile, the stiffness, and weight of the components.

The cam system noise is transmitted from valve train to engine block, then it radiates or transmits to the valve train cover through the air, then radiates from the cover to air.

The measures used to control the noise of cam system include reduce the gap of the valve; increase the manufacturing accuracy of the cam and reduce surface roughness; optimize cam profile; increase the stiffness of cam system; reduce the weight of driving component; employ rubber damper for camshaft.

The sources of the noise of engine oil pumps include pressure pulsation, flow "ripple," trapped oil within teeth, fluid-borne noise throughout system, forces on drive mechanism, and mechanical structure responses. The mechanical excitations sources and pressure pulsation are interactive. The pressure pulsation often produces mechanical noise of the drive mechanism. The pressure of the oil pump and mechanical forces are applied directly on the cover of the pump, which results in vibration radiating noise. Since the cover is usually relatively compliant, the vibration and radiated noise may be quite strong. The common types of oil pump in engines include external gear pumps, gerotor pumps, and vane types. The noise mechanisms associated with engine oil pumps include involutes gear teeth and gear forces, oil transfer in tooth spaces on the periphery, low pressure being "dumped" to high, outlet flow having dynamic "ripple" or fluctuation, ripple decreasing with more teeth, flow ripple created dynamic pressure, and oil trapped in meshing teeth. In oil pump, oil is transferred around the periphery of the gears in the spaces between the teeth. The oil is delivered to the outlet, or discharge, a port in discrete "clumps," producing flow variation or flow "ripple," which gives rise to pressure variation, due to the hydraulic impedance of the oil lubrication system "down-stream" of the pump. The impedance of the system is typically high enough to enable significant pressure ripple that give rise to dynamic forces which produce vibration and noise.

Low-pressure oil from the inlet side could be "dumped" into the high pressure of the discharge port, causing a sudden pressure change. Metering grooves can be placed around the periphery of the gear teeth, connected to the discharge side. As a "pocket" of oil is moved from the inlet to discharge, it encounters the grooves which allow a gradual build-up of oil pressure in the pocket so that its pressure is equal to discharge pressure when it reaches the discharge side. Pressure relief grooves are used to allow trapped oil in the tooth mesh to escape to either the inlet or discharge side. These grooves do not allow a direct leak from discharge to inlet. Without this pressure relief, trapped oil can cause shock and cavitations. Increasing tooth count can divide oil delivery into smaller "pockets" with smaller flow variation. Thus, the pressure ripple and noise are reduced.

The split gear pump reduces ripple by a similar mechanism to increase tooth count without adding teeth.

The generator oil pump is an internal/external gear arrangement, with very compact geometry. The gear teeth are not involutes. Tips of the inner gear teeth seal against the outer gear to form pockets. As the gears rotate, the pockets expand and contract, with a volume that varies approximately sinusoidal.

The expanding pockets are placed over the oil inlet cavity. They "pull" oil from the inlet as they expand. The contracting pockets are placed over the discharge cavity. They "push" oil into it as they contract.

The flow from each pocket gets together to produce the total flow of the pump with harmonics at multiples of $N + 1$ order of the outer gear. Note that with an odd number of pockets, the fundamental order of the total pressure, at $N + 1$ order, is zero amplitude. As such the total waveform does not show any periodic behavior at $N + 1$ pulse per revolution. The 2nd harmonic order ($2N + 2$) is the largest harmonic component.

The inner gear drive frequency is usually used as the reference for representing pump orders. Flow harmonics exist at multiples of $N + 1$ order of outer gear frequency. The outer gear turns at $N/(N + 1)$ times the speed of the inner gear. Therefore, $N + 1$ order of the outer gear is N order of the inner gear. The flow harmonics are present at multiples of the $N + 1$ order of inner gear drive frequency. However, this relationship is only valid for the precise flow waveform of the half-sine wave. The flow in actual pump applications are modified by backflow and trapped oil, thus minimizing the benefit which might be achieved with an odd number of pockets. It is still a good design practice to use an odd number of pockets.

When low-pressure oil from the inlet port is delivered to the high-pressure discharge port, oil from the discharge port compresses it and causes a backflow pulse. The pulse occurs for every pocket $N + 1$ times per revolution. The backflow pulse may modify flow waveform significantly. An $N + 1$ order, the large variation could be introduced. For an odd number of pockets, as in the example, the $N + 1$ order is reintroduced where it was completely absent. Variation of the total flow waveform can be greatly increased. A certain amount of backflow is almost always present in practice. Timing and metering grooves allow controlled backflow which assists the pre-compression from delayed discharge port timing. It is somewhat self-regulating, as higher pressures will create more backflow as is desired for the higher pressure.

Quiet oil pumps should be designed to minimize the noise sources associated with pressure ripple and drive mechanism forces. It is also important to control the structural vibration path from the sources to the noise radiating surfaces of the engine. The integral front cover pump has a very sensitive noise transfer path. Reductions in pump noise can be achieved by isolating the pump from the radiating surfaces. Large noise reductions can be achieved throughout the operating range of the pump. The following experimental facts are helpful for the pump design: timing and metering grooves can attain optimum at specific speeds, pressures, and air entrainment; speed determines the time available for backflow and suction; discharge pressure determines the force acting on the backflow; as speed increases, the relationship of noise and pressure will change.

Timing chain has been applied in camshaft transmission due to its advantageous attributes such as wear durability and high strength, particularly the high-performance engine. The major excitation components of chain noise are meshing impact and polygonal effects. The meshing frequency depends on the tooth number of chain sprocket and shaft rotating speed. The most popular chain noise is whine noise which depends on meshing impact force and the vibration response of engine components and system such as engine front cover and the cover of the camshaft. The polygonal effect causes elevation and drop of chain element thus leads to the transversal and torsional vibration of the chain as well as speed variation. The impact noise of roller generated in roller-sprocket meshing is the dominant component of chain noise.

The impact noise consists of acceleration noise which is air noise due to the acceleration of roller interacts with air and ringing noise which is structure noise due to impact-induced resonance. The employment of grooved rubber ring on two sides of chain sprocket can reduce the chain noise. Moreover, the sprocket design with randomly unsymmetrical tooth profile can reduce noise.

The transmission belts can exhibit many different vibrations patterns. Belt vibrations can be classified as: (a) transversal vibrations; (b) axial (longitudinal vibration); (c) torsional vibrations; and (d) lateral vibration. Usually, noise radiates due to transversal vibrations. The power or motion transmission of the belt can excite some of the above-mentioned vibrations. The axial and torsional vibrations could couple with transversal vibrations, thus exaggerating the transversal vibration and noise radiation. The transversal vibration of the belt can be modeled by using the string model. The transversal vibration of the short belt can be modeled by using the beam model.

Due to the unavoidable difference between the pitch of the toothed belt and sprocket pitch, timing belt creates meshing impact and friction when it meshes with the tooth of the sprocket. The mesh impact is a periodic excitation to the belt.

This periodic meshing excitation force generates transversal vibrations of the belt. The vibration and noise frequency of belt is identical to the meshing frequency of its harmonics. Particularly, when the meshing frequency is identical to one of the natural frequencies of the belt; resonance and severe noise will occur. Moreover, each meshing effect also results in wideband impact noise and high-frequency friction noise.

In applications, the approaches to control timing tooth belt noise include the optimization of the belt and the sprocket to minimize meshing impact and the optimization of system design such as tension and load distribution. Some of the approaches are similar to that for gear noise control, such as the use of optimal tooth parameters and structures.

With the improvement of systems design, many conventional noise issues have been alleviated. It results in that potentially stronger noise paths into a vehicle cabin has become salient such as accessory noise. For instance, the quietness has become a measure of steering system quality. multiple v-ribbed belt has been used for accessory transmission in the engine, which includes pulley at the frontend of the crankshaft, water pump pulley (W/P), tensioner (TEN), alternator (ALT), idler (IDR), power steering pump pulley (P/S), and air-con pulley (A/C).

When belt and pulley had misalignment and or has slippage, belt creates misalignment noise or slip noise. The frequencies of these two-kind noises are in the high-frequency range.

In the belt accessory system of a vehicle engine, the overload occurs due to many reasons such as crankshaft torsional vibration, transition process of start-up, accessory overload, belt wetting, and degradation. An overload could result in the slippage between belt and pulley. When the coefficient of friction is sufficiently high, the slippage belt generates squeal noise [50, 51].

In applications, the approaches to control accessory belt noise include the minimization of misalignment, reduce slippage rate, control coefficient of friction, optimize system design, and attain robust design.

For accessories, such as power steering pump, alternator, water pump, air-con and fan subsystems, the excitation frequencies are usually lower than 300 Hz. There are occasionally the cases that the excitation frequency as high as 3 kHz. The selection of accessories for engine integration should allow each subsystem to have enough low vibration and noise level compared with engine overall noise level, to attain an acceptable noise level for engine system integration. Moreover, in practical designs, the exciting frequencies of respective subsystems should have at least a separation of 10% of the specific frequency to avoid resonance or beat phenomena. Usually, a frequency tuning plot can be used to help analysis, in which the excitation frequencies of respective subsystems vs. speed are plotted together in one figure, the cross and proximity between the lines are tuned to meet the target.

The hydraulic power steering systems have been a noise concern in today's vehicles. The accessory noise has become one of the principle noise sources perceived by the customers, and even sound quality has been required as a primary measure of product quality for the power steering pump.

Figure 3.5 shows the schematic of the power steering pump noise sources.

The sources of the noise of hydraulic steering system include the air-borne noise which results directly from flow fluctuation (ripple) of the pump and the components; fluid-induced noise that is from the pipe and is distributed through the hydraulic system in pressure pulsation form; structure-borne noise that is from hydraulic components and at rack-pinion gear system, also from flow ripple of the pump.

The major noise source is pump flow fluctuation, which allows noise to exhibit as air-borne noise, fluid-based noise and structure-borne noise. When the pulsation wave reflections are built up whenever the incident wave reaches a change in impedance, and if the transmitted and reflected waves combine in phase at any portion of the hose, reinforcement could occur, which establishes up a standing wave leading to resonance. The interaction of the flow fluctuation within the pipe circuit component generates pressure fluctuation in pipe therefore, the fluid-based noise. The pressure pulsations within the pump generate alternating forces which result in structure-borne noise. The resultant pump housing vibration radiates noise.

The measures used to reduce power steering system noise include optimize pumps design to have high NVH performance; minimize noise source levels for all forms of

FIGURE 3.5 Schematic of power steering pump noise sources.

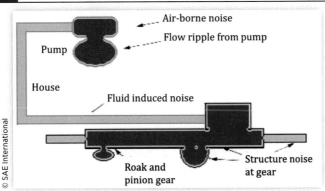

noise, apply noise insulation and absorption materials at the pump and gear mountings [52–54]; deploy structural attachments on the hose located at the anti-resonant points; tune hose design by trial-and-error using high compliance rubber hose; insert devices within hoses design such as silencers and restrictors; modify power steering system design using analytical and measured models for the steering system.

Besides the noise discussed above, other powertrain noise could come from engine fuel injectors, turbocharger systems, fan, etc. Noise from the fuel injector mainly comes from the impact of the moving valve on its seat. Its magnitude is governed by the change in momentum of the moving parts and by the areas of body and inlet manifold which radiate the noise.

If an engine has a turbo system, various noises may occur, which include turbo whine, moan and sigh. Turbocharger whine (due to turbo blade passing) or whistling is an airborne, high frequency, narrowband, or tonal noise. It radiated normally from intake inlet. Turbocharger moan is a structure-borne noise due to bracket resonances. Turbocharger sigh is another noise that may come from exhaust and intake. It is an airborne, broadband noise due to high mean mass flow.

Fan noise is another noise source of the vehicle. Fan noise is a kind of aerodynamic noise, which usually consists of tonal noise and broadband noise. Tonal noise has discrete frequencies and is caused by the interaction of the rotor blades and obstacles in the flow path, which is characterized by the fundamental frequency defined by the rotor speed and by the number of blades and its harmonics. However, broadband noise covers a wide range of frequencies and is caused by vortices around the blades. Broadband noise comes from the vortices behind blades due to changes in the circulation and therefore in the pressure distribution at the aerofoil. This results in sound waves. The acoustic frequencies of broadband noise are identical to the frequencies of the vortex origin, which depend on the air velocity and the thickness of the profile, and change across the height of the blade gives rise to the features of noise spectrum spreading over a wide range of frequencies.

Alternator noise is another source. Some experimental data have shown that the overall sound pressure level (SPL) values from an alternator can be as high as 90 dB(A) with the engine running speed from 4000 to 5000 rpm, and the noise spectra contain both narrow- and broad-band sounds. The rotor is the only component that runs at high speeds and maybe the major noise source. In general, alternator noise is composed of mechanical, aerodynamic, and electromagnetic sources. Test results demonstrate that all these noise sources produce narrow and broadband sounds and their levels increase with the speed. In some cases, rotor cooling fans are the predominant noise sources generated at high speeds, while electromagnetic noise is the most intensive at low rpm. Electromagnetic noise could be due to the excitations from magnetic run out, **rotor** slip, or magnetostriction. Magnetic run out is a variation in the mutual force between rotor and stator during a cycle of rotation. Rotor slip is referred to as the rotating magnetic field revolves faster than the rotor. Magnetostriction is the coupling of material strain and magnetic polarization. Besides fan noise and electromagnetic noise, other mechanical noise sources, such as bearings, brush contacts, structure resonance, unbalance, etc. are of secondary importance. As an example, a typical electrical alternator system has three-phase claw pole alternator. A typical claw pole alternator consists of rotor and stator. There are six pole pairs and the field winding is around the shaft axis. The typical 36 slot stator is typical of conventional three-phase machines. The usual electromagnetic configuration of this machine is like to generate noise at 36 times the shaft speed (36th order), and at idle speed, particularly when there is a high-energy demand after starting up, the resulting noise, typically at around 1000 Hz, can be detectable in the passenger compartment in vehicles in which other noise sources have been reduced.

It is found that a strong electromagnetic generating mechanism exists for 36th order (and harmonics) torque ripple. In a six-phase, six pole pair, 72 slot machine, cogging torque and phase current and voltage-induced additional torque ripple is present in 72nd order harmonics. Some research illustrated that an important noise radiation mechanism in claw pole alternators is the dynamic response of stator and housing excited by rotating electromagnetic forces between the rotor and stator. Some research suggests that machines having a water cooling jacket show little reduction in radiated noise. There exist noise radiation mechanisms in addition to housing dynamic response. The torque ripple and machine and mounting structure response are also considered as an important noise radiation mechanism.

Some engine noise exhibits complicated features and is perceived as unusual noise. Unusual noise and vibration are probably the most annoying to the customer. Unusual vibration happens less frequently than usually annoying noises. Drivers name these unexpected vibration and noises based on their feeling such as a whine, moan, boom [55–57], etc. Engineers correlate these noises with objective measurement based on testing. Among the most occurring and probably most unusual, annoying engine noise is engine tick. Engine tick is a transient phenomenon and is difficult to relate to a specific frequency. Engine tick is particularly perceivable at idle operation. It is also more present in diesel engines. The root cause of the phenomenon can be injector event excitation, fuel pulse generated by the high-pressure pump, fuel pump radiated noise, mechanical excitation of the pump, structural sensitivity (structural amplification and modal alignment), fuel rail fluid resonance (pulse magnification), fuel rail structural resonance (pulse excited), or structural excitation of engine components (cylinder head, intake manifold, engine covers, and oil pan).

Transmission controls the torque-speed relationship for the powertrain. It consists of various gearing structures, shafts, and bearings. Helical gears are used for manual transmission and planetary gears are used for an automatic transmission.

Transmission noises can be separated into two groups. The first group represents transmission noises resulting from gears engaged in the torque flow or meshing noise, the second group is the noises from unloaded transmission components such as loose gears or synchronizer rings bouncing inside their functional backlashes and clearances.

Transmission whine is the most occurring transmission noise, which is in the high-frequency range (2k–15k Hz). This is due to the meshing of teeth. As torque transmitted through these teeth, they will deform leading to variation between the designed profile kinematics requires and actual profile due to deformation. This is referred to as transmission error and is known to be the root cause of many transmission noises.

In reality, gears are not geometrically perfect, not perfectly aligned and not infinitely stiff.

The difference between the actual meshing position and ideal meshing position of the two gears is called transmission error (TE).

The transmission error can be found wherever gear meshing exists such as in transmission, front and rear axles, and transfer case. The transmission error causes meshing excitation, which could result in the bending and torsional vibration of the driveshaft that is transferred to the body through mounts and bearings. The excitation frequency of gear meshing is the product of the first order of the driveline excitation multiplying the numbers of teeth in the driving pinion of the axle. To reduce the excitation caused by transmission error, gear teeth and the gear meshing should be optimally designed. The modal separation between driveshaft's bending and torsional modes also helps reduce the excitation.

Transmission error (TE) could be due to the poor design, manufacturing tolerance, wear and degradation, tooth mesh stiffness variation, and due to impacts (engagement and separation). It can also be due to change of mesh force direction, friction, air and lubricant entrapment, gear tooth geometry, and gear train dynamics.

Gear whine is a typical noise due to TE, which is characterized as a tonal noise with a specific frequency. System resonance could occur when the excitation frequency from the gear pair coincides with the resonance of the system, which may include the individual or coupled torsional, lateral, and axial motion.

The mechanism of whine is the meshing excitation induced torsional vibration and transversal vibration of the gear shaft and the system resonance. This further magnifies dynamic meshing and load, causing the interaction between the bearing and gearbox, and finally cause gearbox and related components to radiate noise.

The attributes of gear transmission are that the meshing point has both rolling motion and sliding motion. The interaction between teeth generates inevitably the impact and friction, thus create meshing vibration and noise that are dependent on rotating speed. The meshing frequency is given by $f = nN/60$, in which N is a number of gear tooth, n is the gear rotating speed (rpm).

Transmission whine is primarily generated by transmission error, which is a deviation of meshing gears from a perfectly conjugate (smooth) motion due to manufacturing tolerances, tooth corrections, and elastic deflection due to transmitted torque. The transmission error excites the geared shaft and bearing system leading to dynamic forces at the bearings, which, in turn, excite the transmission casing causing it to radiate noise. Gear whine can be reduced through a combination of gear parameter optimization and manufacturing process tolerance control directed at reducing the effective transmission error. The process of gear selection and profile modification design is facilitated through the evaluation of the details of the tooth contact analysis through the roll angle, including the effect of gear tooth, gear blank, and shaft deflections under load. The factors influencing gear noise include gear tooth geometric parameters, the accuracy and smoothness of gear, the structure and profile of other gear factors: shaft stiffness, lubricant. The abatement approach of gear noise includes apply gear materials with high inner damping or isolation design, optimization of gear parameters and structures, enhance manufacturing accuracy of gear, edge treatment, and the system dynamics control to increase damping.

Transmission whine occurs (and is probably expected) for most vehicles when they are in reverse.

Low gear whine in partial throttle conditions occurs frequently and is annoying to the drivers. Besides, slowing the vehicle by only friction (no gas or brake) will expose the transmission whine problem.

Vehicle gear meshing noise control usually is implemented through two approaches: one is to reduce transmission errors, include optimize tooth profile and enhance manufacturing accuracy; the second is to control system dynamics, which can be implemented by modifying gear transmission system design.

The impact noise of the wideband spectrum of gear called rattle is due to the impact of gears with slash gap. The onset of impacts from an unloaded gear pair is more likely to occur than from an engaging set. Transmission rattle is caused by engine excitation and backlash in geared transaxles. Backlash yields piecewise nonlinear characteristics in gear mesh stiffness. This slash can create impact, nonlinear vibration [58–60], and noise. The variation of the period load of gear and varied error worsen this dynamic impact load. This dynamic load can make the shaft deformation and generate extra bearing load.

The dynamic bearing load can transmit to the transmission case and engine black/cover, allow shell/plate structure to radiate noise. Gear rattle could be caused by torsional vibrations of the crankshaft due to the engine firing excitation and inertial imbalance, which initiates cyclic angular vibration to be transmitted from engine to transmission input shaft, leading to impacts of gear and gear teeth oscillation within their backlash. The gear vibrations are finally transmitted via the supporting bearings to the transmission case, which radiates noise to air.

The angular velocity fluctuations causing the gear rattle are mainly due to the engine firing order.

The amplitude of the angular crankshaft accelerations depends on the number of cylinders, cycles and the type of fuel (gasoline or diesel). As an example, for a 4-stroke, 4-cylinders engine, the torque engine fluctuation will mainly oscillate with twice frequency of the crank rotational frequency. Although the excitation itself occurs in a well-defined range of low frequencies, the measured rattle noise has a broadband distribution from 1 to 8 kHz.

The measurable rattling noise in the surrounding area of the transmission consists of a broadband frequency spectrum ranging from 1 to 6 kHz. The torsional vibrations excite idle components, such as idler gears, synchronizer rings, and sliding sleeves to vibrate within their functional clearances. Gear rattle is strongly audible inside the vehicle cabin during the run-up at low speed and high load, or with the transmission in neutral and the clutch engaged (drive rattle or idle rattle).

Transmission noise reduction may be obtained acting in different ways: (1) reduction of the exciting source of torsional vibrations; (2) optimizing the design parameters; (3) increase of vehicle damping characteristics. Usually, the third possibility is more complex and economically less advantageous than others. Manual transmission gear rattle could occur when transmission system torsional resonance excited by firing order. The unloaded pair could have an impact that results in rattle, which is transmitted to transmission house through shaft bearing.

Besides whine and rattle, there are other kinds of noise in transmission. The sliding friction between meshing teeth is one of the primary excitations for noise and vibration in geared systems. Existing research suggests that the measured maximum friction coefficients are in the range of 0.05–0.06 subject to load. Friction coefficients increase at low sliding speeds. The reversal of sliding which occurs at the pitch-point does not cause a discontinuity in the friction coefficient, which shows a smooth transition as the friction force reverses direction. The sliding friction has been recognized as a significant source of noise and vibration in gear meshing. It was found that the dynamic friction force could be the same order of magnitude as the forces normal to the tooth profile for certain cases. Friction forces play an important role in load transmitted to bearing and housing in gear meshing, particularly at high torque and low speed. Friction has a predominant effect at higher harmonics of meshing frequency. Both oil viscosity and surface roughness had a large influence on the vibroacoustic behavior. The difference in measured sound spectra for the two lubricants is attributed to the lower values of friction coefficient. A distinct reduction in noise level could be achieved with the high-viscosity lubricant.

A variety of clutches have been extensively used in passenger cars for engaging different gears in automatic transmissions. Clutch system could be excited by torsional and axial flywheel vibrations which are influenced by engine firing orders. There are many vibrations and noise phenomena associated with the clutch system, just to name a few, whoop, tingling, toc-toc, chatter, shuffle, clunk, judder, shudder, squeal, [61–63] etc. Whoop is a tactile foot vibration accompanied by noise. The clutch whoop could occur when the clutch is engaging or disengaging. It causes low-frequency vibrations of the clutch pedal and it also results in noise. Tingling is a strong pedal vibration that is associated with engine harmonic orders. Toc-toc is low-frequency vibration of pedal that is associated with half order that is perceived felt by foot. They depend on engine speed, pedal position and actuation speed, and the temperature of wet clutch and wear.

Shuffle is a vibration of the low frequency of clutch driveline system during load change, which is described in the section of driveline vibration. It is also influenced by the clutch system [74–83].

The chatter could occur when a periodic torque change is created in the sliding clutch, perceived as both vibrations and noise. It could be due to the friction-induced vibrations [84–89].

The clunk of metal impact noise could occur due to rapid clutch engagement or gear shifting. Driveline clunk is a phenomenon that can adversely affect customer perception of vehicle quality. Clunk is created by the impact of two driveline components as they oscillate in response to a torque disturbance in the driveline system. This disturbance is typically caused by an engine torque variation, most severely through a throttle or clutch manipulation. This torque variation excites a torsional vibration from the driveline, amplified by a variety of mechanisms such as resonances of various shafts, housings and axles, clutch oscillations, and gear impacts. Because automotive drivelines are complex systems composed of many rotating components, it is difficult to identify the impacts that cause clunk and evaluating the significant parameters that can affect these phenomena. In the following, we discussed the friction-induced vibration of the clutch.

In starting engagement, fiction clutch is likely to generate judder, which is a low frequency, friction-induced torsional vibration of the driveline. It affects starting smoothness and rides comfort. It is due to the decreasing friction of clutch concerning the slip speed between flywheel and friction disk. Sometimes misalignment of driveline also contributes to it. Take-off clutch judder is another low-frequency vibration of fore and aft type in the frequency range of 5–20 Hz or even higher, due to the torsional vibration of driveline that occurs during the clutch engagement process. Clutch judder generally has two root causes; a particular type of the friction-sliding speed curve and misalignment in the driveline. The beat phenomena and variation in torque-time traces have been observed. It is also noted that clutch judder is affected considerably by the buildup time of the axial load and the friction-sliding speed variation of the facing. Judder is a kind of self-excited vibration of the vehicle driveline, whose frequency is independent of shaft revolution speeds. The vibration could be transmitted to the vehicle body through suspension mount and engine mount. For instance, a front-engine-rear drive vehicle judder could be close to the third mode of driveline torsional vibration. The countermeasure is to improve the clutch friction characteristics and the damping of the driveline system. This will be further elaborated at the end of this section.

Clutch operation or engagement is usually attained through friction effects between driving and driven ends, in which self-excited vibrations could be triggered for some of the systems with improper friction properties and system combinations. Low-frequency torsional vibrations of powertrain called shudder (judder) occasionally occur due to clutch friction during acceleration at low speed. Shudder could exist in dry clutches in a manual transmission or wet clutches in an automatic transmission. Squeal noise could even occur in dry clutch applications.

The slippage of torque converter clutch can excite a low-frequency torsional response of the whole powertrain due to the property of velocity-dependent friction. The shudder severity varies with clutch surface characteristics, temperature, lubrication additives, and vehicle operating conditions. It has been found that some combination of clutch design and transmission fluid can prevent negative damping associated with velocity dependence friction. Besides, some attenuation of problem severity can be obtained with increased damping in the clutch or the torque converter damper.

Most of the wet clutch-related vibrations are attributed to the velocity-dependence properties of friction. A problem area in most wet clutches working at low velocity is the occurrence of shudder associated with the negative slope of the friction-velocity property of lubricated interface. The anti-shudder performance of automatic transmissions has been of primary interest in new designs featuring lock-up and continuous slip torque converter clutches. Generally, stick-slip occurs at low velocities when the static coefficient of friction is higher than the dynamic coefficient of friction. Friction induced instability tales place at higher sliding velocities, due to a negative slope of the friction vs. velocity curve.

Some vibration and noise problems associated with dry friction clutch are attributed to mode-coupling type self-excited vibrations. During the engagement of dry friction clutch in manual transmission, a strong squealing noise could be produced. For this case, in near full engagement, the pressure plate suddenly starts vibrating with a frequency close to the first natural frequency of the rotational sub-system. This problem exhibits typical signs of a dynamic instability associated with a constant friction coefficient. It was observed that the instability of the rigid body wobbling mode is controlled by the friction forces and this mode is also affected by the first bending mode of the pressure plate. A stiffer plate could lead to a design with a reduced tendency to the squealing noise. It has been demonstrated that there exists potential mode-coupling between the pressure plate-wobbling mode and the first elastic deformation mode of the pressure plate. The stability threshold depends on the friction coefficient, the pressure plate geometry, and structural stiffness. Two conditions were observed to generate transient squeal noise. Firstly, the engine speed is within the 1500–2500 rpm range and the slip speed is relatively high. It may assume that a constant coefficient of friction μ corresponding to the threshold of noise. This allows noise to not follow the mechanism of classical stick-slip. Secondly, the clutch pedal motion is in the condition that the clutch is close to the full engagement. Measurements of noise spectra exhibit a dominant frequency f_w around 450 Hz. This frequency has been correlated to the wobbling rigid body motion of the pressure plate (both out-of-plane rotations) from dynamometer tests and the simple math model. The pressure plate and the disk cushion are assumed to be the main components that control the squeal phenomenon. To consider the nonlinear characteristic of the cushion, mathematical modeling yields an asymmetric stiffness matrix characterized by the coefficient of friction and stiffness ratio. The results of complex eigenvalue analysis agree with experimental observations. It is seen that the instability of the rigid body wobbling mode is controlled by the friction forces. This mode may, however, also be affected by the first bending mode of the pressure plate. Therefore, a stiffer plate could lead to a design with a reduced tendency to squeal.

In an engine air intake system, intake noise radiates from the inlet or the open end. The reduction of intake noise is critical for reducing vehicle interior noise and pass-by noise, and for enhancing interior sound quality to meet customer expectations. Conventionally, the priority of intake system design has been the engine volumetric efficiency. More and more efforts have been made to intake system design to attain a lower noise target.

A typical air intake system consists of a dirty side tube, an expansion chamber (normally the air cleaner box), and a clean side duct. The engine manifold system consists of manifolds and surge tank. The air intake system consists of the throttle valve, idle intake pipe, flexible connecter, clean air duct, quarter-wave pipe, air cleaner box, filter, Helmholtz resonator, and dirty side tube (inlet).

The basic need for intake system design is to allow it to have higher engine volumetric efficiency. If an engine has no intake system, the pressure at the throttle valve is ambient, however, after an intake system is applied, the pressure at the throttle valve is higher than the ambient pressure. The difference between the higher pressure due to the application of the intake system and the ambient pressure is called as back-pressure. The existence of back-pressure allows part of the engine's power to be wasted. The larger the cross-section area of the intake duct, the smaller the engine power loss. On the other hand, the smaller the cross-section area of intake duck, the less the noise radiated from the intake system. Therefore, there is a conflict between the requirements of low power loss and low noise on the cross-section area of the intake duct. A trade-off is needed in real design. In addition to the requirements of the low power loss and low noise, there are other requirements also in intake system design, which includes the waterproof as well as vibration reductions. One-dimensional linear acoustic models have been widely used for the evaluation of noise performance of intake and exhaust systems, due to their simplicity and low computational requirements in the description of complex geometries.

In the engine's operation, the intake system works intermittently to allow fresh air to flow into the engine. The interrupted air-flow from the opening and closing of the intake valve gives rise to impulsive pressure variation. This pressure pulse propagates to intake inlet leading to sound radiation.

Certain special volumetric devices called silencers (mufflers) are capable of abating tonal or narrowband noise of specific frequencies in induct system. The silencers are special ducts or pipes or openings/cavity which accommodates the free flow of air while impeding the transmission of sound. There are two kinds of silencers, the resonant (reactive) and resistive silencers.

Reactive silencers include a side-branch resonator (Helmholtz type of resonator or quarter-wave tuner) and expansion chamber. Reactive silencers rely on the mechanism of expansion, contraction, or pipe protrusion to abate noise. The abrupt expansions or contractions of airflow allow the discontinuity which generates reflective sound waves that could result in the attenuation of the incident waves, thus gives rise to a smaller sound waves radiation. The reflection of acoustic waves at the discontinuities and the interaction of these waves can reduce the transmission of sound. The duct with a closed-end acts as an acoustic oscillator which could further result in some attenuation to incident waves. The reactive silencer can be designed to have high attenuation in some specific frequency band. The Helmholtz resonator is a kind of silencer, which has a specified chamber and is widely used to abate noise at specific lower frequencies. Its working principle is similar to vibration dynamic absorber. The air in the chamber serves as spring, and the mass of the air in the chamber reacts like a spring to form a harmonic oscillator system. Change the volume of the air chamber, the silencer can shift the frequency of resonance. Quarter wave tuner is similar to Helmholtz resonator in principle. The expansion chamber is another kind of silencer such as cleaner box that serves as a cleaning component as well as a silencer.

Resistive or dissipative silencers are made by special ducts or pipes or openings/cavity or chamber with its inside walls lined with acoustically absorptive material, which allows for the free flow of air on the surface while impeding the transmission of sound. It attenuates sound by the acoustic energy absorbing action of absorptive material within the silencers. This kind of devices typically provides noise attenuation over a broad frequency range.

Intake noise is one of the major noise sources of a vehicle. Besides the flow pulse induced intake noise, the poor structure design of the intake system with insufficient stiffness also leads to structurally radiated noise. The reduction of intake system noise can significantly reduce vehicle interior noise [90–93] and pass-by noise. On the other hand, the control of intake system acoustic performance helps to tune interior sound quality [94–96].

The first design parameter for intake system noise attenuation is the silencing volume, which is usually referred to as the sum of the volumes of the air cleaner box and Helmholtz silencer. Generally, the larger the silencing volume, the better the noise reduction performance. For expansion silencer, the larger the volume, the larger the transmission loss of noise, and the wider the frequency band for noise abating. For Helmholtz silencer, the larger the volume, the lower the frequency limit for noise attenuation. For 4- and 6-cylinder engines, the volumes of silencing components are usually required up to 10–15 L.

The second parameter for intake system noise control is the cross-section area of duck. The smaller the cross-section area of duck, the larger the expansion ratio of expansion silencer, and therefore, the larger the transmission loss and the better the silencing effects. However, the smaller cross-section area of duck results in higher backpressure and the air friction noise.

The third parameter for intake system design to abate noise is the location of the orifice of intake. As a noise resource, it should be kept away from the compartment. The reallocation of the orifice of intake is also subject to the constraints that need preventing from the entries of water and snow and allows the efficient flow of air.

A manifold system connects an engine cylinder with the intake system.

When airflow reaches the surge tank, the phase difference between the i-th cylinder and the first cylinder consists of two parts: the first part is the difference between the firing timing of different cylinders; the second part is due to the different length of the individual manifold.

In the i-th manifold, the sound pressure is made of the incident sound pressure and reflective sound pressure.

For the other kinds of engine such as 4, 8, and 10 cylinders, if the lengths of manifolds are identical, then the sound pressure in the surge tank has the similar features as that of an above 6-cylinder engine. There only exists firing order and harmonics, the sound pressures of the half order and all of the other orders cancel each other.

Next, we consider a case of the 6-cylinder engine with center feed manifolds. For this case, the sound waves of the half order cancel each other due to structural symmetry. But all of the sound waves of the order with the whole number remain, such as 1st, 2nd, 3rd, etc.

Finally, we consider the case of a 6-cylinder engine with end-feed manifolds. For this case, the sound waves of the half order and whole number order retain.

The configuration and the length of manifold determine the composition of the order number of sound pressures at the surge tank. The order number of sound pressures at the orifice is similar to that of the surge tank. The sound quality of intake sound depends on the order number features; therefore, the design and selection of manifold had a significant influence on the sound quality of the intake system. The manifold design with equal length remains the firing order, it is an ideal option for most of the passenger car, whereas the manifold design with end feed manifold retains half order component, it is an ideal choice for the sportive car.

Besides serving as a cleaner, the air cleaner box also functions as an expansion silencer. The volume of air cleaner box determines the transmission loss and center-specific frequency. Usually, the volume of expansion silencer needs to be at least three times of engine cylinder volume, to attain ideal silencing effects. For many passenger cars, the volume of expansion silencer is in the range of 5–10 L. Basically, the larger the volume, the better the silencing effect.

There are two factors influencing transmission loss; the expansion ratio, m and the length of the box, L. In the design of the ducts and air cleaner box of the intake system, the penetration of ducts into the box also affects transmission loss.

Usually, the expansion ratio is expected to be as large as possible. There are two approaches to increase expansion ratio; one is to reduce the size of the duct, the other is to increase the cross-section area of the air cleaner box. The former is limited by power loss; the latter is constrained by the packaging space.

A portion of the tube can be used to regulate the transmission loss at certain frequencies. The real application of protrusion duct is constrained both by the filter inside air cleaner box and by power loss. The acoustic performance of a silencer can be characterized by the value of insertion loss, which is the difference between the sound pressure levels at the certain specified measurement point with and without employing the silencer. The insertion loss is the noise reduction attained by putting the silencer inside the original system. Insertion loss is a more effective measure than either transmission loss which is the ratio of the incident acoustic power to the transmitted acoustic power, or the noise reduction.

Side branch silencers consist of Helmholtz silencer and quarter-wave tuner, both of which are used to abate narrow-band noise. Helmholtz silencer is more effective than the quarter-wave tuner, as its effective frequency band is wider than that of the quarter-wave tuner. Helmholtz silencer is mainly used to abate the noise of relatively low frequency, whereas a quarter-wave tube is used to abate the noise of relatively high frequency. The installation location of resistive silencer should be at the antinode of the acoustic modes of the system. Certain empirical results show that the proper volume for regulating Helmholtz silencer is given by frequency (Hz) × volume (L) > 300.

Each Helmholtz silencer can only be used to abate one noise peak and the adjacent noise.

The factors influencing the performance of quarter-wave tuner include the length and cross-section area. The specific frequency of transmission loss depends on the length of the quarter-wave tuner, whereas the magnitude of the transmission loss depends on the ratio of the cross-section area of the quarter-wave tuner to that of intake duck.

The configuration of the quarter-wave tuner does not influence the performance of noise abatement, so it can be manufactured to be curved for packaging.

The design of the intake system by using silencer includes the following steps:

1. Characterize the noise property of the engine. As a preliminary analysis, a real straight pipe is used to replacement for the intake duct to be designed. Allow the real straight pipe to connect with engine flexible connecter and measure the orifice noise. Compare the measured noise level with the target noise level to figure out the necessary transmission loss curve for the intake system.

2. Design a manifold system in terms of sound quality requirement.

3. Design air cleaner box, to set sufficient volume.

4. Based on the intake noise level of the intake system after the air cleaner box is included, the silencers are selected. Helmholtz resonator is used to abate the noise of relatively lower frequency. Quarter wave tuner is applied for the noise of high frequency.

Further, we brief an example to illustrate the design process an intake system for a 4-cylinder engine of a passenger car. Its manifold system was designed with equal length type.

Some nonlinear acoustic behavior of intake system cannot be quantified by the above simple linear theory and advanced theory has to be used. To analyze the complicated situation such as area discontinuities, some nonlinear model or three-dimensional linear studies has to be performed.

The intake sound is a source of both unwanted noise and wanted sound. Wanted sound is usually some engine order related sound and can be a good contributor to sound quality. Unwanted noises are detrimental to sound quality. These noises can be controlled by silencer along with the airflow. A typical intake system can have at least one silencer and possibly a handful of them. The diameter of the pipe, characteristics of the air filter box and the intake manifold design are key design parameters to control the intake sound. Many intricate noises such as air whistle can also be due to the intake system. The coupling of shear layer instabilities with the acoustic resonances at the interface of two ducts, the main duct and a connecting side-branch, leads to whistle noise. The shear layer instabilities and the resulting vortices at the interface of the main duct and a connecting side-branch can couple, under discrete flow/geometry conditions, with the acoustic resonances inside the closed side-branches.

Such coupling can generate high-pressure amplitudes in both the side-branch and the main duct, leading to disturbing noise outside the main duct. Inherent in the coupling is the selective mechanism between the discrete vortex modes and the discrete quarter-wave modes, leading to a distinct narrow band, or pure tone noise, called whistle noise. Since the two ends of the intake system are fixed on respectively engine and chassis, it is of significance to optimize intake system to reduce the vibration transmission from engine to body. A flexible connector is usually used in the intake system, the vibrations from the engine can be filtered and its transmission to the body is reduced. The Helmholtz silencer and quarter-wave tuner are respectively installed on two trays that connected with vehicle body through rubber isolators.

The vibration of the silencer shell structure induced by high-speed airflow could radiate strong noise. In design, the natural frequencies of silencers should be higher than that of excitation. For instance, the natural frequencies of the silencer structure should be higher than 300 Hz for the 6-cylinder engine. The natural frequencies of silencer structure can be enhanced by adding rib, avoiding using a large plane, and increasing thickness. The radiation can be reduced by modal optimization or using dual-layer structures. The stiffness of trays should be high enough.

The exhaust system is one of the main exterior noise sources of vehicle. The generation of exhaust noise is mainly due to the periodic pressure pulsation of the burnt gas through the exhaust manifold. The lowest frequency of exhaust noise spectrum is given by the number of exhaust charges per cylinder per second, corresponding to the firing frequency. Combining the exhausts of several cylinders into one exit does provide some cancellation which increases with the number of cylinders. The spectrum of exhaust noise could show a series of harmonics peaking at integral multiples of the firing frequency. Other intermediate peaks may be present which modify the quality of the sound. The most effective method of abating exhaust noise is to use silencers as the treatments in the intake system. Both reactive and dissipative silencers are widely used. The silencers are usually designed to meet twofold needs such as to reduce the overall sound pressure level and to improve the sound quality.

The exhaust system is an assembly of a series of parts from the engine exhaust manifold to the tailpipe.

An exhaust system can perform many vibrations subject to the engine types (in-line or v-type) and applications. There are many configurations of exhaust systems in terms of input and output arrangement, such as one input/one output, one input/two output, two input/one output, two input/two output, and the system with two independent one input/one output. Exhaust system usually has two silencers. The front silencer is a reactive silencer, which is mainly applied to abate noise of specific frequencies. The rear silencer could be either reactive or combined silencer with both reactive and dissipative function which is applied to abate the noise with relative wider frequency band.

Similar to the intake system, an exhaust system has to spend some engine energy to allow exit burnt gas to overcome the backpressure at the tail orifice. The larger the cross-section area of the exhaust pipe, or the smoother the gas flows, the smaller the energy loss. However, the noise reduction needs a smaller cross-section area and thus results in larger resistance to gas flow. Therefore, the design needs to attain the trade-off. The design has also other constraints such as the vibrations transmitted to the vehicle body, air pollution and system reliability, cost, weight, installation, packaging and maintenance, etc.

The operations of engine results in the pressure pulsation of airflow in the exhaust pipe. The propagation of air pressure generates noise.

The outlet of the flow directly generates noise to air due to its pressure pulsation waves. Moreover, in the exhaust process of burnt gas, the flow could exhibit unsteady state and impinge the duct system to cause vibrations, both of which form noise sources. The noise directly from pressure pulsation has discrete frequency components, and the flow-induced noise which is due to turbulent and vortex effect could be both broadband and tonal, and the structural noise due to the gas flow impinging could have correlated with structural modes. The unstable flow in exhaust pipe causes excitation to pipe structures, creating structure vibrations, and flow-induced structure born noise. A large change in cross-section area is likely to cause impact effect. The magnitude of the radiated noise depends on the geometric dimension, configuration, and stiffness of the structures. The frequency of noise radiated from silencer is comparatively low, whereas the frequency of noise radiated from a pipe is high. The approaches to abate noise radiation include to reduce the sound of gas flow and to optimize the characteristics of structures, such as applying sandwich plate and damping materials. Engine vibration can be transmitted to the structures of the exhaust system, thus causing the vibrations of the structures of the exhaust system. The silencers, catalyzer, and pipe are made of shell and plate structure and their vibrations radiate noise.

Besides the above-described pulsation noise and flow friction noise, there are other kinds of exhaust noise source subject to the system. For instance, whistle and flutter noise are two typical cases. The whistle is a disturbing sound with a narrow band high-frequency (above 1000 Hz) aerodynamic sound. There are certain cases such as a flow along a perforation of a tube where whistle takes place, in which whistle becomes audible around 4500 rpm. Flutter is a disturbing sound occurs very frequently in the low rotating speed range (2000 rpm) when the engine is in open throttle run-up operation under full load. It is an intermittent sound due to the very large load at low rpm. This kind of disturbing sound has a frequency range between 300 and 10,000 Hz.

Silencers applied in exhaust system have two types, namely, reactive and resistive. Reactive silencer basically can allow part of the incident sound energy to be reflected toward the sound source, thus suppress the transmitted sound. Resistive silencers make use of acoustic absorptive materials to convert sound energy to thermal energy, thus suppress the transmitted sound. For pure-tone type noise from the engine, the reactive silencer can be effectively used to reflect the incident noise. As used in the intake system, typical resistive silencers such as expansion chambers, Helmholtz resonator, and quarter-wave tuners are also used for the exhaust system. For non-periodic and wideband noise, the resistive silencer is used efficiently or noise abatement through noise absorption.

Unlike the intake system, the exhaust system has a larger temperature gradient along its length direction. At engine' manifold location, exhaust gas temperature could be as high as 700°C or even higher. It may reduce to 300°C under some operational condition. As sound speed is strongly dependent on temperature, the sound frequency and wavelength vary with temperature. Usually, a silencer is designed to abate noise of specific frequency, however, when it is used at a different temperature, its design needs tuning to accommodate the change of application conditions.

Reactive silencer based on impedance mismatch is also called a reflective or passive silencer. There are many kinds of reactive silencers used for the exhaust system. The configuration types of silencers include sudden contraction; sudden expansion; extended outlet; extended inlet; reversal expansion; reversal contraction; concentric tube resonator; cross-flow expansion; cross-flow contraction; perforate reversal expansion; perforate reversal contraction; cross-flow; closed-end three-duct chamber; reverse-flow, closed-end, three-duct chamber; cross-flow, open-end, three-duct chamber, etc.

In the exhaust system, quarter-wave tuners are usually combined with duct (pipe or expansion chamber), due to the limit of installation space.

For perforate silencer, its transmission loss and specific frequency depend on the perforation diameter and total area. If the perforation is small, the silencer works as a Helmholtz resonator. If the perforation is big enough, it behaves like an expansion chamber.

The silencers are based on the principle of conversion of acoustic energy into thermal energy through highly porous or fibrous linings. It is called resistive or dissipative silencers or mufflers.

A resistive silencer is a silencer with installed acoustic absorptive materials. If porous materials or fiber materials are deployed inside the silencer, when sound waves pass these materials, sound energy is absorbed or dissipated by the materials and is transferred to thermal energy. The resistive silencer is capable of abating high-frequency noise and wideband noise.

The noise abatement performance of a silencer depends on the attributes of materials such as structure density, perforating, etc.

For low-frequency noise, the resistive silencer is not effective, and the reactive silencer should be used. The absorption coefficient of material increases with the increase of frequency. The material density has a substantial effect on noise absorption at high frequency.

Usually, tailpipe noise is mainly composed of pulsation noise. In the exhaust process of burnt gas, steady air sound as a plane wave propagates in a pulsation form in pipe. When it reaches tailpipe orifice, the flow disturbs outside as a piston. The real effect is a monopole sound source.

Tailpipe noise may consist of two kinds of noises: air pulsation noise and flow-induced friction noise. Steady flow pulsation creates air pulsation noise and unstable flow generates flow-induced friction noise.

The composition of these two kind noises depends on their respective magnitudes of flow volume and speed. For the case of low flow volume and low speed, air pulsation noise is dominant; for the case of large flow volume and large speed, flow-induced friction noise is dominant.

The application of dual pipes after front silencer (two pipes and two rear silencers) can reduce friction noise significantly.

Due to the acoustic impedance changes between silencer and tailpipe, and between tailpipe and ambient environment, the exhaust flow forms incident wave and reflective wave which finally forms standing waves. When the volume of the silencer is larger enough, the interface between silencer and tailpipe can be treated as an open end. Then the tailpipe can be considered as having free-free boundary condition, and the standing wave frequency can be estimated. The standing wave frequency of free-free end pipe only depends on sound speed and the length of the tailpipe. As sound speed is close to a constant, the frequency of noise radiated by tailpipe can be regulated by adjusting pipe length. In a real application, if the magnitude of a pure-tune noise is larger, it can be abated by regulating the length of the tailpipe.

Moreover, the cross-section area of the inner pipe and tailpipe cannot be designed too small. Otherwise, it causes larger flow friction noise. Typically, the flow speed is controlled in the following range to control flow friction noise: for center pipe, $M_c < 0.35$; for tailpipe, $M_c < 0.25$.

The configuration of the exhaust manifold has a significant influence on the sound quality of tailpipe. The exhaust manifold can be classified as equal length pipe type, center connection type, and tail connection type. The effect of the type of exhaust manifold on sound quality is the same as that of intake. For the V engine, Y pipe has a more significant effect on the sound quality of tailpipe. Next, we discuss the effect of Y pipe of a 6-cylinder engine by assuming that manifold is equal length, and we consider the pipe without using a silencer. Assume the length of sub-pipes of Y pipe are respectively L_1 and L_2, $L_1 = 100$ cm. Consider the designs with different length, $L_1 - L_2 = 0, 20, 40, 60,$ and 80 cm, respectively.

For sound quality design consideration, it is usually to allow the two branch of Y pipe to be identical, to allow the firing order noise to be dominant, and the half order and the other integer order components are very small. This way can allow passengers to have comfort and harmonic sound perception. However, for sportive car design, the half order components need to be magnified, to allow the driver to have exciting and powerful sound perception.

The silencer volume of exhaust system refers to the sum of all silencer volumes. Typically, it equals to the sum of the volumes of front silencer and rear silencer. Usually, for an exhaust system, the larger the silencer volume, the better the noise abatement performance. The tailpipe noise mainly depends on the silencer volume. To attain competitive noise level, silencer volume is usually required to be close to the 10 times of the total volume of a cylinder or even larger.

For silencer system design, the closer the silencer to engine, the better the noise abatement performance. However, due to the packaging limit, it is difficult to install silencer close to the engine. Normally, the front silencer volume (in front of the rear axle) should be at least larger than the cylinder volume by two times.

The design of a silencer of exhaust system includes the following steps: (1) characterize engine parameters, vibration and noise attributes, such as the number of cylinders, exhaust temperature, a stroke of the exhaust valve, pressure wave of the cylinder, firing timing, etc. to determine the noise order, noise magnitudes; (2) for V engine properly chose Y pipe; thus determine the half order component and evaluate the associated sound quality; (3) determine insertion loss and silencer volume; (4) determine the numbers and location of silencers, generally, the more the silencers, the better the noise abatement performance, and the inner pipe length can be shorter thus reducing the effect of standing waves; (5) verify power loss; (6) tune the inner design of silencers, to alter the frequency distribution of transmission loss, to meet the frequency requirement; (7) determine tailpipe design.

Tailpipe noise is determined by the properties of the engine noise source and the insertion loss of the exhaust system. In the design process for noise abatement, tailpipe noise is measured before silencer's installation; then the desired insertion loss is determined by comparing this measured noise value with targeted level. The desired insertion loss can be realized by selecting a system with catalyzer, pipe, reactive silencer, resistive silencer, tailpipes, etc. The desired insertion loss of exhaust system consists of the sum of the insertion losses of the individual components. The design of exhaust systems requires to ensure that the specified back-pressures levels (typically 20–35 for engine <1.4 L; 20–40 for engine <2.5 L; 30–60 for engine >2.5 L) are not exceeded to avoid large power loss, and also meet the other targets such as noise and vibration, weight and volume, and to optimize the overall design.

The transmission loss of silencer is a function of frequency. For the 3rd order noise, the excitation frequency can be determined by f = speed (rpm)/20. Therefore, we can determine the noise level for each specific frequency and then figure out the difference between actual noise and target value at a specific frequency.

Next, we add the second silencer with a specific frequency of 250 kHz to reduce the corresponding peak.

In the actual system, the series type, independent silencers are rarely applied directly. Usually, they are compacted into one enclosure or are used as a combined type.

Tailpipe noise is one of the main contributors to interior noise. On the other hand, we need to reduce it as much as possible. Tailpipe design is expected and sometimes even required to be beneficial to the sound quality of a vehicle. However, the exhaust system can be the source of other unwanted noises. These include order related noises that can be damaging to the sound quality metrics of the vehicle and/or exterior noise requirements. Control of the tailpipe noise is known to be directly related to the silencers, the architecture of the exhaust system (equal length pipe versus unequal length pipe for V6 and V8 engines), internal silencer design and the energy losses. Engineers are usually required to control the noise with limited volumes of silencers, subject to the limit of flow, and power loss.

The engine vibrations can be transmitted to the exhaust system to cause system resonance, which is then transmitted to the vehicle compartment through hangers.

The sources to generate noise in the exhaust system includes the engine vibrations, fluid flow-induced impact, sound wave pulsation excitation and vehicle body vibrations. The engine vibrations can be directly transmitted to the exhaust system; the cylinder-excited flow results in impact excitation to the exhaust system. Particularly the vortex formed around pipe bending position is strong excitation to pipe vibrations. The propagating sound wave in pipe leads to vibrations and resonance of structures. The vehicle body vibration can interact with the exhaust system through hangers.

The vibration modal analysis is necessary for exhaust design. Since the exhaust system relates to engine and vehicle body, the natural modes of exhaust system must be designed to avoid coincidence with the modes of the vehicle body and the exciting frequencies of the engine. From the vibration modal analysis of the exhaust system, the nodes and anti-nodes of the exhaust system can be identified. The location of hangers is usually set at node positions of the exhaust system. Usually, the following several modes need to set targets: the first-order vertical bending modes; the first order horizontal bending modes; the first order horizontal torsional modes and the mode density. The first order vertical bending modes and the first-order horizontal bending modes are the two modes that are most susceptible to engine excitations. And these two kinds of vibrations are easy to be transmitted to the vehicle body to cause resonance. In four-wheel drive and all-wheel drive (AWD) cars, the exhaust system may share identical bracket/fixture with transmission shaft. Therefore, the natural modes of the exhaust system also need to avoid the coincidence with that of the shaft system.

Engine's idle speed is typically 600 rpm. For 4-cylinder engines, the firing frequency corresponding to 600 rpm is 20 Hz; and for 6 cylinder engine, the firing frequency corresponding to 600 rpm is 30 Hz. For a typical exhaust system, the modes with a frequency of fewer than 20 Hz are unlikely to be excited, whereas the dynamic response with frequencies higher than 250 Hz is usually small. As such, the typical frequency range for modal analysis is from 20 to 250 Hz. Usually, the design should allow the system to have as fewer natural modes as possible, to reduce the likelihood of resonance. The pipe should be arranged as straight as possible. Moreover, vibration response analysis is also needed to evaluate the system vibration and the force transmitted to the vehicle body. In the calculation, the engine force acted on the exhaust system needs to be estimated, which could be obtained from either measurement or calculations.

For a typical car, the peak of the hanger-transmitted force is in the range from 2 to 10 N. When this force is too larger, the vibrations transmitted from hanger structures and the noise affect customer's perception. The hanger of the exhaust system is usually connected to a vehicle body through isolator. The transmitted force can be measured, but the direct measurement is quite difficult. The practical procedure is to measure the accelerations on both sides of the isolator, then to figure out the displacements of two sides.

The transmitted force is equal to the product of difference of the displacements and isolator stiffness. The transmission rate is usually used to evaluate the effect of the isolator. The transmission rate is referred to as the ratio of the vibration magnitudes of the driving side to that of the driven side. The larger the transmission rate, the better the isolation effect of the isolator. Usually, when the transmission rate is larger than 20 dB, the isolator can be assumed as acceptable, which means that the vibrations need to decay 10 times.

The force transmitted to the body depends on both transmission rate and the vibrations of the exhaust system, If transmission rate has reached 20 dB, the transmission force has not been suppressed below target, then further effort needs to be made to reduce the vibrations of the exhaust system, such as using flexible pipe (transversal arrangement, with crankshaft axle, is perpendicular to exhaust pipe), or using a tuned mass damper. The hanger/bracket stiffness is very crucial to attain the vibration isolation effect.

Sometimes, the exhaust system can cause structure born noise of compartment interior such as the low-frequency boom and moan due to the vibrations of the front exhaust system (hot end).

The front portion of the exhaust system (hot end) relates to the engine on one side, and with the flexible connector on the other side. It behaves like a cantilever. This cantilever could perform "swaying" motion under the excitation of the engine. When the vibration is high enough, it leads to one or several resonances, which is likely to be transmitted into the compartment. The countermeasures include applying tuned mass damper (dynamic absorber) or adding bracket/hanger.

3.3 **Driveline NVH**

A driveline usually consists of the drive shaft, half shafts, mounted gear systems, bearings, and brackets as well as other components. For rear-wheel drive (RWD) and AWD vehicles, the driveline includes a front driveshaft, a rear driveshaft (if more than one shaft is used), a front axle, a rear axle, half shafts, and other components. In front-wheel drive (FWD) applications, half shaft defines most of the driveline system.

As an example, a typical driveline system used for a 4-wheel drive system. It consists of transmission, differentiator, front half shaft, rear half shaft, front and rear axles, support bearing and universal joints, etc.

Mounted gears include rear axles in RWD vehicles (typically with north-south engine configurations) or all-wheel-drive versions of originally FWD vehicles; transfer case which is a mechanism that exists in the AWD version of FWD vehicles. It consists of gears and a chain system; the power take-off unit. This is a bevel gear that exists in AWD versions of a FWD vehicle. It is close to the transmission.

In the analysis of power plant vibration isolation, engine and transmission are usually considered as a whole block. On the other hand, in the analysis of the driveline system, the transmission is also included for completeness. The driveshaft relates to the transmission which influences the vibrations and noise of the driveline system.

The vibration excitations to driveline could come from the transmission, differentiation, driveshaft, axles, half shaft, and universal joint. The transmission consists of a lot of gears. The imperfect meshing of gears results in vibrations. This could occur for drive axles and differentiation as well. Driveshaft and half shaft are rotation parts with a larger moment of inertia. When the center of gravity does not fall in the rotational centerline, centrifugal force will be generated, which excites the whole system. The system of the driveshaft and half shaft have bending and torsional vibration modes. When they are identical to the engine' excitation frequency, system resonance takes place. The application of universal joint could cause the speed variation of the driveshaft, which leads to the secondary oscillations of the system. Moreover, the gaps in gear joints could result in impact, vibrations and rattle noise. There are a lot of gears applied in transmission, drive axles, and differentiation. The operation of imperfect gears directly generates meshing noise and impact noise.

Driveline system is connected with frame or vehicle body through bearings and isolators. Bearings installed on shafts are used to support shafts and to enhance system natural frequencies. Isolators/mounts of axles, transmission, and power differentiation are used to support the respective system and to reduce vibrations.

As bearings and isolators are connected with driveline and vehicle body, the vibrations of driveline are transmitted to body interior through bearings and isolators. The usually perceived noises in vehicle interior include several types: boom due to the excitations related to driveline order; moan due to the excitation related to engine firing order, pure tone type gear meshing noise, and some squeaks and rattles [97–104]. The perceived vibrations in the interior include the vibrations of the floor, driving steer wheel and seat.

Usually, the magnitudes of vibrations and noise of driveline are larger in the rear-drive vehicle than that of front-drive vehicle, as the rear-drive vehicle uses longer driveshaft. The front-drive vehicle uses relatively shorter shaft, or even without shaft, which renders it to have relatively higher natural modes that are hard to be excited. The vibrations and noise of driveline in the whole wheel drive vehicle have the characteristics of vibrations that appear both in front and RWD vehicles.

In shaft design, the bending and torsional natural frequencies should be higher than the specific frequencies corresponding to the maximum rotation speed of shafts, to prevent the resonance caused by the rotating unbalance force. The secondary motion of universal joints which is proportional to the angle between driving and driven shafts affects system vibration. The natural mode frequencies of subsystems, such as shafts, half shafts, axles and transmission, bearings, isolators, and body acoustic modes should be designed or tuned to avoid coincidences.

The transmission uses gears to transmission power and motion. The transmission error of transmission gears may result in shaft vibrations and the vibration of transmission. The transmission error of the gears in axles and differentiation case may result in shaft vibrations and may be transmitted to half shaft and cases. The vibrations are finally transmitted to the body through shaft bearings and isolators. The meshing error also directly leads to tonic noise which transmitted to the body through the case of transmission and axles. This is usually perceived as whine noise as discussed in the preceding section.

In this section we discuss the effect of unbalance, run out, second-order excitation, unbalance and critical speed of the shaft.

The driveline unbalances forces are the results of individual unbalance effects and are related to the driveline rotation speed. The first order excitations are the results of driveline unbalance. Driveline would not produce disturbances to its bearing supports when a driveline is an ideal axis-symmetric structure and its rotation axis coincides with the principal axis. However, this ideal case is usually unachievable. The following reasons may cause the unbalance of shaft: the driveshaft and other rotating components are not axis-symmetric; driveshaft is not supported on its supporting members; driveshaft is not straight; driveshaft has deformation (as a result of weight); interface unbalance (run out), etc., some of them are detailed as follows.

The following factors may allow a rotating shaft to create run out and to create an exciting force at supporting points: the gravity center is not identical to the rotation center; the geometric center is not identical to rotation center; shaft has geometric unsymmetrical; bending deformation due to the elasticity of shaft; tolerance and gap variations of shafts. The centrifugal force due to run out is an excitation acting on bearings, which is finally transmitted to the body. The run out could be due to the misalignment of connected shafts. When power is transmitted by several shafts that are connected by joints, the rotation center of different shafts could be non-co-axial, which leads to run out. There are two kinds of misalignments; one is the in-line misalignment, the other is an inclined misalignment in which two shafts axis form an inclined angle. Usually, the run out of shafts of a passenger car is required to be controlled below 25 g-cm.

At first, we consider the case in which the geometric center is identical to the rotation center, only gravity center has deviation. Assume that the shaft mass is m, the deviation of gravity center from rotation center is r, then the centrifugal force is

$$F = mr\omega^2 \qquad (3.3)$$

The secondary excitation due to the universal joint is briefed as follows.

The first order vibration of shafts are mainly due to the excitation of unbalance of shafts. The secondary order vibration are due to the excitation of a universal joint (unequal speed).

In one rotating cycle, driven shaft speed could change twice between maximum and minimum. This speed variation leads to the parametrically excited vibrations of the system. The induced vibrations are referred to as secondary vibrations. The magnitude of the speed variation depends on the angle between the two shafts. The larger the angle, the larger the speed variation. The secondary vibrations are salient at vehicle start-up process.

Radical run out is one of the main factors leading to shaft vibrations. The real shaft system usually needs balance treatment. For an unbalance on a disk, it can be identified by synchronizing the measurements of the acceleration of support bearing and the measurements of reference position on disk using an optical sensor. Once the unbalance is identified, a counteract mass can be added in the point opposite to the unbalance concerning the rotation center.

A real shaft usually has distributed unbalance along the longitudinal direction. Therefore, multiple plane balance treatment is needed. The practical approach is to assume the total unbalance of the system to be decomposed on two reference planes by using the principle of total unbalance equivalence. In applications, the counter mass to be added for balancing will be only applied to the two reference planes. For single shaft driveline, the unbalance can be evaluated by using accelerometer mounted on transmission case and axle's case. The balance masses are usually added on the surfaces of the shafts at two locations close to the two ends.

For the whole wheel and four-wheel-drive system, the balance masses are usually added at the locations close to front transmission, front axle, rear transmission, inner bearing, and rear axles.

The procedures for balancing a shaft are as follows: measure the unbalance forces on two reference plane; add small mass on the first plane, then measure the unbalance force on two planes; remove the added small mass from the first plane, and add a small mass on the second plane, then measure the unbalance forces on the two plane; then figure out the two needed masses to attain balance, and add the two small masses on the corresponding locations. Usually, in design, manufacturing and installation of the driveshaft, the unbalance should be reduced as much as possible. The balancing treatment or even deploying dynamic absorber is also required if necessary.

The modal analysis of driveline system includes the modes of transmission shaft, the bending modes of the whole driveline system, the modes of the bracket of center bearing, the modes of axles, etc. These modes need to be verified to avoid mode coincidence and to have enough separation from vehicle body modes.

In operation, the shafts undergo the excitations of both torque modulation and unbalance, thus both torsional and bending vibrations are important. For torsional vibrations, the analysis procedure is like that of an engine crankshaft system.

When the rotation speed of a shaft is sufficiently high, the unbalance excitation frequency could be identical to the bending natural frequency of shaft. Under this situation, resonance occurs, the corresponding specific speed is referred to as critical speed. Operation at critical speed is likely to lead to failures of shaft and system.

The first order critical speed can be calculated using a formula. There exist its harmonics. If unequal velocity universal joints are used, the secondary excitation should also be considered. The maximum speed of shaft depends on driving speed instead of engine speed. From the maximum driving speed, we can derive out the maximum speed of the shaft. In real design, driveshaft natural frequency should be higher than the specific frequency of critical speed by 15% or above. This is usually attained by following approaches: increasing stiffness, or reducing mass, or applying two or multiple shaft designs to enhance natural frequency; improving balance; separating bending and torsional modes by at least 15%.

The vibrations of the driveline system and mode decoupling are briefed as follows.

In addition to the bending and torsional modes of individual parts and subsystem, the bending modes of the whole driveline system are also critical to determine the system's vibrations.

The bending mode frequency of power plant is required to be higher than the first-order natural frequency of the shaft system by 10%, otherwise, the unbalance of driveshaft could lead to the vibrations of the whole system. This is likely to occur when speed is high.

The structures and systems connecting with driveline system include body/frames, power plant, suspension systems, and exhaust systems. Usually, there are isolators installed between these systems and driveline to reduce the vibrations transmitted to these systems. In the driveline system, adjacent parts should have their mode with sufficient separation. For instance, the modes of axle case should be separated from that of the driveshaft, which requires that the axle case has sufficient stiffness; the axle case modes should be separated from that of body and suspension system. The modes needed to be taken into account include the bending and torsional modes of driveshaft; the bending and torsional modes of driveline system; power plant bending modes; vertical modes of center bearing; the modes of axles; the modes of the body; the modes of frames; the modes of body acoustic; the vertical and lateral modes of steering wheel-column, etc.

One of the approaches to reduce vibration and noise of the body is to change the vibration transmission path to attain minimum transmission. Conventionally, the axle of driveshaft uses three mounts. The improved design with four mounts could improve vibration transmission. The application of hydraulic mount could further improve the performance.

In the driveline system, the above-described excitation sources causing a variety of vibrations and noise. Some of these vibrations and noises associated with a different part of driveline are schematically illustrated in the **Figure 3.6**, which includes shaking, rattle, shuffle, shudder, clunk, boom, moan, and whine.

The gear whine due to meshing referred to as axle whine (when it originates at the rear axle). Axle whine occurs when forcing functions due to transmission error resonates with any of the natural frequencies (driveline case modes, axle case, etc.). Torsional vibration in the driveline system can cause unusual and unwanted rattle noise. The clearance between the teeth in the gear systems in idle condition causes rattles.

One of the most problematic issues with driveline systems is the critical speed of the shaft. If the shafts are driven at their bending frequency, resonance will occur, which leads to shaft failure and possible detachment from the vehicle. This can cause rollover and safety concern. Thus shafts are designed such that their first bending frequency is higher than critical speed by at least 15%–20%. Another phenomenon that frequently occurs is powertrain bending, which is fundamentally driven by shaft bending. It is known to cause vibration fatigue and durability concerns and higher warranty cost. Powertrain bending mode frequency is required to exceed the firing frequency by a certain margin to avoid durability problems [105]. There are other problems in the driveline system which are called driveline boom and driveline shudder. During torque reversal, driveline components accelerate through various lash zones and the impacts across such lashes can cause high-frequency noise (300–6000 Hz). The clunk is perceived as short duration jerk and may be accompanied by a low-frequency boom (20–50 Hz). The uncomfortable vehicle shuffle is due to the sudden load changes (pedal tip in or out), which is related to the first-order natural frequency of driveline and is of 2–8 Hz.

FIGURE 3.6   Schematic of vibrations and noises originated in the driveline.

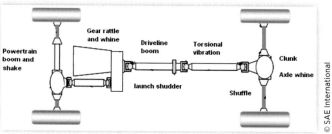

© SAE International

FIGURE 3.7   Plot of the specific frequency range of vibration and noise in driveline.

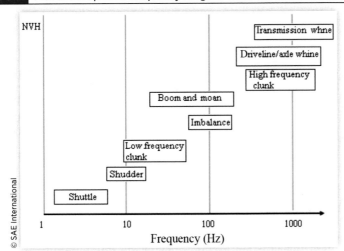

TABLE 3.2 Summary of the driveline vibration and noise phenomenon, their frequency ranges and possible root causes.

NVH phenomena	Frequency range	Root cause
Boom	30–80 Hz	Imbalance
Roughness	20–80 Hz	Imbalance
Moan	80–300 Hz	Imbalance/driveline angles
Whine	300–2000 Hz	Gear meshing forces
Take-off shudder	10–30 Hz	Suspension/axle wind-up Excessive driveline angles
Shuffle	2–8 Hz	Load change/resonance
Clunk	Wideband	Vibroimpact
Idle shake (FWD)	20–40 (idle firing order)	Engine firing

The dynamic meshing of a rear axle leads to whine in the frequency range of 300–1000 Hz. The transmission whine goes up to 3–4 kHz. **Figure 3.7** is the plot of the frequency range of major driveline vibration and noise.

Table 3.2 summarizes the driveline related vibration and noise phenomenon in a typical application, together with their frequency ranges and possible root cause.

3.4 **Vibration Isolation of Powertrain System**

The common approach to reducing power plant excitation to the vehicle in isolation. To evaluate power plant isolation, power plant and supporting mounts system is usually modeled as a six-degree-of-freedom system. The model integrates many parameters such as power plant moment of the inertial, mass, center of gravity, mount stiffness, and locations. The locations of mounts are constrained by packaging limitations; thus they must be chosen in the early phase of vehicle development. The isolation system must be designed to satisfy following requirements such as mounts should be rigid enough to statically support powertrain; mounts should be resilient enough to isolate power plant vibration from being transmitted to the body; the mounts should have the resistant capability to counteract the torque of power plant output and inertia unbalance; the mounts should have the resistant capability to counteract the dynamic forces during acceleration and deceleration process due to starting, stopping, cornering, road inputs, truck/rail shipment, etc.; the mounts should have shock resistance capability to suppress high-frequency vibration resulting from power plant structural elements; and the mounts should have proper damping capability to absorb vibrations from the road and tire/wheel excitation.

The requirements are usually in conflict with each other. For instance, high mount stiffness is required for shock isolation at low frequency, but low stiffness is required for vibration isolation at high frequency. The dynamic stiffness of mounts usually varies with frequency. Conventional rubber mounts may not be able to realize the ideal design. Well-designed hydraulic, semi-active, and active mounts can be used for satisfying both conditions.

If an engine has been chosen for a vehicle, and the idle condition has been specified in the development, then the lowest firing frequency is known. Then isolation design can be conducted based on the excitation frequency and the system natural frequencies. The first objective in design is to determine rigid modes of the power plant. The roll frequency is determined and roll vibration is isolated first as the roll motion is comparatively more important than the rest. The firing order excitation torque usually acts on the engine block along the axis of roll direction. Consider an isolation system of single-degree-of-freedom.

The vibration can be reduced only in the region where the ratio of the excitation frequency to natural frequency is greater than $\sqrt{2}$.

If the roll frequency, f_{roll}, satisfies $f_{roll} < f_{firing}/\sqrt{2}$, the roll vibration will be reduced significantly. Once the firing frequency is specified, it is desirable to design the roll mode frequency as low as possible. The lower the roll frequency, the lower the transmissibility attained. However, the very soft mounting system is impractical, this is because the mounting system has to satisfy other requirements: support power plant weight; react output torque; sustain dynamic acceleration and deceleration. To effectively isolate power plant roll excitation, in engineering practice, the frequency ratio is usually chosen to be in the range from 2 to 3.

The isolation evaluation using six-degree-of-freedom model is presented as follows.

If a system has been effectively isolated at the minimal firing frequency, it should satisfy the isolation condition for higher-order excitation. The roll mode frequency of an isolation system is desirable to be minimized sufficiently, otherwise higher vibration and noise levels will be induced. According to some experiences, 1 Hz increase in roll mode frequency of an isolation system could cause vibration increase at occupants perceived points by 4–6 dB.

For example, if the idle speed for a 6-cylinder engine is 600 rpm, the firing order is 3rd and the firing frequency is 30 Hz, i.e., $f_{firing} = 30$ Hz. In the next, we elaborate on how to determine the natural frequency of the power plant roll mode to allow it to meet the isolation requirement.

According to Equation (3.1), if firing frequency is twice of the natural frequency, the roll frequency should be f_{roll} = 0.5 Hz, f_{firing} = 15 Hz. If the firing frequency is three times of the natural frequency, the roll frequency should be $f_{roll} = f_{firing}/3$ = 10 Hz. Therefore, the roll mode frequency can be chosen between 10 and 15 Hz. The roll mode frequency of a 6-cylinder (east-west layout) engine is usually set between 10 and 12 Hz in engineering practice.

Besides the roll mode frequency requirement, acceleration reduction ratio, η, is used to evaluate the isolation effectiveness of mounts. The acceleration attenuation is defined as follows.

$$\eta = \frac{a_{active}}{a_{passive}} \tag{3.4}$$

where

a_{active} is the acceleration on the power plant side, or the active side

$a_{passive}$ is the acceleration on the side of vehicle frame or subframe, or the passive side

For a good isolation system, the ratio should be larger than 10, i.e., the vibration transmitted to the vehicle should be less than one-tenth of the power plant vibration. The criterion can be expressed in dB as follows,

$$20\log\eta = 20\log\left(\frac{a_{active}}{a_{passive}}\right) > 20\,dB \tag{3.5}$$

The roll mode frequency depends not only on the required transmissibility rate but also on the locations of the mounts. Poor location design could introduce extra load to the system. To effectively design the isolation system, the mount locations should be chosen to allow the power plant to move around its free-free roll axis of torque. The free-free roll axis is the roll axis of a rotating object for which there are no external restraints to constrain its motion. This axis always goes through the Center of Gravity (C.G.) and depends only on the axis of the applied torque and the inertial properties of the object. If the mounts are chosen to allow the power plant to move about the free-free roll axis, no additional forces will be generated within the mounting system to disturb its motion.

In many isolation designs, the power plant is supported on a subframe by three mounts. The mounts are not located at roll axis. Thus, the extra load will be induced in addition to the roll torque. In a typical mount system, power plant moves around its free-free roll axis of torque and two mounts pass through the roll axis. The extra load is avoided.

The upper mount is usually fixed on the vehicle body, such as on shock tower. This system is called the pendulum mounting system, which is the desirable system in mount design. However, it is not easy to secure the desired pendulum mounting design because of package limitations and as other requirements.

The power plant is usually supported on mounts. Mounts are usually installed on subframe/frame which connects with the body. Mounts design is critical to reducing engine vibration and noise and its transmission to the compartment. Its function includes (1) Sustain the weight of power plant which could be up to several 100 kg and should be equally supported by mounts; (2) When the engine is in operation, it transmits power to the driveline and it also creates force and torque to mounts. The mounts should be able to react this torques properly; (3) Besides, serving for vibration isolation, the mount should be able to reduce high-frequency structural vibration transmission; (4) When the vehicle runs in the transition process, such as acceleration, deceleration, cornering, on a rough road, or run over a bump, the vehicle undergoes impact excitation. The mount should have a proper function to reduce this impact. Some of these requirements are conflicting. Generally, the mount stiffness, particularly vertical stiffness is required to be high enough, to sustain weight and resist engine torque and impact from acceleration/deceleration. However, the mount stiffness should also be small enough to attain the vibration isolation function.

The factors influencing mount design includes (1) Body and frame structures; (2) Power plant gravity center, the inertia of moment, the axis of rotation torque, the output of engine force and torque as well as its frequency characteristics; (3) Isolation system model: power plant is usually modeled as a multiple-degree-of-freedom system, system natural mode frequencies and the mode-decoupling level should be taken into account for mount system target setting; (4) The number, layout and types of mounts (rubber or hydraulic). After having determined these factors, the location and stiffness of mounts can be preliminarily determined to attain target through testing or calculation analysis. Then the actual effects can be evaluated and validated, and further consideration is extended to reliability and robustness. Usually, the design consists of two steps: the optimal design of mount structures stiffness and powerplant-whole vehicle design.

In the phase of optimal design of stiffness structures of the power plant, the number and location of mounts, the stiffness and damping of mounts, and the installation of mounts will be determined. In the modeling of the vibration of the power plant–mount system, the power plant is treated as a rigid body. Each mount has one end connecting with a rigid body and the other end being fixed. This system needs to meet the requirements of mode decoupling, transversal rotation frequency target, and isolation transmission rate targets.

In the phase of power plant-whole vehicle design, the performance of the isolation system is evaluated in the context of the vehicle. When power plant is installed on subframe/body through mounts, the boundary conditions are changed, as the vehicle subframe/body is elastic instead of rigid. The natural frequencies of the power plant system will have some changes. In the whole vehicle response analysis, this should be taken into account.

There are several ways to arrange mounts. Mounts can be installed at different locations on the power plant. The total number of mounts used for different power plant is usually from 2 to 5.

For a passenger car, the installation location could be at the bottom of power plant or at the top of the power plant. Although the bottom installation has a relatively large space available, the isolation effect is not desirable. Consequently, the top installation has a relatively smaller space to be used, however, the isolation effect is desirable. Besides mounts, usually, there are one to two roll restrictors to create constraints to the transversal rotation motion of power plant. Its effect is along the longitudinal direction of the roll restrictor.

When an excitation acts on power plant to cause system vibrations, the vibrations usually consist of the contributions from different natural modes. In many cases, the vibrations from two or more modes could be coupled with each other. If the vibrations

FIGURE 3.8   Schematic of a powerplant-mount system of six-degree-of-freedom.

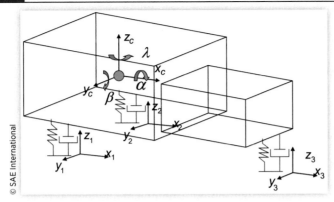

© SAE International

from different modes are coupled, it will be difficult to isolate the vibration in a specific direction, for example, the vertical direction. In design, it is desired to decouple the six natural modes of the power plant isolation system. The decoupling process is to make each mode motion to be independent of the other modes. A complete decoupling may not be possible and as such the practical objective is to minimize the coupling. The decoupling of power plant modes depends on the mount numbers, locations, CG, and moment of inertia of the system. Decoupling is another critical index to evaluate power plant isolation in addition to the roll mode frequency. Moreover, mode energy is used to measure the strength of the coupling between the modes.

Consider a power plant modeled as a rigid body and supported by mounts, as shown in **Figure 3.8**. The system has six-degree-of-freedom: three translation motions and three rotation motions. The equations of motion of the powerplant-mount system can be expressed as follows,

$$\begin{cases} M\ddot{x}_c = F_x \\ M\ddot{y}_c = F_y \\ M\ddot{z}_c = F_z \\ I_{xx}\ddot{\alpha}_c - I_{xy}\ddot{\beta}_c - I_{xz}\ddot{\gamma}_c = M_x \\ I_{xy}\ddot{\alpha}_c - I_{yy}\ddot{\beta}_c - I_{yz}\ddot{\gamma}_c = M_y \\ I_{xz}\ddot{\alpha}_c - I_{yz}\ddot{\beta}_c - I_{zz}\ddot{\gamma}_c = M_z \end{cases} \tag{3.6}$$

where

M is the power plant mass

$I_{xx}, I_{yy},$ and I_{zz} are the moment of inertia

$I_{xy}, I_{xz},$ and I_{yz} are the cross effect coefficients of inertia

$x_c, y_c,$ and z_c are the coordinates of the power plant gravitational center

α_c, β_c and γ_c are the rotational angular displacement around the x-axis, y-axis and z-axis, respectively

$F_x, F_y,$ and F_z are the summation of all forces acting on the body in x, y and z directions, respectively

$M_x, M_y,$ and M_z are the torques acting on system concerning the x-axis, y-axis and z-axis, respectively

Considering free vibration of the system, assuming a free harmonic motion, the motion equations will yield characteristic equation of eigenvalue,

$$\left(\left[K\right]-\omega^2\left[M\right]\right)\Phi=\left\{0\right\} \qquad (3.7)$$

where

[K] and [M] are respectively the stiffness and mass matrices of the mounting system
Φ is the eigenvector of the six-degree-of-freedom
ω is the natural frequency

Setting the determinant of Equation (3.7) to zero, the natural frequency and modal shape of the mounting system can be solved. The matrix of actual mode Φ can be expressed in terms of Ψ using target modes $\hat{\Phi}$ as a basis,

$$\Psi=\Phi^T M\Phi \qquad (3.8)$$

The modal energy matrix E is defined as the matrix of modal energies expressed as a percentage as follows,

$$E=100\Psi\Psi \qquad (3.9)$$

E has the property that the sum of each column or row is identical to the unit or 100%. The percentage of modal energy for each mode represents its coupling or decoupling levels with other modes. For example, if a roll mode's energy is 95% at its roll frequency, the roll mode can be considered as almost decoupled with the other modes. If the energy of a specific mode is only 50% at the corresponding specific frequency, then the other modes have the rest 50% modal energy at this frequency, thus this specific mode is not well decoupled.

Next, we describe the solution procedures on determining the modal frequencies and modal energy distribution.

Substitute the parameters in this example into the above mentioned modal equation and energy equation; the natural frequency, modal shape, and modal energy can be obtained.

Mount is used to connecting power plant and subframe/body through two brackets. If the brackets' stiffness is comparatively low, their function equals to add two springs to the system. The equivalent stiffness for the series connection of the three springs can be evaluated.

If the bracket stiffnesses are too low, the equivalent stiffness of the assembly will be different from the designed stiffness of the mount K_i. Thus the desirable requirements on roll mode frequency and mode decoupling cannot be met. The flexible bracket in a mounting attachment may result in resonance and isolation failure. Therefore, brackets on both mount sides must be rigid enough.

There are two rules that are used for bracket design guideline. The first is that the bracket stiffness should be 6–10 times of the mount stiffness. The second one is that 1st natural frequency of the brackets should be above a certain value (such as 500 Hz). The mount location should make the natural frequency within the effective frequency range for vibration isolation and make the power plant modes decoupled. The mount brackets must possess enough stiffness. Besides these requirements, the following conditions should be considered when the mount locations are determined; locations should be chosen to allow the power plant to move around an axis as close as possible to its free-free torque roll axis, mounts should be located at the points of low powertrain vibration, mounts should be located at points of low body sensitivity, mounts should be located near suspension attachment points for great absorption of suspension disturbances, and power plant vibration modes should be decoupled.

In isolation design, power plant system is usually modeled as a six-degree-of-freedom-system. This system can be decoupled to get six independent coordinates by using theoretical modal analysis. On the other hand, if a mount design is properly chosen, the system mode can be decoupled, and each coordinate's vibration can be treated separately, thus the problem is attributed to be the isolation problem of a single-degree-of-freedom system, which was discussed in the last section.

To some extent, the effect of power plant mount system on the whole vehicle is also similar to a dynamic vibration absorber: when the vehicle undergoes road excitation, the power plant-mount system can react as a dynamic vibration absorber concerning the rest part of the vehicle, which should have a positive effect on the reduction of whole vehicle vibrations. However, in the real design, usually, only the isolation function of mounts is considered.

Mount design should balance many requirements. Vehicle drivers receive undesirable vibrations through one of two possible excitation sources. The first source, from engine eccentricity, typically contains frequencies in the range of 25–200 Hz with amplitudes generally less than 3 mm. The second source of excitation originates from road inputs and engine torque during harsh accelerations. Road inconsistencies cause disturbances to the vehicle frame via the suspension system, whereas force accelerations cause excessive engine torque and motion at the mounts. Excitations of this nature are typically under 30 Hz and have amplitudes greater than 0–3 mm. During low-frequency high-amplitude vibrations, the ideal mount should exhibit large stiffness and damping characteristics to reduce relative displacement transmissibility whereas for high-frequency low-amplitude vibrations the ideal mount should have low stiffness and damping characteristics. These conflicting properties indicate that an ideal mount system has stiffness and damping characteristics dependent on the amplitude of excitation. Hydraulic engine mounts have been developed to address the conflicting amplitude- and frequency-dependent characteristics desired for automotive applications.

For ideal mounts, its stiffness should large enough at low frequency, and smaller enough at high frequency, to sustain weight and impact and attain ideal isolation at high frequency. This is also valid for ideal mount damping.

There are many types of mounts available, such as elastic mounts, hydraulic mounts, semi-active mounts and active mounts, etc.

In most elastic mounts, the elastic element is rubber. Thermosetting elastomeric synthetic rubbers have been applied to fabricate mounts. Elastomers for mount applications cover a wide range from extruded silicone synthetic rubber to Ethylene Propylene Diene Monomer (FEAD) EPDM. Usually, a rubber isolator can be represented by an elastic linear spring and a dashpot with linear viscous damping.

Hydraulically-damped rubber mounts (HDM) having frequency- and amplitude-dependent dynamic characteristics superior to conventional rubber mount have become an effective mount to attenuate vibrations transmitting between powertrain and body/chassis and to reduce the interior noise of vehicle compartment. HDM is widely equipped not only in powertrain isolation system of the vehicle but also in body isolation system. Hydraulic mounts have a much better impact on isolation performance.

The coupled hydraulic mount is mainly composed of a rubber spring, two fluid chambers, a fluid track, and a coupler membrane. Rubber spring possessing high elasticity serves as the main support of powertrain load and as a piston to pump fluid to flow between upper and lower chambers through the fluid track. The main physical property of rubber spring is characterized by its elastic stiffness and damping. The fluid track plays a role as tuned isolator damper. Subsystem consisting of fluid inertia in fluid track and volumetric elasticity of upper chamber results in nonlinear frequency- and amplitude-dependent isolation performances of HDM, which provides large dynamic stiffness and higher damping to isolate low-frequency and large-amplitude vibration effectively. Coupler membrane helps to increase volumetric elasticity of the upper chamber to achieve good isolation for high-frequency and small-amplitude vibration when there is less fluid flowing between chambers. When system under the impact, the inertia block has no motion and the mount behaves like a non-coupled mount when under high-frequency motion, coupling block involves motion and reduced liquid motion, which gives rise to low damping and stiffness at the high-frequency range.

Semi-active and active mounts may offer even better performance than that of hydraulic mounts. Most semi-active and active mounts are based on hydraulic mounts by adding a feedback control system. Active mounts not only consist of a control system but also have an energy provider to supply force to counteract system vibrations. Semi-active mounts consist of a control system, but it does not have extra energy provider, and the counteracting force comes from the mounting system itself.

Mount is usually installed along the free rotation axis which passes the gravity center of the power plant. If the mounts are installed on the free rotation axis, mounts only sustain engine's torque and force. Otherwise installed, it has additional force and torque besides engine's torque and force.

Besides the considerations of mounts installing on the free rotation axis, the following factors need to take into account: mounts should be installed on the location of the engine where it has smaller vibrations, such as vibration nodes, at least it should be far away from antinodes; the mount should also be installed on the nodes of frame/body. The installation location should allow the six natural modes to be uncoupled, thus allow the system to have six independent modes. Under this situation, each mode can be treated separately.

On the other hand, consider that the power plant-mount system behaves as a dynamic absorbing system for whole vehicle system, the location close to suspension could be an ideal location for mounts, which has high stiffness and is usually close to the node of first-order bending modes of body. But it is usually not practical to allow all the above conditions to be satisfied.

The concept of transmission rate is applicable for one-degree-of-freedom system analysis. For six-degree-of-freedom-system, more design index is needed for analysis. In the real design, each mount's stiffness, damping, the excitation characteristics, inertial parameters, and space are all needed to be taken into account. The evaluation index includes the transmission rate, the modal decoupling degree of the power plant isolation system, and the rotation frequency of power plant concerning the free rotation axis.

The excitation magnitude and frequency of power plant depend on the engine's structure and attributes; the response of power plant depends on the excitation and the system of power plant and mounts. In practice, the transmission rate of mounts is determined either by measurement or FEM calculations. In the determination of transmission rate through measurement, the accelerations at two supporting brackets are measured.

Since the vibration excitation source of the engine is concerned with rotation speed and frequency, the transmission rate is also dependent on speed and frequency. The above-mentioned criteria for a transmission rate of mount should be valid in whole engine operating speed range.

Based on Equation (3.6), the power plant isolation system is modeled as a six-degree-of-freedom system. Assume the energy of a specific vibration can be represented as E_{ij}, in which the subscript i refers to the mode coordinates $i = 1, 2, ..., 6$; j refers to the specific vibration frequency f_j, $j = 1, 2, ..., 6$. For the vibration under specific frequency, the total mode energy under this specific frequency equals to the sum of individual mode energies.

For this specific frequency the relative magnitude of energy of a specific mode can be quantified by the ratio of the energy of a specific mode to the sum of all individual mode energy.

Under a specific frequency, if single-mode energy accounts for up to 85%–90% of total mode energy, this single-mode vibration is the dominant vibration and its mode decoupling level is considered to be acceptable.

The mode decoupling is a critical index to evaluate mounts design. One of the major targets of isolation design is to allow the six modes decoupled as much as possible, particularly the coupling between transversal rotation mode and the other. Usually, under specific frequency, if mode energy accounts for up to 85%, it's decoupling with the rest is considered satisfied. For transversal rotation mode, its decoupling requires 90%.

In Equation (3.6), the engine crankshaft rotates in the direction of x, therefore, the transversal rotation modes are most likely to be excited. If the transversal excitation frequency is identical to the natural frequency of transversal rotation mode, the system resonance occurs. The natural frequency of transversal rotation mode, the system resonance should be designed to avoid coincidence with the transversal rotation excitation frequency. Assume the transversal rotation mode has decoupled with the rest, and then the treatment of transversal rotation mode can be done as a single-degree-of-freedom system. In the consideration of the design transversal rotation mode, the engine excitation frequency should be known. The engine excitation frequency depends on engine rotation speed and orders.

If the lowest firing frequency is known as f_e then the transversal rotation frequency f_{rn} should satisfy the isolation condition for frequency. When engine speed increase, the excitation frequency increases as well and they are larger than the lowest firing frequency. If the transversal rotation frequency should satisfy both isolation conditions for the lowest firing frequency and the higher-order excitations. The ratio of excitation frequency to transversal rotation frequency is usually designed in the range of 2–3, to attain the trade-off between the requirements of both vibration isolation and impact isolation which requires mounts to be stiffer.

For instance, an engine has the lowest idle speed of 600 rpm or 30 Hz. If the ratio of excitation frequency to transversal rotation frequency is taken as 2.5, then the transversal rotation frequency is 12 Hz. Mounts can be preliminarily designed and arranged in terms of this value. If a design cannot reach it, sometimes the idle speed can be increased, for instance, 600–700 rpm.

In the above three evaluation index, the transmission ratio is mainly used to evaluate mounts, the model decoupling and transversal rotation frequency are used to evaluate the overall performance of the power plant-mount system. The model decoupling and transversal rotation frequency depend on system inertia parameters such as mass, gravity center, the moment of inertia, and free rotation axis. They also depend on the numbers of mounts, the installation location, and the stiffness. Once power plant and transmission are decided, system inertia parameters are determined, as such, to acquire ideal isolation performance, the numbers, the installation location, and the stiffness of mount should be selected properly.

The optimization of isolation design can further help attain ideal mount design, optimal model decoupling, and optimal transversal rotation frequency.

3.5 Road/Tire-Induced NVH

The noise and vibration [106–109] transmission paths from external excitation to vehicle compartment interior include door/quarter panels, floor, roof, dash, door window strip, glass window strip, glass, vents and ducts, engine/transmission, mounts, exhaust hangers, suspension, control cables, body structural resonances, and cavity acoustic resonances, etc. There exist many excitations to the body, which include the imbalanced dynamic force in tire/wheel, road excitation, impact due to engine combustion, rotating and reciprocal inertia force and torque of the engine, as well as the motions of other systems. These excitations and response transmit to the body directly or indirectly, which causes body structure vibration radiating noise into the compartment. The compartment noise due to this kind of mechanism is referred to as structure-borne noise.

In contrast to the structure-borne noise, airborne noise is referred to as the compartment interior noise transmitted into the interior through opens or orifice from the noise sources outside the body which includes the tire/road noise. The compartment of airborne noise usually ranges from several hundred to several thousand Hz or even higher.

Next, we will discuss the excitations of tire/road to the body and their transmission and attenuation.

The excitation generated by rough road decreases rapidly with the increases of frequency. In the ideal situation, tire and wheel should only play a role of isolation. However, in a real situation, on one hand, tire acts as an isolator to attenuate the excitation energy, on the other hand, it creates specific excitation due to its inherent properties, exhibiting as a new excitation source.

Tire and wheel assembly consists of tire, wheel, hub, and brake assembly system.

The nonuniformity of tire material and the inaccuracy of geometry dimension allow a rotating tire to generate radial excitation force. Even on a flat road surface, this type of excitation exists. The bias or non-concentricity of tire and wheel generate unbalance excitation force. If in the assembly process the unbalance of tire and wheel are allowed to be offset each other, and the balance has been done by adding extra mass, the static unbalance can be eliminated. However, the dynamic unbalance excitation (torque) of tire/wheel still exists, and it may result in the vibrations of the steering system. Some vehicle's steering wheel shiver vibrations when the vehicle travels at high speed is due to the excitation of dynamic unbalance of tire/wheel.

The road forces acted on the tire comes from two aspects: one is the continuous press and releases action by road through contact patch, which generates vertical excitation force; the other one is the continuous roll-press and releases action from the road at the contact surface, which generates longitudinal excitation force [110, 111]. The rough surface usually generates larger excitation force than smooth surface does.

The excitation forces of tire/road are transmitted to tire, and then transmitted to the axle through the coupling system of tire air chamber and hub, thus forming the vertical force, longitudinal force, lateral force, and torque acted on an axle or steering joint.

Finite element analysis has been used to analyze the forces and the transmission of tire/road, which usually includes tire and road model, air pressure effect and thermal effect. Finite element analysis has been used to quantify the effect of stiffness, pressure, temperature, to estimate natural mode and force response. Normally, the fundamental natural frequency of tire is in the low-frequency range, the up limit is usually not beyond 150 Hz.

With the increase of frequency, the effect of air chamber gradually becomes strong. **Figure 3.9** illustrates the vertical acceleration spectrum of a loaded tire at the end of the axle, in which peak A corresponds to the first-order mode of tire structure. Peak C corresponds to the higher-order mode. Peak B is not due to the natural vibration mode of tire structure, it is due to the resonance of the coupling between structure and the acoustic mode of the air chamber. Its natural frequency is about 240 Hz. If these modes are excited, it could cause the acceleration response of the end of the axle, thus cause the excitation to suspension [112, 113]. Finally, the excitations are transmitted to the body through control arm and suspension and cause body vibration and noise.

When the above tire is pressed against with road and generates deformation, the acoustic mode of air chamber exhibits two peaks at the frequency close to 230 Hz.

FIGURE 3.9   Vertical acceleration spectrum of a loaded tire at the end of the axle.

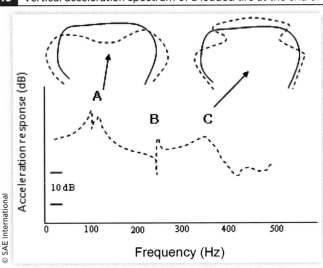

© SAE International

FIGURE 3.10 Schematic of the transfer path.

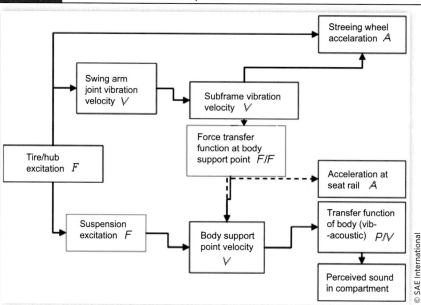

© SAE International

The higher mode is likely to cause vertical vibrations of the end of the axle. The lower mode usually causes the horizontal vibrations of the end of the axle.

The tire excitation is transmitted to the body through a suspension system through many paths. **Figure 3.10** is the block chart of the forces transmitted to the body from tire-road excitation for a vehicle. For the widely used, independent suspension and loaded body, there are two major transmission paths; one is to transmit to the body through suspension spring and shocker; the other is to transmit to the body through subframe. The tire/wheel excitation force and torque could be transmitted to the steering arm, then transmitted to body and steering column and steering wheel through steering mechanism installed on the subframe. If a vehicle structure consists of the longitudinal push rod, then the third path is that the force is transmitted from pushrod to subframe or body. Suspension parts are usually connected flexibly. They are connected through rubber isolators such as a rubber ring or bush. As shown in **Figure 3.10**, the performance of this kind of isolators has a direct influence on force transmission from road excitation to the body.

For unloaded body structure, the excitation force of tire/wheel is usually transmitted to frame, then transmitted to the body through multiple rubber isolators installed on the frame. Since unloaded body structure has one more layer of isolation treatment than that of loaded body structure, for the same tire excitation and same the suspension design, the force transmitted to the body of an unloaded body vehicle is smaller.

For non-independent suspension, the force of tire excitation is transmitted to axle first. Then one path is that the force is transmitted to the body (or suspension) through suspension spring and shocker. The other path that the force is transmitted to the body (or suspension) through longitudinal and lateral push rod and central support.

Besides being transmitted to the body, road excitation could be directly transmitted to the steering system and finally causing the vibrations of the steering wheel.

FIGURE 3.11 Sound pressure levels of interior noise of a vehicle traveling on rough and flat roads.

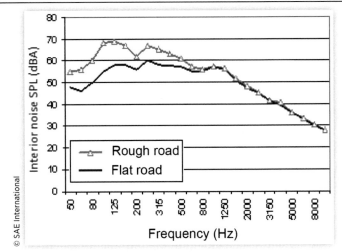

In the following, we simply present the design points on the attenuation of interior noise from road excitations.

The varied surface of road results in different interior noise, particularly in the low and middle-frequency range. **Figure 3.11** is the sound pressure levels of interior noise of a vehicle traveling on rough and flat roads respectively. In the frequency range 50-800 Hz, The interior noise of the vehicle on the rough road is higher than that on the flat road by 10 dB. In the high frequency above 1000 Hz, the effect of roughness or unevenness of road on the car's interior noise is very small.

In **Figure 3.11**, the interior noise of car on the rough road has a peak at about 250 Hz. This is due to the strong mode coupling between wheel mode and the first order circular acoustic mode of tire air chamber under the excitation of road. For P205/60R14 tire, the first order acoustic mode of tire air is about 240 Hz. The method to suppress this peak is to change the mode frequency of the wheel hub vibration to allow it to have a separation with the first-order acoustic mode of the tire air chamber. Some experiences show that aluminum wheel has higher first-order natural frequency than that of steel wheel, which is effective to abate this peak. On the other hand, the analysis on the coupling of tire structure and the tire air chamber shows that the increase in tire structure damping has a very small effect on axle vibrations, and the tuning of natural mode frequency of tire structural also has a very small effect on axle vibrations. Certain control approaches for tire air chamber have been investigated, including adds Polyurethane sponges [114], display regularly distributed insulation curtain within the tire chamber, etc.

A more frequently used method for abating vibrations of road excitation is to insert vibration isolation components in the vibration transmission paths. One of the vibration isolation components is rubber bush, whose stiffness is different in longitudinal, vertical, and lateral directions. It has relatively high stiffness in longitudinal and lateral directions to acquire good handling performance, and it has relatively low stiffness in a vertical direction to attain better isolation performance. In practical applications, the stiffness of isolator at different directions are usually coupled due to rubber's properties, thus an independent design is not feasible. An improved design is to separate the function of vibration control from its function of handling.

In the frame of unloaded body structure, or the front and rear subframe as well as body's front and rear strengthening parts, the brackets of mounts (isolators) need to be designed with enough stiffness.

The acceptance of the subframe exhibits the attributes of acceptance of a static stiffness which is close to the acceptance of a pure spring. The design of subframe should consider structure stiffness of support point at varied directions. The general principle is that the local stiffness of support point is higher than that of the component of vibration reduction by one order magnitude. Sometimes, the first-order frequency of subframe is used to evaluate the acceptance performance of subframe. The higher the first-order natural frequency, the better the acceptance performance. The lower the first-order natural frequency (for instance, lower than 50 Hz), the worse the acceptance performance. In some applications, there exist some cases in which the high first-order natural frequency does not guarantee the acceptance performance being good at all directions. The detailed design method of isolators has been described in the last chapter.

The transfer path approach has been used to conduct system analysis of road-tire vibrations and interior noise [115–117]. Its strength is that it can pinpoint the sensitive transmission path, which is usually the path where the excitation force is larger at its support point. In the sensitive transmission path, the suspension isolator or subframe isolator has large magnitudes; whereas the support point on suspension or subframe exhibits comparatively low stiffness at sensitive frequency range, thus allowing larger dynamic forces being transmitted to the body easily.

The countermeasure to structure-borne noise include well-balancing engine shafts, and using internal counterweighing, using light-weight components; using low-frequency mounts; using hydraulic mounts; putting exhaust hanger at nodal points; using exhaust flex coupler; allowing Front End Accessory Drive (FEAD) bracket resonance to be above engine firing frequency; isolating cooling fan and shroud resonances; allowing power plant bending mode target above exciting frequencies and 2nd order of engine; and using mass tuned dampers.

The noise sources outside body radiate in all directions. Some of the sounds radiate to ambient air and some reaches to the vehicle body. The incident sound impinging on the body is partially reflected and partially absorbed by body structure and its insulation layer, and the rest propagates into the inside of the body that is referred to as air-born noise. The seal quality directly affects the sound propagating into the body. Poor seal and orifice, as well as openings, lead to the coupling of air inside and outside of the body. Besides seals, the panels and windows of the body also influence the transmission of sound from outside to inside. The transmission of air sound includes several paths, which include the coupling of an outside air pressure wave with the structure of the body, the transmission of the excited vibrations of the body, the coupling of body structure vibrations and the inside acoustic field. The coupling of an outside air pressure wave with the structure of the body can be considered as the inverse process of sound radiation from structure vibrations. The strength of this coupling is weaker than the coupling between structure and structure in terms of response level; however, sometimes it still results in effective sound transmission. In many cases, the acoustic insulation

FIGURE 3.12 Decomposed interior noise spectra of a passenger car traveling at low-middle speed.

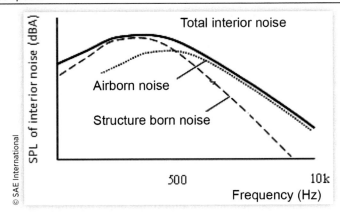

of a thin panel of the body is not sufficient, additional interior acoustic treatment inside the body is needed. Some of this acoustic treatment is implemented by attaching acoustic materials on the panel of the body, and some of it is integrated with either panel or trim. This treatment helps to allow the interior noise level meets the design target. **Figure 3.12** schematically illustrates the decomposed interior noise concerning frequency for a passenger car traveling at low-middle speed. In the frequency range higher than 500 Hz, the body air-born sound exceeds the structure-born sound to become a dominant interior noise. With the increase of traveling speed of the car, wind noise could substitute airborne sound to become the dominant sound, which will be discussed in the next session.

Road and tire noise mechanism are briefed as follows. Tire/road noise is another major noise source of body air-born noise. Tire noise is generated by several mechanisms. For a modern tire, noise due to wall vibration, air pumping and air resonant radiation are all considered to be important. The interaction between tire and road constitutes the dominant noise source for a vehicle traveling at middle speeds (above 50 km/h). Vibrations of the tire structure (belt and sidewall) are the combination of several different wave types which appear at different frequencies. In a low-frequency range, tire behaves like an elastically supported beam. Above middle-frequency range (i.e., 300 Hz), which is the transition point from one dimensional to two-dimensional waveguide properties of the tire, a cylindrical shell model can be used to analyze flexural waves propagation. Most of the tires are inefficient sound radiators since their wavenumbers are larger than the acoustic wave number.

The design and application of tires directly influence vibrations and noise performance as well as power handling/safety. The grooves and treads of tire need sophisticated design. There are many design targets for groove configuration and geometry such as water rejection, insertion proof, etc. The design of surface and wrinkle of the tire has a direct influence on the noise generation.

For tires, the air pumping noise of grooves, the impact noise of tread blocks and the vibration noise of tire body have been considered as the main noise sources. There are also several other noise mechanisms.

1. Before the tread blocks or patch of tire get into a roll-press state with the road, it gets impact with road first and it causes the vibrations of the patch. This kind of impact is intermittent and has a period which is dependent on the number of patches and the vehicle speed. The impact of patch results in vibrations and noise, the waveform of the sound is approximately close to sinusoidal function that may consist of third-order harmonic components. This impact effect behaves more obviously on the smooth surface than that on the rough surface road. This kind of impact effects has been alleviated by modern improved designs.

2. When the patch of the tire comes into the roll-press state, some portion could have slip motion concerning road surface thus causing friction noise. The roughness of non-smooth road could resist the slip and tends to reduce friction noise. On the other hand, the rough surface tends to excite the contacted patch thus causes the related noise. When a tire rolls on road, it flattens in the contact patch. The changing radial deflection produces tangential forces between the tire and road. These forces are resisted by friction and tire stiffness. The residual forces are dissipated by a slip of the tread material over the road surface. Friction between the tread and the surface can be divided into hysteresis and adhesion components. The adhesion component has its origin at a molecular level and is governed to a large extent by the small-scale roughness characteristics or micro-texture of the road surface. During relative sliding between tire and road surface, the adhesion bonds that have been formed between tire and road surface begin to rupture and break apart, so that contact is effectively lost and the rubber is then free to slip across the road surface. Contact may be regained as these residual forces are dissipated. The hysteresis force is due to a bulk phenomenon which also acts at the sliding patch. In the contact zone, tread rubber drapes around asperities in the road surface and the pressure distribution about each asperity is roughly symmetrical in the case of no slip. When slip occurs, tread rubber tends to accumulate at the leading edges of these surface irregularities and begins to break contact on the downward slope of the surface profile. This gives rise to an asymmetric pressure distribution and a net force which opposes the sliding motion. At high speeds, this mechanism is largely responsible for the tread element regaining contact with the road surface. The hysteresis component of tire surface friction is largely controlled by the surface macro-texture which comprises texture wavelengths corresponding to the size of the aggregate used in the surface material. It is not the slippage of the tread elements alone that gives rise to vibrational excitation of the tire. It is the combination of the slip of the tread elements in the contact patch and the hysteresis effect due to the deformation of the tread that gives rise to a "slip-stick" process in the contact patch, which excited the vibrations the tire. Tire vibration and noise generated by this mechanism have been related to the slip velocity of the tread elements. The highest velocities tend to be found to the rear of the contact patch and may contribute to blocking "snap out" effects.

3. The tube air noise of the circular groove is due to the resonance of air in the tube formed by the tire groove and road surface. This kind of noise could be strong. But for rough surface, it is not significant. The air pumping noise is due to the ejection of air from the small chamber formed by tire tread and road surface.

While tire rotates, in the moment of the formation of the small compressed chamber, the air in the small chamber erupts out due to compression; in the time when the compressed patch leaves road surface and stretched, the compressed cavity get volume expanded thus form a vacuum to a certain extent, which pulls in the surrounding air. This kind of effect causing pressure increase and decrease is similar to that of pumping, thus it is named as pumping noise. The frequency is from 800 Hz to several thousand Hz. The motion of tire tread could create lateral, front and rear noise radiation at the lateral, and front and rear grooves. If the road surface is porous, the air could be pumped into pores thus pumping noise is reduced significantly. The air noise may be generated by more complex aerodynamic phenomena such as turbulence or vortex shedding.

4. The rebound of compressed tire tread block after rotate-press contact state causes structure impact vibration and radiate noise. The lateral tread structure of tire in the proximity of the contact area undergoes deformation which generates structural noise radiation. The first order acoustic mode frequency of tire inner chamber is about 240 Hz, and it is the circular mode. The tire acoustic vibration has a lateral mode of 1100–1200 Hz, which is likely to get coupled with other structural modes of vehicle, this is also one of the important sources to generate noise. The other effects include the amplifying effect of "broadcast" formed by the angle between tire and road, the aerodynamics effect, etc.

Pumping noise has a wide band spectrum and attains the maximum at the rotating frequency of tread block. The circumferential frequency of tire tread depends on speed, the dimension of the tire and the number of tread block. For most cars at the speed of 60 km/h, this frequency is usually in the range of 800–1200 Hz. The overall sound pressure level of this kind of noise increases with the increase in vehicle speed. The sound pressure level increases about 10–20 dB when speed doubles. It is also affected by load, the sound pressure level increases about 8–10 dB when the load is increased by 50%. This is because the intensity of the impact noise of tread block is proportional to the traction of the vehicle.

There are many parameters affecting tire noise. Hayden once gave a formula for tire noise.

$$SPL(r) = 68.5 + 20\log\left(\frac{gw}{s}\right) + 10\log(n) + 2 - \log(fc) + 40\log(V) - 20\log(r) \quad (3.10)$$

where

r is the distance

g, w, s, n are respectively the depth, width, length of grooves, and the number of groove column

fc is the groove volume changes due to deformation

V is the tire speed

In tire tread block design, the longitudinal tread block (along the circular direction) design has minimum noise; transversal tread block design has the maximum noise. The mixed tread block of 45° design has advantageous. To avoid resonant peaks in the proximity of rotation frequency of tread block, modern design allows the distribution of tread groove and block along the circular direction to be not uniform, instead, to have some random variation. This random variation of tread block along circular direction helps to flatten the peaks around the rotation frequency of tread block and reduce response magnitudes. In some contemporary tire design, the dimension of the wheel and tire get larger and larger. The larger dimension of wheel and tire do not necessarily create high noise. The tread block or tread pattern design has been the dominant factor influencing noise. The sound pressure level of noise of tires with different tread pattern design could vary up to tens dB. The increase of tire dimension could affect its structure noise.

Besides tire itself, the road surface flatness has a larger effect on noise. The tire noise on a rough surface is substantially higher than that at the smooth surface at low-frequency range. In the frequency range above the rotation frequency of tread pattern, the tire noise on a rough surface is compatible with that at a smooth surface. For the same road, the tire noise on a wet road is higher than that on a dry case by 8–10 dB in the high-frequency range. Moreover, the texture and density of the road have larger effects on tire noise. Some road installed special construction materials with sound absorption function can effectively abate tire pumping noise.

The calibration of tire noise sources is usually conducted in the outside field.

In the aspect of tire/road noise modeling and analysis, finite element analysis, statistical energy method and boundary method have been used to model and simulate tire vibrations and noise and are used as a supplement means for design and development.

3.6 **Wind Noise**

Wind noise is the major noise source of a vehicle traveling at high speed. Traveling vehicle interacts with airflow around it, and airflow disturbance acts on the surface of the vehicle thus directly or indirectly affect vehicle interior noise. This section describes the mechanisms of wind noise. The related testing and analysis technology of wind noise is also covered [118–122].

To quantify wind noise, vehicle level tests can be made to vehicles traveling at high speeds (>90 km/h) where the wind is the dominant source of the noise. To isolate wind noise from another noise source, the test needs to be conducted in a wind tunnel. Moreover, to prevent the wind noise coming from the turbulent air around the vehicle from entering the vehicle through certain "leaks," special air leakage tests is developed. On the other hand, to model wind noise, computational fluid dynamics has been used.

When a vehicle is in idle operation, both road and wind noises do not exist and the only powertrain is excitation source. As the vehicle accelerates, road noise and vibration pick up at a higher rate than that of the powertrain and may exceed the powertrain noise and vibration on rough roads, and become the dominant source of noise and vibration in mid-speed ranges. At higher speeds, wind noise can become the dominant source of the noise [123–127].

Wind noise is correlated with many other attributes, which includes styling, vehicle performance, and vehicle fuel economy.

Wind noise mechanism and testing are presented as follows.

When the vehicle travels at low and middle speed, the contribution of wind noise to interior noise can be ignored. However, when the vehicle travels at high speed, for instance, 90 km/h, wind noise becomes the dominant component in vehicle interior noise.

The wind noise can be simply classified as noise leak, aspiration, buffeting and wind rush which have different generation mechanisms.

In terms of Lighthill's acoustic theory of aerodynamics, the noise of aerodynamics can be attributed as following three kinds of sound sources of linear acoustics.

The first one is a monopole, which exhibits as the variation of system volume or mass concerning time. The sound intensity of monopole is influenced by flow velocity and is proportional to the 4th order of power of velocity.

The second one is a dual pole, whose variation of momentum varies with time. It exhibits the variation of surface pressure. A dual pole can be considered as the coexistence of two monopoles with the same vibration magnitude and opposite phase, which have a small distance. The sound intensity of the dual pole is influenced by flow velocity and is proportional to the 6th order of power of velocity.

The third kind of sound source is quarter-pole, which exhibits as coacts of multiple dynamic forces and forming shear to flow liquid. A quarter-pole can be considered as the co-acts of two dual-poles with same vibration magnitude and opposite phase, which are kept with a small distance. The sound intensity of quarter-pole is highly influenced by flow velocity and is proportional to the 8th order of power of velocity. For vehicle under typical travelling speed, the flow field velocity is far smaller than sound speed ($M_c \ll 1$), a quarter-pole source has small sound radiation efficiency. The sound radiation efficiency of dual-pole is larger than that of quarter-pole but smaller than that of the monopole.

The perceived wind noise inside the body could come from monopole sources, such as the noise transmitted from door and window's seal, called leaks. Another interior noise from monopole type source of wind noise is the aerodynamic pressure outside of the body which causes the local negative pressures at the seals of doors and windows, thus leading to the deformation of the seals and finally transmitting the outside noise into the inside. This kind of noise is called aspiration noise.

The perceived wind noise inside the body could also come from dual-pole sources, such as the unsteady flow pulsation on the vehicle surface. This kind of unsteady flow pulsation exhibits random properties in time and space. It can be modeled as that the vehicle surface is excited by numerous microphone and exciter arrays that have statistical relationship and coherence, thus causing noise to be transmitted into the body, and allowing body vibrations to radiate noise into the body. The generation of unsteady flow pulsation or disturbance depends on the profile of the vehicle surface. **Figure 3.13** illustrates the schematic of several unsteady flow profile when air flows over blocks with different profiles: (a) flow generates aerodynamic separation due to front right angle profile; (b) the aerodynamic separation is reduced due to rounded right angle profile, but an adhesion region is generated; (c) the aerodynamic separation is eliminated due to obtuse angle profile design, and the adhesion region is also avoided. In body profile design, it is usually required to reduce the aerodynamic separation and to control airflow adhesion.

FIGURE 3.13 Effect of block profile on flow field: (a) aerodynamic separation; (b) aerodynamic separation and adhesive flow; (c) ideal case.

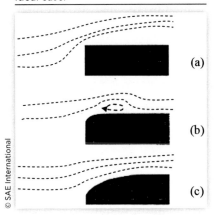

© SAE International

The unsteady flow disturbance at the different location has a different effect on interior noise.

The unsteady flow disturbance on vehicle body surface can also be classified as two types in terms of the excitation frequency. One is the interaction between the aerodynamic flow disturbance and the air inside the body chamber through a roof window or side windows, which generates low-frequency flutter or buffeting. The other is that the aerodynamic flow disturbance transmits noise into the body through body structure, which has a frequency of about 500 Hz, called wind rush.

Another case of interior noise due to dual-pole excitation of the wind noise source is that airflow across the protrusion components on the vehicle, such as antenna, rack of case, etc., which generates vortex tune noise.

There rarely have the case that the interior noise is from the quarter-pole source. Some air/liquid pipe/tube in the vehicle (for instance, the exhaust tailpipe) may generate quarter-pole type noise source.

The wind noise can also be classified from the perspective of design: the first is the wind noise due to poor seal; the second is the wind noise caused by groove, seam, or another non-flat transition zone on vehicle surface; the third category is the wind noise due to the protrusive objects such as rearview mirror, windshield wiper, antenna, a rack for suitcase, etc.

The wind noise test is usually conducted in a wind tunnel to avoid the effect of other noise sources such as tire noise and engine noise. Usually, the automotive wind tunnel is primarily used to test wind resistance coefficient and other aerodynamic properties of the vehicle. Unlike a typical wind tunnel, the wind tunnel for noise tests needs special acoustic design or treatment to attain stringent requirements on low background noise. The wall of the testing chamber needs sophisticated acoustic design.

Figure 3.14 illustrates the variation of the environmental noise of the test chamber concerning wind speed for three wind tunnels.

FIGURE 3.14 Variation of environmental noise of test chamber concerning wind speeds for three wind tunnels (1 – typical wind tunnel without acoustic design/treatment; 2 – acoustic wind tunnel, with good acoustic treatment; 3 – acoustic wind tunnel, with excellent acoustic design/treatment).

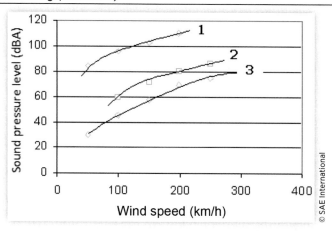

© SAE International

The wind tunnel tests usually include the following items: (1) the visualization of the distribution of flow field outside the vehicle; (2) the measurement of aerodynamic pressure, speed and the sound intensity of specific points outside of the vehicle in the flow field; (3) the measurement of body vibrations, aerodynamic pressure and their space-time statistical coherence; (4) the measurement the pressure, the sound intensity of the vehicle interior noise.

To observe the distribution of flow field outside the vehicle is an effective way to understand the physical mechanisms of flow-vehicle interaction. **Figure 3.15** shows the measured static pressure distribution on the front side window of a vehicle, in which the pressure shows negative value in the region behind A (front)-pillar where the flow is separated. This approach can qualitatively indicate the vortex zone after aerodynamic separation and the re-adhesion zones.

To calibrate the attributes of the flow field around the vehicle in wind tunnel test, the aerodynamic pressure, velocity, and sound intensity at specific points in the flow field need are measured. The sensors should be mounted below the body surface. The aerodynamic pressure of vehicle surface p_s consists of two parts, the average pressure of the flow field $\bar{p}$ and the pressure disturbance of the unsteady flow field p_r.

$$p_s = \bar{p} + p_r \tag{3.11}$$

The average pressure of the flow field directly influences aspiration noise by generating negative pressure on door seals, which may change the actual seal pressure thus malfunction the door seal and allowing noise to enter the body. Although the door seal pressure can be increased by design, its magnitude has constraints due to the limited force applied to close the door. Therefore, the negative pressure generated by the flow field should be controlled under a certain value. The generated negative pressure magnitude is usually proportional to the square of flow velocity.

The pressure pulsation of the unsteady flow field has a significant effect on vehicle interior noise. To quantify the effect of pressure disturbance of the unsteady flow field on interior noise, the characterization of pressure disturbance of the unsteady flow field is necessary. The related parameters need to be measured include the following: (1) the radiated sound pressure from pressure disturbance of unsteady flow field, which can be obtained by sound intensity method or near field holographic method; (2) the vehicle

FIGURE 3.15 Measured static pressure distribution on the front side window of a vehicle.

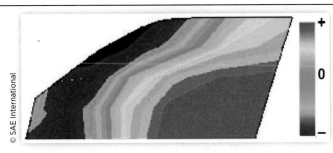

© SAE International

structural vibrations due to pressure disturbance of unsteady flow field, which can be measured by using accelerometer installed inside the vehicle or non-contact sensors like laser Doppler vibrometer; (3) the coherence spectrum of the pressure pulsations of unsteady flow field measured at different locations, which can be used to derive space-time coherence of the pressure disturbance of unsteady flow field. The coherence spectrum of the structural vibrations are measured at different locations; (4) the vehicle interior noise pressure or sound intensity.

Some testing procedures are described as follows: (1) Preparation stage: use tape to seal whole vehicle interior including the doors and windows seam, sometimes attached mass layer as insulation. The objective of the seal is to avoid sound propagation to body interior. The seal and package of outside vehicle body include door hander, mirror, A-pillar and windows, etc. The purpose of the outside seal is to eliminate the grooves and all non-flatness on vehicle surface; (2) To measure interior noise at certain vehicle speed. The measured sound represents the transmitted noise due to pressure pulsations of unsteady flow; the characterized noise is referred to as "profile-structure transmission noise;" (3) To remove all of the interior tape seals, repeat the above step 2. The measured sound pressure is higher than that previously measurement from step 2. The value of the pressure increment can be used to evaluate the seal performance and the design of the doors and windows seals. The characterized noise is referred to as "boundary leak noise rise"; (4) To remove all of the exterior seals, repeat the above step 2. The measured sound pressure would be further higher. The difference can be used to evaluate the surface groove, discontinuity, non-flatness on the interior noise. The characterized noise is referred to as "surface discontinuity noise rise." The total wind noise is the sum of the above three components.

The modeling and analysis of wind noise are briefed as follows.

The modeling and calculation of wind noise have made large progress over the last decade by using the latest technology of computational fluid dynamics [128–130]. In principle, aerodynamic noise can be predicated by solving the compressible fluid NS equation. However, the direct calculation needs a large number of volume elements to model local attributes of the large fluid field including sound source and response points. This is constrained by the calculation capability and time. Thus, it is only applicable to very small fluid zone and low-frequency acoustics. In practice, the indirect method has been used, which divides the zone to be calculated as the inner zone and outer zone and uses two steps to calculate noise.

The first step is to calculate the fluid attributes in the inner zone, under small M_c case, ($M_c < 0.3$), the inner zone enclosing structure encompasses sound source and is a zone with a relatively small thickness. The fluid in the inner zone can be considered as being unaffected by sound field thus, the sound field and flow field coupling can be ignored, and the incompressible assumption can be used. It is attributed to be as a problem to solve unsteady and compressible fluid NS equation.

The second step is to calculate sound based on the unsteady pressure disturbance of structure surface obtained from step one. Lighthill's equation has been used for the treatment of this step,

$$\frac{1}{c^2}\frac{\partial^2 p}{\partial t^2} - \nabla^2 p = \frac{\partial^2 T_{ij}}{\partial x_i \partial x_j} \tag{3.12}$$

$$T_{ij} = \rho\delta_{ij} - \tau_{ij} + \rho u_i u_j - c^2 \rho'\delta_{ij} \tag{3.13}$$

The solution of the equation is,

$$p(\vec{x},t) = \frac{\partial^2}{\partial x_i \partial x_j} \int_V \frac{T_{ij}\left(\vec{y},t - \frac{|\vec{x} - \vec{y}|}{c}\right)}{4\pi|\vec{x} - \vec{y}|} d^3\vec{y} \tag{3.14}$$

where

    V is the fluid volume consisting sound source

    $\vec{x}$ is the location vector of sound response point

    $\vec{y}$ is the location vector of the sound source

    $t - \dfrac{|\vec{x} - \vec{y}|}{c}$ is a time delay

If the above solution is applied to the location behind the obstacle of fluid, Lighthill-Curle solution is obtained,

$$p(\vec{x},t) = \frac{\partial^2}{\partial x_i \partial x_j} \int_V \frac{T_{ij}\left(\vec{y},t - \frac{|\vec{x} - \vec{y}|}{c}\right)}{4\pi|\vec{x} - \vec{y}|} d^3\vec{y} - \frac{\partial}{\partial x_i} \int_S \frac{T_{ij}\left(\vec{y},t - \frac{|\vec{x} - \vec{y}|}{c}\right)}{4\pi|\vec{x} - \vec{y}|} d^2\vec{y}$$

$$+ \frac{\partial}{\partial t} \int_S \frac{\rho\vec{u}\left(\vec{y},t - \frac{|\vec{x} - \vec{y}|}{c}\right)\vec{n}}{4\pi|\vec{x} - \vec{y}|} d^2\vec{y} \tag{3.15}$$

In which the first term is quarter-pole, the second term is a dipole and the third term is a monopole. In real applications, more simplification assumption will be introduced. For instance, if "profile structure transmission noise" is considered, the dominant sound source is dual-pole, and the far-field sound pressure is given by,

$$p(\vec{x},t) = \frac{1}{4\pi c} \int_S \frac{\vec{n},\vec{r}}{r^2} \frac{\partial P_j}{\partial t} dS \tag{3.16}$$

From Equation (3.16), the spectrum of sound pressure can be derived by using FFT. Moreover, the above approach has been used to calculate wind-induced vibrations (flutters). When aerodynamic flow pass opens on vehicle body such as roof window, it creates a vortex which could couple with the acoustic mode of air in the body chamber, thus leading to self-excited vibrations. It usually occurs at low-frequency range of 20–50 Hz when the vehicle speed is at 50–70 km/h.

    Besides the computational fluid dynamics approach, the approach combined statistical energy analysis and the experimental method has also been explored for wind noise analysis and design.

    The statistical energy analysis is a feasible method to analysis vehicle interior noise. The interested frequency range of wind noise is like that of air noise that can

FIGURE 3.16   An input model of wind noise by using the statistical energy method.

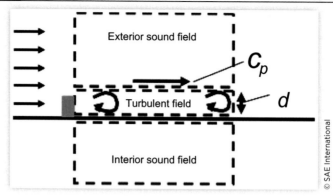

be handled by using statistical energy analysis. The statistical energy analysis uses the concept of virtual pressure to deal with the wind noise problem. The airflow is separated as aerodynamic wave (turbulent excitation) and scatter waves (propagation excitation). The aerodynamic wave is generated from convective flow, which couples with structure modes and transmits energy and leads to structure vibrations and sound radiation. Scatter wave directly transmits sound through the structure. From the perspective of body interior, the contribution of the aerodynamic wave is similar to that of the structure-born noise, whereas scatter waves are similar to that of the air-born noise source. **Figure 3.16** illustrates an input model of wind noise by using the statistical energy method.

The aerodynamic wave and scatter wave have different contributions to interior noise. **Figure 3.17a** shows the vibration accelerations of lateral window glass under the excitation of flow with speed 130 km/h; the contributions of aerodynamic wave and scattered wave from SEA estimations are also plotted in the same figure. It can be seen that the vibrations of lateral window glass are mainly due to aerodynamic wave. The scatter wave's effective influence on the glass vibrations occurs at a frequency higher than 3000 Hz. On the other hand, their effects on interior noise are quite different.

FIGURE 3.17   (a) Vibration accelerations of lateral window class under the excitation of flow with speed 130 km/h, and the contributions of aerodynamic wave and scatter waves from SEA estimations; (b) interior noise of the vehicle, and the contributions by an aerodynamic wave and scatter waves from SEA estimations.

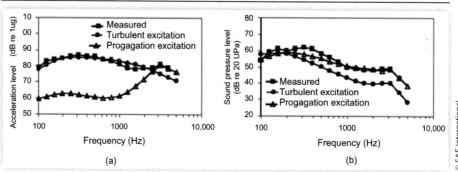

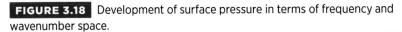

FIGURE 3.18 Development of surface pressure in terms of frequency and wavenumber space.

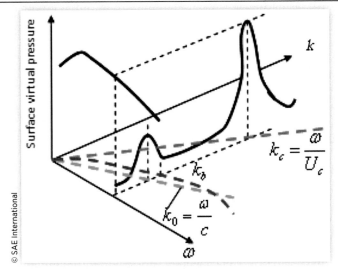

© SAE International

Figure 3.17b shows that the interior noise is mainly contributed by scattering waves over the wide frequency range, the aerodynamic wave's effect is obvious only at low frequency.

Even though the magnitude of scattering waves (propagation excitation) is usually much smaller than that of aerodynamic wave (turbulent excitation), in other words, it is only a small portion of the total pressure of surface, it can have a significant effect on interior noise. The measured magnitude of surface total pressure cannot be used to effectively predict vehicle interior noise, sometimes even cannot be used for qualitative interpretation.

Figure 3.18 elaborates the physical meanings of Figure 3.18. After the FFT operation of surface pressure spectrum in the flow direction, its projection on the plan (k) exhibits aerodynamic wave (k_a) and scatter waves (k_c). The convective propagation speed of the aerodynamic wave U_c is usually 70% of the flow speed of the adjacent local fluid and is lower than sound speed. Thus, it is a slow wave, with wavenumber being larger than that of sound. In the same figure, the wavenumber of the elastic bending wave of glass is plotted as k_b, which is proportional to the root square of the frequency. At high frequency (about 3000 Hz), it transferred from slow-wave to fast wave. The difficulty of the application of SEA lies in the modeling of complex wind noise excitation and the energy input.

The wind noise abatement is discussed as follows.

There has been a lack of a unified approach to evaluate vehicle wind noise. Generally, the A-weighted overall sound pressure level is not always appropriate. This is because the vehicle wind noise spectrum usually decreases with the increase of frequency, and the dominant component of wind noise is centered at the frequency range above 500 Hz.

If A-weighted overall sound pressure level is used, the influence of the component below 500 Hz is strong, which allows the sophisticated difference in design is not distinguishable. In contrast to this, AI as an index to evaluate wind noise is superior to the A-weighted overall

TABLE 3.3 Varied evaluation methods for different wind noise category/perceptions.

Customer report	SPL	Pressure pulsation	AI	Orientation	N	S	R
Perception exterior noisy	O		O		O		
Perception of unsealing	O		O		O		
Unable to talk			O				
Ear uncomfortable		O	O			O	
Rearview mirror noise perceived				O			
Wind flutter perceived	O	O					
Sound roughness feeling							O

N – loudness; S – sharpness; R – roughness.

© SAE International

sound pressure level. The application frequency range of Articulation Index (AI) is in the range of 200–6300 Hz. Besides, loudness is also used to evaluate wind noise. Another approach is to select a reprehensive vehicle wind noise spectrum as a standard reference for the spectrum of wind noise. The measured wind noise of a vehicle is then compared with the standard and finally, a new single value evaluation index can be derived [131–133].

The evaluation of wind noise includes the single value evaluation index for steady front wind flow and steady lateral wind flow (usually 10° inclined angle). The transient wind noise is important but there is no simple evaluation approach. **Table 3.3** lists the varied evaluation methods for different wind noise category/perceptions.

The design considerations for low wind noise should be included in the preliminary design and development stage of the vehicle. The basic configuration, A-pillar design, etc. have a significant effect on vehicle wind noise. The streamline form and streamline shape design helps to reduce wind resistance; to prevent aerodynamic separations of air at A-pillar, front beam of roof and other structures. The non-smooth corners should be avoided in design. For example, the transition from A-pillar to the front beam of the roof should not have abrupt geometry change, as illustrated in **Figure 3.19**.

The A-pillar design is crucial. There is a step between A-pillar and windshield, which is required for the installation of the windshield and is also served as a barrier to avoid the rainwater or cleaning liquid from windshield blade to flow onto the lateral windows [134], as illustrated in **Figure 3.20**. The step usually has a height requirement, which should not be higher than 5 mm; otherwise, it will affect interior perceived wind noise.

FIGURE 3.19 Sensitive design locations of vehicle profile for wind noise abatement.

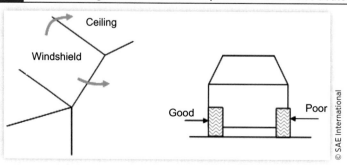

© SAE International

FIGURE 3.20 (a) A-pillar design; (b) wiper blade's rest position for reducing interior noise from wind noise excitation.

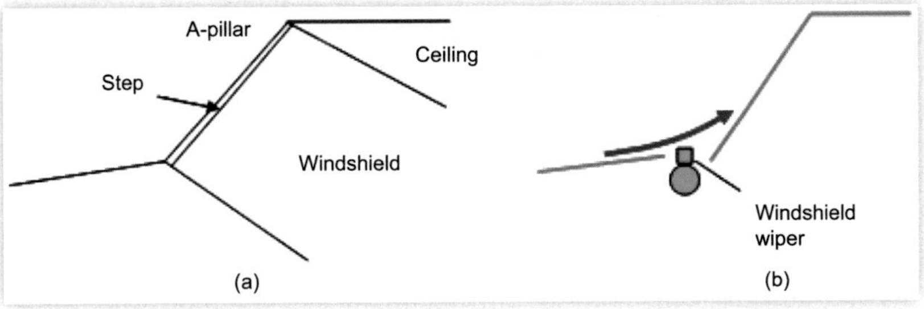

In preliminary design, the wiper blade's rest place also should not be exposed on the flow field to black flow path; instead, it should be hidden in a predesigned groove.

The other related design considerations for low wind noise are described as follows: the smoothness of the cover surface of mirror affects wind noise; the mirror with cantilever beam to keep a distance from a window may help suppress wind noise; if the mirror is integrated with the body, special modeled plastic connection is needed; the seals of windows should allow windows to have a smooth and uniform connection with the body, and allow body surface not to generate strong non-flatness; the seal of door and body must be used and installed either on doors or body; the exterior antenna should have as small dimension as possible, and the constant cross-section area should be avoided and the spiral type should be applied, to avoid tune noise; in the front of roof window, an automatic fender can be used for reducing flow disturbance, which can change the energy feedback process of wind flutter in the opened roof window.

3.7 Whole Vehicle Vibrations under Road Excitation

In the next, we discuss whole vehicle vibration under road excitation. The road unevenness is the major outside vibration source. The vibrations are transmitted to occupants through suspension and seating systems.

3.7.1 Vibration Sources from Road

Vibration excitation from tire-road is due to the interaction between tire tread block and the unevenness of road surface. Road excitation level depends on the road profile, vehicle speed, and tire dynamic properties.

Road surface unevenness is one of the major excitation sources to ride dynamics. It has been found that the power spectral density of the interior vibration is roughly proportional to the power spectral density of the road surface profiles.

Speed is another important factor in vehicle vibration. Increasing vehicle speed biases, the vibration spectrum to higher frequencies, and usually increases overall vibration level.

The tire dynamic properties include tire stiffness and pressure, tire patch/tread length and shape, tire width, and size. These properties also influence the ride characteristics. The vibration levels increase as tire pressure or stiffness increase. Shortening the tire tread block length biases the vibration spectrum to higher frequencies. A wider tire typically involves a shorter tire tread block, which will tend to bias frequencies upwards.

Next, we describe the input model of road, the power spectral density in spatial frequency.

A typical road profile can be represented by a random function. Usually, a road profile is a wideband random function. The instantaneous values for a random function cannot be expressed by deterministic methods. Thus, statistical analysis is used to describe road vibration input. When a vehicle travels on a road in the x-direction, the roughness, waviness and shape profile change of road causes the vehicle vibration in the z-direction. A road profile includes many wave components. Spatial frequency, Ω, stands for the wavelength number per meter of the road and the unit is m^{-1}. The spatial frequency is the inverse of wavelength. Roughness is defined as a road elevation profile. Roughness is used to evaluate road smoothness. The power spectral density of road profile is a useful parameter to describe the random function, $W_z(\Omega)$, which is defined as the mean-square response of an ideal narrow-band filter divided by the frequency bandwidth of the filter $\Delta\Omega$ as $\Delta\Omega \to 0$ at frequency Ω, as follows:

$$W_z(\Omega) = \lim_{\Delta\Omega \to 0} \frac{\overline{z_{\Delta\Omega}^2}}{\Delta\Omega} \tag{3.17}$$

where $\overline{z_{\Delta\Omega}^2}$ is the mean-square value within $\Delta\Omega$. Then, the mean-square value for the whole frequency range is,

$$\overline{z^2} = \int_0^\infty W_z(\Omega)\,d\Omega \tag{3.18}$$

Or the mean-square value during the wavelength L_i can be expressed as follows,

$$\overline{z_{L_i}^2} = \frac{1}{L_i} \int_0^{L_i} z_i^2\,dx \tag{3.19}$$

The power spectral density is a function of spatial frequency. Based on many road test data, a relation between the power spectral density of road profile and spatial frequency was proposed by ISO as follows,

$$W_z(\Omega) = W_z(\Omega_0)\left(\frac{\Omega}{\Omega_0}\right)^{-N} \tag{3.20}$$

where

Ω_0 is reference spatial frequency

$\Omega_0 = \dfrac{1}{2\pi}\,m^{-1}$

$W_z(\Omega_0)$ is the profile spectral density corresponding to the reference spatial frequency

$W_z(\Omega_0)$ is also called profile roughness coefficient

N is the frequency index, which determines the frequency structure of the spectral density and slope of log-log curves

TABLE 3.4 Roughness coefficients of road profile.

| Road class | Roughness coefficient $W_z(\Omega_0) \times 10^{-6}$ | |
	Range	Geometric mean
A	8–32	16
B	32–128	64
C	128–512	256
D	512–2048	1024
E	2048–8192	4096
F	8192–32,768	16,384
G	32,768–131,072	65,536
H	131,072–524,288	262,144

The road roughness is divided into 8 levels from A to H in terms of the ISO standard. Table 3.4 lists the range and geometric mean values of the roughness coefficients.

The frequency index, N, can be chosen as several values for different road wavelength. For most cases, N can be chosen as 2. For this case, Equation (3.20) can be written as,

$$W_z(\Omega) = \frac{W_z(\Omega_0)\Omega_0^2}{\Omega^2}$$

(3.21)

The power spectral density is inversely proportional to the square of spatial frequency or proportional to the square of the wavelength. According to Equation (3.21), the power spectral density of road profile for the different levels is plotted in **Figure 3.21**.

Usually, the elevation of the road surface where a vehicle travels along is used to describe the road roughness, which means that the vertical displacement input to the vehicle is the roughness. On the other hand, vibration is commonly expressed by acceleration and velocity. Hence, acceleration and velocity and their power spectral densities will be used as vibration input for vehicle ride dynamics analysis. The relation between

FIGURE 3.21 Power spectral density for different road profiles.

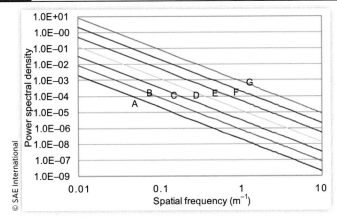

displacement and acceleration can be obtained by harmonic analysis. Assume that roughness displacement z of a road profile during a wavelength L can be expressed as

$$z(x) = A \sin\left(\frac{2\pi x}{L}\right) = A \sin(\Omega x) \tag{3.22}$$

where A is the amplitude of road profile during the wavelength. The corresponding velocity and acceleration are given by,

$$\dot{z}(x) = \frac{2\pi A}{L} \cos\left(\frac{2\pi x}{L}\right) = A\Omega \cos(\Omega x) \tag{3.23}$$

$$\ddot{z}(x) = -\left(\frac{2\pi}{L}\right)^2 A \sin\left(\frac{2\pi x}{L}\right) = -A\Omega^2 \sin(\Omega x) \tag{3.24}$$

The mean square value of $\dot{z}(x)$ during the wavelength is obtained,

$$\overline{\dot{z}^2} = (2\pi\Omega)^2 \overline{z^2} \tag{3.25}$$

The mean square value $\ddot{z}(x)$ is given by,

$$\overline{\ddot{z}^2} = (2\pi\Omega)^2 \overline{z^2} \tag{3.26}$$

Substitute Equations (3.24) and (3.25) into Equation (3.26), the velocity and acceleration power spectral density can be obtained as follows,

$$W_{\dot{z}}(\Omega) = (2\pi\Omega)^2 W_z(\Omega) \tag{3.27}$$

$$W_{\ddot{z}}(\Omega) = (2\pi\Omega)^4 W_z(\Omega) \tag{3.28}$$

For the special case where $N = 2$, substitute Equation (3.21) into Equation (3.28), the velocity spectral density will be,

$$W_{\dot{z}}(\Omega) = (2\pi\Omega)^2 \frac{W_z(\Omega_0)\Omega_0^2}{\Omega^2} = 4\pi^2 W_z(\Omega_0)\Omega_0^2 \tag{3.29}$$

The velocity spectral density is constant and is independent of spatial frequency. That is, the spectral density is "white noise."

The vehicle vibration is usually described by temporal frequency instead of spatial frequency. Thus, it is more convenient to express the spectral density of surface profiles using temporal frequency in Hertz than using spatial frequency in m^{-1}. The temporal frequency relates to vehicle speed and spatial frequency. The relation between the temporal frequency and spatial frequency can be expressed as follows,

$$f = V * \Omega \tag{3.30}$$

where

f is the temporal frequency

V is the vehicle speed

From Equation (3.30), it can be seen that the temporal frequency is proportional to vehicle speed for a given spatial frequency or a certain wavelength.

The statistical data shows that the spatial frequency for most road profiles is between 0.011 and 2.83 m^{-1}. The temporal frequency range is between 0.33 and 28.3 Hz for the corresponding spatial frequencies during vehicle normal operating speed range ($V = 10$–30 m/s). Thus, the frequencies for vehicle sprung mass (1–2 Hz) and the frequencies for unsprung mass (10–15 Hz) fall into the road excitation frequency range.

Similar to Equation (3.21), the power spectral density in temporal frequency can be expressed as follows,

$$W_z(f) = \lim_{\Delta f \to 0} \frac{\overline{z_{\Delta f}^2}}{\Delta f} \tag{3.31}$$

where Δf is the temporal frequency range. Substituting Equation (3.30) into above equation, relation between power spectral density in spatial frequency and that in temporal density can be obtained:

$$W_z(f) = \frac{1}{V} W_z(\Omega) \tag{3.32}$$

Substitute Equation (3.21) into Equation (3.32), the power spectral density in temporal frequency can be rewritten as,

$$W_z(f) = \frac{W_z(\Omega_0)}{V} \left(\frac{\Omega}{\Omega_0} \right)^{-N} \tag{3.33}$$

For the special case where $N = 2$, the density is,

$$W_z(f) = \frac{W_z(\Omega_0)}{V} \left(\frac{\Omega}{\Omega_0} \right)^{-2} = \frac{\Omega_0^2 W_z(\Omega_0)}{V\Omega^2} = \frac{\Omega_0^2 W_z(\Omega_0)V}{f^2} \tag{3.34}$$

Performing similar manipulation on Equations (3.27) and (3.28), the velocity power spectral density and acceleration power spectral densities are:

$$W_{\dot{z}}(f) = (2\pi f)^2 W_z(f) = 4\pi^2 W_z(\Omega_0)\Omega_0^2 V \tag{3.35}$$

$$W_{\ddot{z}}(f) = (2\pi f)^4 W_z(f) = 16\pi^4 W_z(\Omega_0)\Omega_0^2 V f^2 \tag{3.36}$$

Figure 3.22 is the power density spectra of road profile, velocity and acceleration for road class A and 10 m/s vehicle velocity.

The velocity density spectrum is constant for a given vehicle velocity. The road profile power spectral density decreases as the frequency increases. However, the acceleration power spectral density increases as the frequency increases. All three density spectra are proportional to vehicle speed. The spectral densities increase as the speed increases. The acceleration density increases with the square of temporal frequency, i.e., the vibration transferred to the vehicle at high frequency are more severe than that transferred at low frequency.

FIGURE 3.22 Power density spectra of road profile, velocity and acceleration.

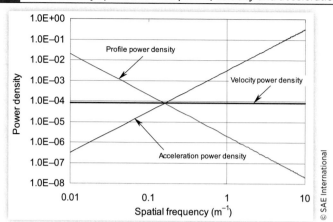

© SAE International

3.7.2 Vehicle Ride Model: Quarter Car Model, Pitch-Bounce Model, and Complex Models

A vehicle model needs to be considered to analyze ride dynamics. The vehicle can be modeled to be a multiple-degree-of-freedom system in many ways according to the type of response. For example, in automobile engineering, finite element model is usually used to calculate vehicle body modes and dynamic response. However, building and running the finite element model is complicated, time-consuming, and costly. In ride quality analysis, only a few low frequencies are considered [135–142]. Approximate solutions are enough to determine vehicle parameters, such as suspension damping and stiffness. Thus, simplified models will satisfy the requirements.

In ride dynamic analysis, a vehicle can be treated as a rigid body installed on its suspension system that is supported by tire and axles. The body and suspension constitute the sprung mass. The tire and its axles constitute the unsprung mass. The two simplified models used in this chapter are the quarter car model and the bounce-pitch model. The models will be described in detail.

The purpose of ride dynamics is to find the relation between the response at the occupant's touch points and vibration sources. In this section, vehicle body response will be the focus. The vehicle response will be covered in later sections. As discussed in the previous section, there are two major vibration sources input to the body. One is from the road uneven profile. The other is from the vehicle itself, such as the engine, driveline, and exhaust. Thus, the relation between the body response and the road excitation and relation between the body response and the vehicle excitation will be analyzed in this section.

A vehicle is typically a symmetric structure around the vehicle x-axis. The vehicle body is supported by two front wheels and two rear wheels. The body motions are usually coupled with the front and rear wheels. However, for many cases, the front and rear wheel motion are independent of each other. Thus, a quarter model, shown in **Figure 3.23a**, has been used to analyze the ride characteristics. Only vertical vibration is analyzed in this model. **Figure 3.23b** is a more complicated model. The complicated model can be obtained by including spring surge, the resonance of the shock absorber, and the frequency dependence of bushing stiffness, etc., the degrees-of-freedom could be up to tens.

FIGURE 3.23 A simple quarter car model and a multi-degree-of-freedom model.

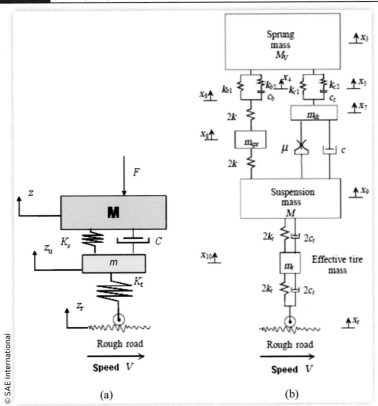

© SAE International

(a)   (b)

The quarter car model is a two-degrees-of-freedom system with two masses. The upper mass represents vehicle body and suspension mass, called sprung mass M. The lower mass is the weight of the tire and its axle, called unsprung mass, m. The quarter car model has two frequencies associated with a vehicle suspension system.

The motion equations of the quarter model are given by,

$$M\dot{z} + C\ddot{z} + K_s z = K_s z_u + C\dot{z}_u + f \qquad (3.37a)$$

$$m\ddot{z}_u + C\dot{z}_u + \left(K_s + K_t\right)z_u = K_s z + C\dot{z} + K_t z_r \qquad (3.37b)$$

where
    M is the sprung mass
    m is the unsprung mass
    C is the suspension damping coefficient
    K_s is the suspension stiffness
    K_t is the tire stiffness
    z is the sprung mass displacement
    z_u is the unspring mass displacement
    z_r road displacement input
    f is the vehicle internal excitation force

The excitation force and the corresponding displacement solutions can be expressed,

$$f = Fe^{j\omega t}, \quad z_r = Z_r e^{j\omega t}$$

$$z = Ze^{j\omega t}, \quad z_u = Z_u e^{j\omega t} \tag{3.38}$$

Substitute Equation (3.37) into Equations (3.35) and (3.36), the equations are transferred into the frequency domain, as follows,

$$\left(K_s - M\omega^2 + jC\omega\right)Z = \left(K_s + jC\omega\right)Z_u + F \tag{3.39a}$$

$$\left(K_s + K_t - m\omega^2 + jC\omega\right)Z_u = \left(K_s + jC\omega\right)Z + K_t Z_r \tag{3.39b}$$

For a passenger car, the sprung mass is higher than the unsprung mass by an order of magnitude while the sprung stiffness is lower than the unsprung stiffness by an order of magnitude. The unsprung mass can be neglected and the two springs are connected in series. The approximate frequency, ω_s, for the sprung system can be obtained by the following:

$$\omega_s = \sqrt{\frac{K_s K_t}{\left(K_s + K_t\right)M}} \tag{3.40}$$

Usually, the sprung mass is about 10 time of unsprung mass, i.e., $M \approx 10$ m and the unsprung stiffness is about 9 times of the sprung stiffness, i.e., $K_t \approx 9K_s$. Thus, Equation (3.40) can be further simplified as

$$\omega_s = \sqrt{\frac{K_s K_t}{\left(K_s + K_t\right)M}} \approx \sqrt{\frac{K_s}{M}} = \omega_s' \tag{3.41}$$

where ω_s' is the approximate values of ω_s.

The approximate frequency of the unsprung system can be calculated based on the assumption that the vehicle body is fixed and only the tire moves. The frequency is:

$$\omega_u = \sqrt{\frac{K_s + K_t}{m}} \tag{3.42}$$

Usually, the sprung system frequency is around 1 Hz, while the unsprung system frequency is around 10 Hz. The two frequencies are so widely separated that the approximate frequencies are useful to understand the system dynamic characteristics. The reason that the sprung system is designed with 1 Hz natural frequency is that this frequency is close to walking pace frequency which generally excites the natural modes of the human body.

The damping ratio provided by shocker absorber is about 0.2–0.4. Thus, the damped natural frequency has little difference from the undamped natural frequency. To investigate the system natural frequencies and modal shapes, damping is neglected, and external excitation is not included. Equations (3.8) and (3.9) can be rewritten as

$$\left(\omega_s'^2 - \omega^2\right)Z - \omega_s'^2 Z_u = 0 \tag{3.43}$$

$$-\frac{K_s}{m}Z + \left(\omega_u^2 - \omega^2\right)Z_u = 0 \qquad (3.44)$$

The characteristic equation of the above equations can be written as,

$$\omega^4 - \left(\omega_s'^2 + \omega_u^2\right)\omega^2 + \omega_s'^2\omega_u^2 - \omega_s'^2\frac{K_s}{m} = 0 \qquad (3.45)$$

The two real characteristic values of the solutions of the above equation are the frequencies of the system, they are,

$$\omega_{1,2}^2 = \frac{\omega_s'^2 + \omega_u^2}{2} \pm \sqrt{\frac{1}{4}\left(\omega_s'^2 - \omega_u^2\right)^2 + \omega_s'^2\frac{K_s}{m}} \qquad (3.46)$$

According to the relationship between sprung mass and unsprung mass, and the relationship between sprung stiffness and unsprung stiffness, the approximate formula between sprung frequency, and unsprung frequency can be obtained as

$$\omega_u = \sqrt{\frac{K_s + K_t}{m}} = \sqrt{\frac{K_s + 9K_s}{M/10}} = 10\sqrt{\frac{K_s}{M}} = 10\omega_s' \qquad (3.47)$$

Submit Equation (3.47) into Equation (3.46), the two frequencies of the system are:

$$\omega_1 = 0.95\omega_s'$$

$$\omega_2 = 10\omega_s' \qquad (3.48)$$

The results show that the first frequency of the system is close to sprung system frequency and the second frequency is close to the unsprung system frequency.

The mode shapes corresponding to the above two natural frequencies can be obtained by substituting Equation (3.48) into Equation (3.44), the relationship is,

$$\frac{Z}{Z_u} = \frac{\omega_s'^2}{\omega_s'^2 - \omega^2} \qquad (3.49)$$

Substitute Equation (3.48) into Equation (3.49), the modal shapes of the two-degree-of-freedom system are obtained.

- The 1st modal shape: $\left(\dfrac{Z}{Z_u}\right)_1 = \dfrac{\omega_s'^2}{\omega_s'^2 - \left(0.95\omega_s'\right)^2} = 10.26$

- The 2nd modal shape: $\left(\dfrac{Z}{Z_u}\right)_2 = \dfrac{\omega_s'^2}{\omega_s'^2 - \left(10\omega_s'\right)^2} = -0.01$ or $\left(\dfrac{Z_u}{Z}\right)_2 = -100$

When the excitation frequency ω is close to the first natural frequency ω_1, the system will vibrate at a frequency ω_1 and the mode shape is the 1st order mode shape. The body vibration is about 10 times larger than the tire's. The apparent motion comes from the body, thus the tire's vibration can be neglected. The first mode is called body mode. On the other hand, if the excitation frequency is close to the second natural frequency ω_2, the tire' amplitude is about 100 times larger than the body's motion. The body is almost at a static position. The mode is called tire mode. The separation of the two modes is significant for vibration isolation. One benefit is that the two modes can be controlled independently. The other benefit is that the body can be effectively isolated if the vehicle hits a bump and oscillates at unsprung frequency.

When a vehicle travels over an undulating road, the excitation includes multiple frequency input. As mentioned above, the high frequency can be effectively isolated by the suspension. However, the low-frequency excitation will be transferred to the body. It may even be amplified. Thus, the dynamic response must be analyzed over a wide frequency spectrum. The vibration sources are either from road excitation or from vehicle excitation. The body responses caused by the two sources need to be analyzed. Transmissibility between body response and road input, and transmissibility between the body response and vehicle excitation will be described.

References

1. Williams, R., Henderson, F., Allman-Ward, M., Dunne, G. et al., "Using an Interactive NVH Simulator for Target Setting and Concept Evaluation in a New Vehicle Programme," SAE Technical Paper 2005-01-2479, 2005, doi:https://doi.org/10.4271/2005-01-2479.

2. ISO 2631-1:1997, "International Organization for Standardization. Mechanical Vibration and Shock-Evaluation of Human Exposure to Whole-Body Vibration-Part 1: General Requirements. International Standard 72," ISO 2631-1:1997, International Organization for Standardization, Geneva, Switzerland, 1997.

3. ISO532B, "Acoustics-Method for Calculating Loudness Level, International Standard ISO 532B:1975," International Organization for Standardization, Geneva, Switzerland, 1975.

4. Abe, T., Cheng, M., Na, L., Schwalm, B. et al., "The Ford Motor Company Spin-Torsional NVH Test Facility-2," SAE Technical Paper 2003-01-1684, 2003, doi:https://doi.org/10.4271/2003-01-1684.

5. Abe, T., Obourn, L., Cheng, M., Maskill, M. et al., "The Ford Motor Company Spin-Torsional NVH Test Facility," SAE Technical Paper 1999-01-1837, 1999, doi:https://doi.org/10.4271/1999-01-1837.

6. Okamura, H., Naganuma, T., and Morita, T., "Influences of Torsional Damper Temperature and Vibration Amplitude on the Tree-Dimensional Vibrations of the Crankshaft-Cylinder Block System under Firing Conditions," SAE Technical Paper 1999-01-1775, 1999, doi:https://doi.org/10.4271/1999-01-1775.

7. Taraza, D., "Quantifying Relationships Between the Crankshaft's Speed Variation and the Gas Pressure Torque," SAE Technical Paper 2001-01-1007, 2001, doi:https://doi.org/10.4271/2001-01-1007.

8. Aoki, K., Shikata, T., Hyoudou, Y., Hirade, T. et al., "Application of an Active Control Mount (ACM) for Improved Diesel Engine Vehicle Quietness," SAE Technical Paper 1999-01-0832, 1999, doi:https://doi.org/10.4271/1999-01-0832.

9. Ernster, S., Tudor, J., and Kathawate, G., "Acoustical Advantages of a New Polypropylene Absorbing Material," SAE Technical Paper 1999-01-1969, 1999, doi:https://doi.org/10.4271/1999-01-1969.

10. Watanabe, K., Minemura, Y., Nemoto, K., and Sugawara, H., "Development of High-Performance All-Polyester Sound-Absorbing Materials," *JSAE Review* 20(3):357-362, 1999, doi:10.1016/S0389-4304(99)00014-4.

11. He, H., Zhang, Q., and Fridrich, R., "Vehicle Panel Vibro-Acoustic Behavior and Damping," SAE Technical Paper 2003-01-1406, 2003, doi:https://doi.org/10.4271/2003-01-1406.

12. He, H. and Huang, Q.B., "Damping of a Vehicle Floor System," presented at *the InterNoise02*, Dearborn MI, Aug. 19, 2002.

13. Nagaoka, H., Ohashi, Y., and Suzuki, H., "Development of New Sound Insulator Damping Coat," SAE Technical Paper 2003-01-0232, 2003, doi:https://doi.org/10.4271/2003-01-0232.

14. Saha, P. and Hussaini, A., "A Graduated Assessment of a Sprayable Waterborne Damping Material as a Viable Acoustical Treatment," SAE Technical Paper 2003-01-1588, 2003, doi:https://doi.org/10.4271/2003-01-1588.

15. Subramanian, S., Surampudi, R., Thomson, K., and Vallurupalli, S., "Optimization of Damping Treatment for Structure Borne Noise Reduction," SAE Technical Paper 2003-01-1592, 2003, doi:https://doi.org/10.4271/2003-01-1592.

16. Symietz, D., "High Performance Damping by a New Generation of Spray-On Coatings," SAE Technical Paper 2003-01-1581, 2003, doi:https://doi.org/10.4271/2003-01-1581.

17. Guo, R., Zhang, L.J., Zhao, J., and Zhou, H., "Interior Structure-Borne Noise Reduction by Controlling the Automotive Body Panel Vibration," *Proceedings of the Institution of Mechanical Engineers, Part D: Journal of Automobile Engineering* 226(7):943-956, 2012, doi:10.1177/0954407011433119.

18. Lewitzke, C. and Lee, P., "Application of Elastomeric Components for Noise and Vibration Isolation in the Automotive Industry," SAE Technical Paper 2001-01-1447, 2001, doi:https://doi.org/10.4271/2001-01-1447.

19. Matsumoto, Y. and Griffin, M.J., "Movement of the Upper-Body of Seated Subjects Exposed to Vertical Whole-Body Vibration at the Principal Resonance Frequency," *Journal of Sound and Vibration* 215(4):743-762, 1998, doi:10.1006/jsvi.1998.1595.

20. Mignery, L., "Designing Automotive Dash Panels with Laminated Metal," SAE Technical Paper 1999-01-3201, 1999, doi:https://doi.org/10.4271/1999-01-3201.

21. Brach, R., "Automotive Powerplant Isolation Strategies," SAE Technical Paper 971942, 1997, doi:https://doi.org/10.4271/971942.

22. Liu, S. and Seybert, A.F., "Acoustic Attenuation Analysis of a Resonator Muffler Element Using Boundary Element Method," in: *Proceedings of the 6th International Pacific Conference on Automotive Engineering*, Seoul, Korea, 1991, 1073.

23. Madjlesi, R., Khajepour, A., Ismail, F., Mihalic, J. et al., "Advance Noise Path Analysis, A Robust Engine Mount Optimization Tool," SAE Technical Paper 2003-01-3117, 2003, doi:https://doi.org/10.4271/2003-01-3117.

24. Yu, Y., Peelamedu, S.M., Naganathan, N.G., and Dukkipati, R.V., "Automotive Vehicle Engine Mounting System: A Survey," *Journal of Dynamic Systems Measurement and Control* 123(2):186-194, 2001, doi:10.1115/1.1369361.

25. Ford, D., "An Analysis and Application of a Decoupled Engine Mount System for Idle Isolation," SAE Technical Paper 850976, 1985, doi:https://doi.org/10.4271/850976.

26. Johnson, S. and Subhedar, J., "Computer Optimization of Engine Mounting Systems," SAE Technical Paper 790974, 1979, doi:https://doi.org/10.4271/790974.

27. Kadomatsu, K., "Hydraulic Engine Mount for Shock Isolation at Acceleration on the FWD Cars," SAE Technical Paper 891138, 1989, doi:https://doi.org/10.4271/891138.

28. Liu, C., "A Computerized Optimization Method of Engine Mounting System," SAE Technical Paper 2003-01-1461, 2003, doi:https://doi.org/10.4271/2003-01-1461.

29. Lee, N., Lee, M., Kim, H., and Kim, J., "Design of Engine Mount Using Finite Element Method and Optimization Technique," SAE Technical Paper 980379, 1998, doi:https://doi.org/10.4271/980379.

30. Lee, D., Hwang, W., and Kim, C., "Noise Sensitivity Analysis of an Engine Mount System Using the Transfer Function Synthesis Method," SAE Technical Paper 2001-01-1532, 2001, doi:https://doi.org/10.4271/2001-01-1532.

31. Morita, T., Tsuna, Y., and Okada, A., "Influence of Dynamic Damper Pulley Design on Engine Front Noise," SAE Technical Paper 2001-01-1419, 2001, doi:https://doi.org/10.4271/2001-01-1419.

32. Nakahara, K. and Ohta, K., "Dynamic Characteristics of Cylindrical Hydraulic Engine Mount with Simple Construction Utilizing Air Compressibility," SAE Technical Paper 2000-01-0036, 2000, doi:https://doi.org/10.4271/2000-01-0036.

33. Onorati, A., "Numerical Simulation of Exhaust Flows and Tailpipe Noise of a Small Single Cylinder Diesel Engine," SAE Technical Paper 951755, 1995, doi:https://doi.org/10.4271/951755.

34. Sheng, G., Chen, Z.Y., and Li, L.F., "Torsional Vibrations of Engine Crankshaft System with Variable Moment of Inertia," *Chinese ICE Transaction* 9(2):143, 1991.

35. Togashi, C. and Ichiryu, K., "Study on Hydraulic Active Engine Mount," SAE Technical Paper 2003-01-1418, 2003, doi:https://doi.org/10.4271/2003-01-1418.

36. Athavale, S. and Sajanpawar, P., "Analytical Studies on Influence of Crankshaft Vibrations on Engine Noise Using Integrated Parametric Finite Element Model: Quick Assessment Tool," SAE Technical Paper 1999-01-1769, 1999, doi:https://doi.org/10.4271/1999-01-1769.

37. Geck, P. and Patton, R., "Front Wheel Drive Engine Mount Optimization," SAE Technical Paper 840736, 1984, doi:https://doi.org/10.4271/840736.

38. Jameson, R. and Hodgins, P., "Improvement of the Torque Characteristics of a Small, High-Speed Engine Through the Design of Helmholtz-Tuned Manifolding," SAE Technical Paper 900680, 1990, doi:https://doi.org/10.4271/900680.

39. Jones, A.D., "Modelling the Exhaust Noise Radiated from Reciprocating Internal Combustion Engines - A Literature Review," *Noise Control Eng J* 23(1):12-31, 1984, doi:10.3397/1.2827635.

40. Kuroda, O. and Fujii, Y., "An Approach to Improve Engine Sound," SAE Technical Paper 880083, 1988, doi:https://doi.org/10.4271/880083.

41. Lacin, S., Lopes, E., and Bazzi, B., "An Experimental (In-Vehicle) Study of a Dual Mode Crankshaft Damper for an Engine Crankshaft," SAE Technical Paper 2003-01-1676, 2003, doi:https://doi.org/10.4271/2003-01-1676.

42. Ma, Z. and Perkins, N., "Efficient Engine Models Using Recursive Formulation of Multibody Dynamics," SAE Technical Paper 2001-01-1594, 2001, doi:https://doi.org/10.4271/2001-01-1594.

43. Mancò, S., Nervegna, N., Rundo, M., Armenio, G. et al., "Gerotor Lubricating Oil Pump for IC Engines," SAE Technical Paper 982689, 1998, doi:https://doi.org/10.4271/982689.

44. Li, Y., Kamioka, T., Nouzawa, T., Nakamura, T. et al., "Evaluation of Aerodynamic Noise Generated in Production Vehicle Using Experiment and Numerical Simulation," SAE Technical Paper 2003-01-1314, 2003, doi:https://doi.org/10.4271/2003-01-1314.

45. Iida, N., "Measurement and Evaluation of Aerodynamic Noise," SAE Technical Paper 1999-01-1124, 1999, doi:https://doi.org/10.4271/1999-01-1124.

46. Linden, P.J.G. and Wyckaert, K., "Modular Vehicle Noise and Vibration Development," SAE Technical Paper 1999-01-1689, 1999, doi:https://doi.org/10.4271/1999-01-1689.

47. Diemer, P., Hueser, M., Govindswamy, K., and D'Anna, T., "Aspects of Powerplant Integration With Emphasis on Mount and Bracket Optimization," SAE Technical Paper 2003-01-1468, 2003, doi:https://doi.org/10.4271/2003-01-1468.

48. Widmann, U., Lippold, R., and Fastl, H., "A Computer Program Simulating Post-Masking for Applications in Sound Analysis Systems," *Proceedings of the Noise Conference 1998*, Ypsilanti, MI, 1998.

49. Ushijima, T., Takano, K., and Kojima, H., "High Performance Hydraulic Mount for Improving Vehicle Noise and Vibration," SAE Technical Paper 880073, 1988, doi:https://doi.org/10.4271/880073.

50. Munjal, M.L., *Acoustics of Ducts and Mufflers with Application to Exhaust and Ventilation System Design* (New York: John Wiley & Sons, 1987).

51. Sheng, G., Liu, K., Otremba, J. et al., "A Model and Experimental Investigation of Belt Noise in Automotive Accessory Belt Drive System," *International Journal of Vehicle Noise and Vibration* 1(1-2):68-82, 2004, doi:10.1504/IJVNV.2004.004068.

52. Moeller, M., Thomas, R., Maruvada, H., Chandra, N. et al., "An Assessment of a FEA NVH CAE Body Model for Design Capability," SAE Technical Paper 2001-01-1401, 2001, doi:https://doi.org/10.4271/2001-01-1401.

53. Nakazato, K. and Fukudome, H., "Development of Sprayable Sound Deadening Material for the Floor Panel," *JSAE Review* 18(4):418-421, 1997, doi:10.1016/S0389-4304(97)00040-4.

54. Onsay, T., Akanda, A., and Goetchius, G., "Vibro-Acoustic Behavior of Bead-Stiffened Flat Panels: FEA, SEA, and Experimental Analysis," SAE Technical Paper 1999-01-1698, 1999, doi:https://doi.org/10.4271/1999-01-1698.

55. Du, H., Frederiksen, M., and Happel, S., "Vibration Modeling and Correlation of Driveline Boom for TFWD/AWD Crossover Vehicles," SAE Technical Paper 2003-01-1495, 2003, doi:https://doi.org/10.4271/2003-01-1495.

56. Hatano, S. and Hashimoto, T., "Booming Index as a Measure for Evaluating Booming Sensation," in *Proceedings of INTER-NOISE 2000*, Nice, France, Aug. 1-5, 2000.

57. Matsuyama, S. and Maruyama, S., "Booming Noise Analysis Method Based on Acoustic Excitation Test," SAE Technical Paper 980588, 1998, doi:https://doi.org/10.4271/980588.

58. Pang, J., Dukkipati, R.V., Qatu, M., and Sheng, G., "Nonlinear Seat Cushion and Human Body Model," *International Journal of Vehicle Noise and Vibration* 1(3/4):194, 2005, doi:10.1504/IJVNV.2005.007523.

59. Pang, J., Dukkipati, R.V., Qatu, M., and Sheng, G., "Automotive Seat Cushion Nonlinear Phenomenon: Experimental and Theoretical Evaluation," *International Journal of Vehicle Autonomous Systems* 1(3/4):421-435, 2004, doi:10.1504/IJVAS.2003.004381.

60. Pang, J., Qatu, M., Dukkipati, R.V., Sheng, G. et al., "Model Identification for Nonlinear Automotive Seat Cushion Structure," *International Journal of Vehicle Noise and Vibration* 1(1/2):142-157, 2004, doi:10.1504/IJVNV.2004.004078.

61. Zhang, L., Ning, G., and Yu, Z., "Brake Judder Induced Steering Wheel Vibration: Experiment, Simulation and Analysis," SAE Technical Paper 2007-01-3966, 2007, doi:https://doi.org/10.4271/2007-01-3966.

62. Jiang, W., Yu, Z., and Zhang, L., "Integrated Chassis Control System for Improving Vehicle Stability," *2006 IEEE ICVES*, Shanghai, China, Oct. 2006.

63. Zhang, L., "The Experimental Study of Braking Feel for Some Electric Vehicle," *2006 IEEE ICVES*, Shanghai, China, Oct. 2006.

64. Dom, S., Riefe, M., and Shi, T., "Brake Squeal Noise Testing and Analysis Correlation," SAE Technical Paper 2003-01-1616, 2003, doi:https://doi.org/10.4271/2003-01-1616.

65. Eriksson, M., *Friction and Contact Phenomena of Disc Brakes Related to Squeal,* Comprehensive Summaries of Uppsala Dissertations from the Faculty of Science and Technology, 537 (Uppsala: Acta Universitatis Upsaliensis, 2003), ISBN:91-554-4716-3.

66. Eriksson, M., Bergman, F., and Jacobson, S., "On the Nature of Tribiological Contact in Automotive Brakes," *Wear* 252(1-2):26-36, 2002, doi:10.1016/S0043-1648(01)00849-3.

67. Eriksson, M., Bergman, F., and Jacobson, S., "Surface Characterization of Brake Pads after Running under Silent and Squealing Conditions," *Wear* 232(2):163-167, 1999, doi:10.1016/S0043-1648(99)00141-6.

68. Misra, H., Nack, W., Kowalski, T., Komzsik, L. et al., "Brake Analysis and NVH Optimization Using MSC.NASTRAN," *MSC Conference*, 2002.

69. Papinniemi, A., Lai, J.C.S., Zhao, J., and Loader, L., "Brake Squeal: A Literature Review," *Applied Acoustics* 63(4):391-400, 2002, doi:10.1016/S0003-682X(01)00043-3.

70. Reeves, M., Taylor, N., Edwards, C., Williams, D. et al., "A Study of Disk Modal Behavior during Squeal Generation Using High-Speed Electronic Speckle Pattern Interferometry and Near-Field Sound Pressure Measurements," *Proceedings of the Institution of Mechanical Engineers Part D Journal of Automobile Engineering* 214(3):285-296, 2000, doi:10.1243/0954407001527420.

71. Rhee, S.K., Tsang, P.H.S., and Wang, Y.S., "Friction-Induced Noise and Vibration of Disc Brakes," *Wear* 133(1):39-45, 1989, doi:10.1016/0043-1648(89)90111-7.

72. Tirovic, M. and Day, A.J., "Disc Brake Interface Pressure Distributions," *Proceedings of the Institution of Mechanical Engineers Part D: Journal of Automobile Engineering* 205(24):137-146, 1991, doi:10.1243/PIME_PROC_1991_205_162_02.

73. Watany, M., Abouel-Seoud, S., Saad, A., and Abdel-Gawad, I., "Brake Squeal Generation," SAE Technical Paper 1999-01-1735, 1999, doi:https://doi.org/10.4271/1999-01-1735.

74. Walter, J., Duell, E., Martindale, B., Arnette, S. et al., "The DaimlerChrysler Full-Scale Aeroacoustic Wind Tunnel," SAE Technical Paper 2003-01-0426, 2003, doi:https://doi.org/10.4271/2003-01-0426.

75. Wickern, G. and Lindener, N., "The Audi Aeroacoustic Wind Tunnel: Final Design and First Operational Experience," SAE Technical Paper 2000-01-0868, 2000, doi:https://doi.org/10.4271/2000-01-0868.

76. Zou, T., Mourelatos, Z., and Mahadevan, S., "Simulation-Based Reliability Analysis of Automotive Wind Noise Quality," SAE Technical Paper 2004-01-0238, 2004, doi:https://doi.org/10.4271/2004-01-0238.

77. Ahmadian, M. and Huang, W., "A Numerical Evaluation of the Suspension and Driveline Dynamic Coupling in Heavy Trucks," SAE Technical Paper 2004-01-2711, 2004, doi:https://doi.org/10.4271/2004-01-2711.

78. Bunne, J. and Jable, R., "Air Suspension Factors in Driveline Vibration," SAE Technical Paper 962207, 1996, doi:https://doi.org/10.4271/962207.

79. Capitani, R., Delogu, M., and Pilo, L., "Analysis of the Influence of a Vehicle's Driveline Dynamic Behaviour Regarding the Performance Perception at Low Frequencies," SAE Technical Paper 2001-01-3333, 2001, doi:https://doi.org/10.4271/2001-01-3333.

80. Gilbert, D., O'Leary, M., and Rayce, J., "Integrating Test and Analytical Methods for the Quantification and Identification of Manual Transmission Driveline Clunk," SAE Technical Paper 2001-01-1502, 2001, doi:https://doi.org/10.4271/2001-01-1502.

81. Kazemi, R., Hamedi, B., and Izadkhah, M., "The Vibrational Improvement of the Two-Piece Driveline of the Passenger Car," SAE Technical Paper 2002-01-1320, 2002, doi:https://doi.org/10.4271/2002-01-1320.

82. Krishna, M., Yoshioka, J., and Sharma, M., "A Web Based Finite Element Frequency Analysis Program for Designing a Driveline System," SAE Technical Paper 2004-01-2724, 2004, doi:https://doi.org/10.4271/2004-01-2724.

83. Lee, Y. and Kocer, F., "Minimize Driveline Gear Noise by Optimization Technique," SAE Technical Paper 2003-01-1482, 2003, doi:https://doi.org/10.4271/2003-01-1482.

84. Pang, J., Dukkipati, R., and Patten, W.N., "Model Recovery Technique from Reduced Model," *International Journal of Vehicle Design* 29(4):317, 2002, doi:10.1504/ IJVD.2002.002016.

85. Pang, J., Dukkipati, R., Patten, W.N., and Sheng, G., "Comparative Analysis of Model-Reduction Methods," *International Journal of Heavy Vehicle Systems* 10(3):224, 2003, doi:10.1504/IJHVS.2003.003208.

86. Parrett, A., Zhang, Q., Wang, C., and He, H., "SEA in Vehicle Development Part I: Balancing of Path Contribution for Multiple Operating Conditions," SAE Technical Paper 2003-01-1546, 2003, doi:https://doi.org/10.4271/2003-01-1546.

87. Pielemeier, W., Greenberg, J., Jeyabalan, V., and van Niekerk, J., "The Estimation of SEAT Values from Transmissibility Data," SAE Technical Paper 2001-01-0392, 2001, https://doi.org/10.4271/2001-01-0392.

88. Trapp, M., McNulty, P., and Chu, J., "Frictional and Acoustic Behavior of Automotive Interior Polymeric Material Pairs Under Environmental Conditions," SAE Technical Paper 2001-01-1550, 2001, doi:https://doi.org/10.4271/2001-01-1550.

89. Trapp, M.A., Karpenko, Y., Qatu, M.S. et al., "An Evaluation of Friction-and Impact-Induced Acoustic Behaviour of Selected Automotive Materials, Part I: Friction-Induced Acoustics," *International Journal of Vehicle Noise and Vibration* 3(4):355-369, 2007, doi:10.1504/IJVNV.2007.016398.

90. Qian, Y. and Vanbuskirk, J., "Sound Absorption Composites and Their Use in Automotive Interior Sound Control," SAE Technical Paper 951244, 1995, doi:https://doi.org/10.4271/951244.

91. Sakai, T., Terada, M., Ono, S., Kamimura, N. et al., "Development Procedure for Interior Noise Performance by Virtual Vehicle Refinement, Combining Experimental and Numerical Component Models," SAE Technical Paper 2001-01-1538, 2001, doi:https://doi.org/10.4271/2001-01-1538.

92. Salaani, M., Chrstos, J., and Guenther, D., "Parameter Measurement and Development of a NADSdyna Validation Data Set for a 1994 Ford Taurus," SAE Technical Paper 970564, 1997, doi:https://doi.org/10.4271/970564.

93. Uchida, H. and Ueda, K., "Detection of Transient Noise of Car Interior Using Non-Stationary Signal Analysis," SAE Technical Paper 980589, 1998, doi:https://doi.org/10.4271/980589.

94. Bray, W.R., Blommer, M., and Lake, S., *Sound Quality 2005 Workshop, SAE Noise & Vibration Conference*, Traverse City, 2005 (Warrendale, PA: SAE International).

95. Moeller, M., Thomas, R., Chen, S., Chandra, N. et al., "NVH CAE Quality Metrics," SAE Technical Paper 1999-01-1791, 1999, doi:https://doi.org/10.4271/1999-01-1791.

96. Lee, S., Kim, B., Chae, H., Park, D. et al., "Sound Quality Analysis of a Passenger Car Based on Rumbling Index," SAE Technical Paper 2005-01-2481, 2005, doi:https://doi.org/10.4271/2005-01-2481.

97. Brines, R., Weiss, L., and Peterson, E., "The Application of Direct Body Excitation Toward Developing a Full Vehicle Objective Squeak and Rattle Metric," SAE Technical Paper 2001-01-1554, 2001, doi:https://doi.org/10.4271/2001-01-1554.

98.   Cerrato-Jay, G., Gabiniewicz, J., Gatt, J., and Pickering, D., "Automatic Detection of Buzz, Squeak and Rattle Events," SAE Technical Paper 2001-01-1479, 2001, doi:https://doi.org/10.4271/2001-01-1479.

99.   Juneja, V., Rediers, B., Kavarana, F., and Kimball, J., "Squeak Studies on Material Pairs," SAE Technical Paper 1999-01-1727, 1999, doi:https://doi.org/10.4271/1999-01-1727.

100. Kavarana, F. and Rediers, B., "Squeak and Rattle - State of the Art and Beyond," *Sound and Vibration Magazine* 35(4):56-65, 2001, doi:10.4271/1999-01-1728.

101. Peterson, C., Wieslander, C., and Eiss, N., "Squeak and Rattle Properties of Polymeric Materials," SAE Technical Paper 1999-01-1860, 1999, doi:https://doi.org/10.4271/1999-01-1860.

102. Hsieh, S., Borowski, V., Her, J., and Shaw, S., "A CAE Methodology for Reducing Rattle in Structural Components," SAE Technical Paper 972057, 1997, doi:https://doi.org/10.4271/972057.

103. Trapp, M. and Pierzecki, R., "Squeak and Rattle Behavior of Elastomers and Plastics: Effect of Normal Load, Sliding Velocity, and Environment," SAE Technical Paper 2003-01-1521, 2003, doi:https://doi.org/10.4271/2003-01-1521.

104. Yang, B. and Rediers, B., "Experimental Investigation of a Friction-Induced Squeak Between Head Restraint Fabric and Glass," SAE Technical Paper 951269, 1995, doi:https://doi.org/10.4271/951269.

105. Casati, F., Berthevas, P., Herrington, R., and Miyazaki, Y., "The Contribution of Molded Polyurethane Foam Characteristics to Comfort and Durability of Car Seats," SAE Technical Paper 1999-01-0585, 1999, doi:https://doi.org/10.4271/1999-01-0585.

106. Donavan, P. and Rymer, B., "Assessment of Highway Pavements for Tire/Road Noise Generation," SAE Technical Paper 2003-01-1536, 2003, doi:https://doi.org/10.4271/2003-01-1536.

107. Park, S., Cheung, W., Cho, Y., and Yoon, Y., "Dynamic Ride Quality Investigation for Passenger Car," SAE Technical Paper 980660, 1998, doi:https://doi.org/10.4271/980660.

108. Zhu, J., Roggenkamp, T., and Yan, D., "Lab-to-Lab Correlation for Tire Noise Load Cases," SAE Technical Paper 2003-01-1533, 2003, doi:https://doi.org/10.4271/2003-01-1533.

109. Harriaon, M., *Vehicle Refinement: Controlling Noise and Vibration in Road Vehicle* (London, UK: Elsevier Butterworth-Heinemann, 2004), doi:10.1016/B978-0-7506-6129-4.X5000-7.

110. Kim, B.S., Kim, G.J., and Lee, T.K., "The Identification of Sound Generating Mechanisms of Tires," *Applied Acoustics* 68(1):114-133, 2007, doi:10.1016/j.apacoust.2006.05.019.

111. Minakawa, M., Nakahara, J., Ninomiya, J., and Orimoto, Y., "Method for Measuring Force Transmitted from Road Surface to Tires and Its Applications," *JSAE Review* 20(4):479-485, 1999, doi:10.1016/S0389-4304(99)00054-5.

112. El-Demerdash, S., Selim, A., and Crolla, D., "Vehicle Body Attitude Control Using an Electronically Controlled Active Suspension," SAE Technical Paper 1999-01-0724, 1999, doi:https://doi.org/10.4271/1999-01-0724.

113. Elliott, A. and Wheeler, G., "Validation of ADAMS® Models of Two USMC Heavy Logistic Vehicle Design Variants," SAE Technical Paper 2001-01-2734, 2001, doi:https://doi.org/10.4271/2001-01-2734.

114. Kinkelaar, M., Neal, B., and Crocco, G., "The Influence of Polyurethane Foam Dynamics on the Vibration Isolation Character of Full Foam Seats," SAE Technical Paper 980657, 1998, doi:https://doi.org/10.4271/980657.

115. Guo, R., Qiu, S., Yu, Q.L., Zhou, H. et al., "Transfer Path Analysis and Control of Vehicle Structure-Borne Noise Induced by the Powertrain," *Proceedings of the Institution of*

Mechanical Engineers, Part D: Journal of Automobile Engineering 226(8):1100-1109, 2012, doi:10.1177/0954407012438501.

116.   Lee, S., Park, K., Lee, M., Rho, K. et al., "Vibrational Power Flow and Its Application to a Passenger Car for Identification of Vibration Transmission Path," SAE Technical Paper 2001-01-1451, 2001, doi:https://doi.org/10.4271/2001-01-1451.

117.   Magalhães, M., Arruda, F., and Filho, J., "Driveline Structure-Borne Vibration and Noise Path Analysis of an AWD Vehicle Using Finite Elements," SAE Technical Paper 2003-01-3641, 2003, doi:https://doi.org/10.4271/2003-01-3641.

118.   DeJong, R., Bharj, T., and Lee, J., "Vehicle Wind Noise Analysis Using a SEA Model with Measured Source Levels," SAE Technical Paper 2001-01-1629, 2001, doi:https://doi.org/10.4271/2001-01-1629.

119.   Chen, K., Johnson, J., Dietschi, U., and Khalighi, B., "Wind Noise Measurements for Automotive Mirrors," *SAE Int. J. Passeng. Cars - Mech. Syst.* 2(1):419-433, 2009, doi:https://doi.org/10.4271/2009-01-0184.

120.   Dobrzynski, W. and Soja, H., "Effect on Passenger Car Wind Noise Sources of Different A-Post Configurations," presented at *the Inter-Noise Conference 94*, Yokohama, Japan, Aug. 29-31, 1994.

121.   Hoshino, H. and Katoh, H., "Evaluation of Wind Noise in Passenger Car Compartment in Consideration of Auditory Masking and Sound Localization," SAE Technical Paper 1999-01-1125, 1999, doi:https://doi.org/10.4271/1999-01-1125.

122.   Hoshino, H. and Kato, H., "An Objective Evaluation Method of Wind Noise in a Car Based on a Model of Subjective Evaluation Process," *Japanese Society of Automotive Engineers, Annual Congress*, September 12, 2000, doi:10.3397/1.2827853.

123.   Zhang, L., Yang, S., and Yu, Z., "Elementary Investigation into Road Simulation Experiment of Powertrain and Components of Fuel Cell Passenger Car," *Journal of Nursing Care Quality* 28(2):147-152, 2013, doi:10.1097/NCQ.0b013e318277e874.

124.   Qatu, M., Sirafi, M., and Johns, F., "Robustness of Powertrain Mount System for Noise, Vibration and Harshness at Idle," *Proceedings of the Institution of Mechanical Engineers, Part D: Journal of Automobile Engineering* 216(10):805-810, 2002, doi:10.1177/095440700221600103.

125.   Singh, V., Wani, N., Monkaba, V., Blough, J. et al., "Powertrain Transfer Path Analysis of a Truck," SAE Technical Paper 2001-01-2817, 2001, doi:https://doi.org/10.4271/2001-01-2817.

126.   Hwang, S., Stout, J., and Ling, C., "Modeling and Analysis of Powertrain Torsional Response," SAE Technical Paper 980276, 1998, doi:https://doi.org/10.4271/980276.

127.   Takeuchi, Y., Tsukahara, H., and Sato, M., "Prediction of Excitation Forces and Powerplant Vibration in a Horizontally Opposed Engine," SAE Technical Paper 980283, 1998, doi:https://doi.org/10.4271/980283.

128.   Blanchet, D. and Cunningham, A., "Building 3D SEA Models from Templates - New Developments," SAE Technical Paper 2003-01-1541, 2003, doi:https://doi.org/10.4271/2003-01-1541.

129.   Bray, W.R., "The 'Relative Approach' for Direct Measurement of Noise Patterns," *Sound & Vibration* 38(9):20-23, 2004.

130.   Bremner, P. and Zhu, M., "Recent Progress Using SEA and CFD to Predict Interior Wind Noise," SAE Technical Paper 2003-01-1705, 2003, doi:https://doi.org/10.4271/2003-01-1705.

131.   Data, S., Pascali, L., and Santi, C., "Handling Objective Evaluation Using a Parametric Driver Model for ISO Lane Change Simulation," SAE Technical Paper 2002-01-1569, 2002, doi:https://doi.org/10.4271/2002-01-1569.

132. Weisch, G., Stücklschwaiger, W., de Mendonca, A., Monteiro, N. et al., "The Creation of a Car Interior Noise Quality Index for the Evaluation of Rattle Phenomena," SAE Technical Paper 972018, 1997, https://doi.org/10.4271/972018.

133. Namba, S., Kuwano, S., Kinoshita, A. et al., "Psychological Evaluation of Noise in Passenger Cars - The Effect of Visual Monitoring and the Measurement of Habituation," *Journal of Sound and Vibration* 205(4):427-433, 1997, doi:10.1006/jsvi.1997.1008.

134. Blewett, J. and Brouckaert, R., "The Importance of Sealing Pass-Through Locations Via the Front of Dash Barrier Assembly," SAE Technical Paper 1999-01-1802, 1999, doi:https://doi.org/10.4271/1999-01-1802.

135. Kazemi, R., Hamedi, B., and Javadi, B., "Improving the Ride & Handling Qualities of a Passenger Car via Modification of Its Rear Suspension Mechanism," SAE Technical Paper 2000-01-1630, 2000, doi:https://doi.org/10.4271/2000-01-1630.

136. Kim, T., Cho, Y., Yoon, Y., and Park, S., "Dynamic Ride Quality Investigation and DB of Ride Values for Passenger and RV Cars," SAE Technical Paper 2001-01-0384, 2001, doi:https://doi.org/10.4271/2001-01-0384.

137. Sheng, G., Jiang, C.H., and Yang, S.Z., "Dynamic Modeling of Pass-By Vehicle on Bridge," *Journal of Vibration, Measurement & Diagnosis* 12(2):20-26, 1992.

138. Yu, L. and Khameneh, K., "Automotive Seating Foam: Subjective Dynamic Comfort Study," SAE Technical Paper 1999-01-0588, 1999, doi:https://doi.org/10.4271/1999-01-0588.

139. Foumani, M., Khajepour, A., and Durali, M., "Application of Shape Memory Alloys to a New Adaptive Hydraulic Mount," SAE Technical Paper 2002-01-2163, 2002, doi:https://doi.org/10.4271/2002-01-2163.

140. Sheng, G., Chen, Z.Y., and Li, L.F., "Torsional Vibrations of Engine Crankshaft System with Variable Moment of Inertia," *Chinese ICE Transaction* 9(2):143, 1991.

141. Iqbal, J. and Qatu, M., "Robustness of Axle Mount System for Driveline NVH," SAE Technical Paper 2003-01-1485, 2003, doi:https://doi.org/10.4271/2003-01-1485.

142. Marjoram, R., "Pressurized Hydraulic Mounts for Improved Isolation of Vehicle Cabs," SAE Technical Paper 852349, 1985, doi:https://doi.org/10.4271/852349.

Electric Motor Vibrations and Noise

4.1 Motors for BEV/HEV/FCEV

Electrified vehicles are vehicles wholly or partially driven by electric motors. Specifically, they are battery-powered electric vehicles (BEVs), fuel cell electric vehicles (FCEVs), and hybrid electric vehicles (HEVs). Induction machines (IM), switched reluctance machines (SRM), and permanent magnet synchronous machines (PMSM) are standard motor topologies in current BEVs/FCEVs/HEVs [1–9].

More electric cars also mean the more extensive utilization of different electric motor systems for drive safety, comfort, communications, etc. Electric drives have been used as a solution for attaining a higher comfort, including drive-by-wire or break-by-wire systems, electric windows, electric seats, heating and ventilation systems, and so on.

In electric drive systems, motors consist of following major components: rotor is the moving part which turns the shaft to output mechanical power. It usually has conductors, laid into it which carry currents that interact with the magnetic field of the stator to generate the forces that drive the shaft. Some rotor has permanent magnets and the stator holds the conductors. Stator is the stationary part, usually has either winding or permanent magnets. Air gap is kept to provide the space between the rotor and the stator. It is an important factor to the power generation and wave form quality. In general, it is put as small as possible as a large gap has a strong negative effect on the performance of the motor. Windings are wires laid in coils and usually wrapped around a laminated soft iron magnetic core so as to form the magnetic poles when energized with current. The rotor is supported by bearings which allow the rotor to rotate on its axis. The bearings are supported by the motor housing. The motor shaft extends through the bearings to the outside of the motor, where the load is applied. The motor's armature current is supplied through the stationary brushes in contact with the revolving commutator, which results required current reversal and applies powers to the machine in an optimal manner as the rotor rotates from pole to pole. Endcap assembly consists of a brush holder, endcap, brush, brush leaf, and terminals. It is used to transfer current from source to commutator.

The automotive electric motors applications require a big working range in speed, supply frequency, and torque capability. The most suitable motor types for this application are the SRM, IM, and PMSM. In some research, the IM is illustrated to have a better overall score in the automotive mass-produced scale market where the concerns of noise, cost, and fault tolerance are essential.

Due to the very important characteristics of high power density and high efficiency, PMSMs have become the preferred choice in various electric drives applications.

For example, the uses of IM in Tesla Motors Model S, Roadster, Toyota RAV4 EV, and Mercedes B-Class electric drive prove that IM technology is very well-suited for high-end and high-performance sports cars, as well as for SUVs and compact city cars. On the other hand, the efficiency of PMSM is higher than IM and SRM. Moreover, since the control of IM is more complex and the torque ripple of SRM is more conspicuous, PMSM has become a common solution for EV motors.

The Toyota 2004 model has been widely used as a benchmark for electric drives, whose torque-speed envelop is shown in **Figure 4.1**. Its peak torque is 300 Nm up to base speed of 1500 rpm. It has torque of 60 Nm at maximum speed of 6000 rpm. Its electric motor type is permanent magnet synchronous motor.

The structures of the three kinds of motor have been widely investigated [11–15]. Typical structures and topologies are shown in **Figure 4.2**, and their comparison is given in **Table 4.1**.

FIGURE 4.1 Toyota Prius motor's torque performance [10].

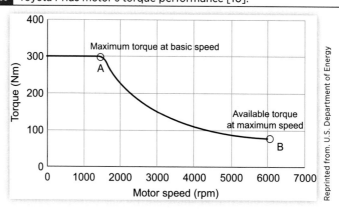

Reprinted from. U.S. Department of Energy

FIGURE 4.2 Three popular motors in EVs: Switched reluctance machines (SRM), induction machines (IM), and permanent magnet synchronous machines (PMSM).

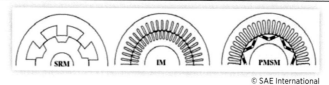

© SAE International

TABLE 4.1 Comparison of motors used in HEVs.

Category	PMSM	Induction	SRM
Power density	High	Medium	Medium
Torque ripple	Low	Medium	High
Acoustic noise	Low	Medium	High
PM material	Yes	No	No
Motor cost	High	Moderate	Low
Motor efficiency	High	Medium	Medium
Reliability	Medium	High	Medium
Size	Small	Large	Small
Maximum revolution	High	Medium	High
Overload capacity	Medium	High	Low
Temperature admission	Low	Medium	High
Construction	Complex	Complex	Simple

4.2 **Motor-Induced NVH**

In terms of the magnitudes of generated acoustic noise and torque ripple, PMSM is usually low, induction motor is medium, and SRM is high.

Electric motors transfer electric energy to kinetic energy and generate noise and vibrations as a side effect. The magnetic flux crossing the air gap between the stator and rotor results in radial and tangential forces. Radial forces are the main source exciting the stator housing to have vibration and noise. Vibrations of the stator housing could be transferred to other structure or create airborne sound. Mechanical noise is caused by torque ripple that excites gears and couplings connected to motor. The torque ripple is coupled to motor orders that depend on the combination of slots and poles in the motor [16].

In the next, we present the basic principles of noise and vibration generation from electric motors, introduce the analysis of motors, and present countermeasures of motor noise and vibration. The following observations have been drawn for motors used in typical EVs/HEVs [16]:

- Motors are used for power conversion, i.e., converting electrical energy to mechanical energy.
- Motor and generator are operated under the same principle, i.e., Maxwell theory. Therefore, a reversing operation of a motor becomes a generator and vice versa. This is applied in regenerative braking of many EV/HEV.
- The interaction between the stator magnetic field flux and the rotor current produces a force on the rotor coil, called "Lorentz Force," which generates the rotating torque.
- The higher magnetic flux density in the air gap produces higher vibration and noise.
- With the increase power density of motors for EV/HEV and more environmental requirements, the prediction of noise/vibration of motors at the early stage of design has become a very critical issue.

- The prediction of noise is rather difficult due to the complexity of the structure and only very small amount of energy is converted into noise and vibration.

Generally, vibration and noise produced by electrical machines can be divided into three categories [16]:

- Electromagnetic vibration and noise associated with parasitic effects due to higher space and time harmonics, eccentricity, phase unbalance, slot openings, magnetic saturation, and magnetostrictive expansion of the core laminations;
- Mechanical vibration and noise associated with the mechanical assembly, such as bearings;
- Aerodynamic vibration and noise associated with the flow of ventilating air through or over the motor.

Figure 4.3 shows the various sources of acoustic noise in electric motors: aerodynamic, mechanical, and electromagnetic.

The electromagnetic excitation is generally the dominating source of noise and vibration in low-to-medium power rated machines (for example, a rated power lower than 15 kW and rotational speed lower than 1500 rpm), which includes cogging torque, torque ripple, and magnetic radial forces.

The electromagnetic force acting on the cores of the stator and rotor produces noise and vibration, especially when the frequencies of the exciting forces are equal to the natural frequencies of the machine system. In most of the practical cases, the predominant acoustic noise is produced by the radial force.

FIGURE 4.3 Noise sources and transmission in electrical motors.

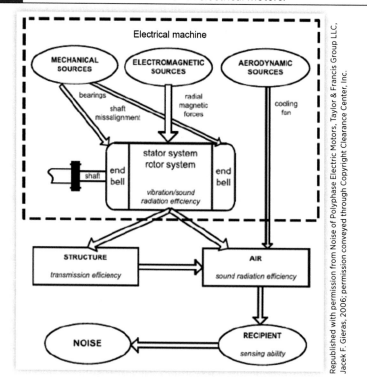

Republished with permission from Noise of Polyphase Electric Motors, Taylor & Francis Group LLC, Jacek F. Gieras, 2006; permission conveyed through Copyright Clearance Center, Inc.

The electromagnetic force is the main cause of noise and vibration of motor, rather than the torque ripple and cogging torque and it is also found that the root cause of noise and vibration is the radial forces not the torque ripples. The acoustic noise in electric machines has a close correlation with the resonant frequencies of stator and the harmonics of normal force.

The ripple torque is produced from the harmonic content of the current and voltage waveforms in the electric machine. In the case of multiphase motors, torque ripple is a crucial factor in the performance and operation because it determines the magnitude of the machine's vibration and acoustic noise. The pulse-width modulation (PWM) strategies add many harmonics to the air-gap Maxwell forces spectrum, leading to possibly harmful noise and vibrations. Stator windings induce in rotor currents additional time harmonics which can also significantly enrich the electromagnetic forces spectrum, especially when running at high slip. Audible electromagnetic noise spectrum, therefore, results from a complex combination of both PWM time harmonics and winding space harmonics.

The cogging torque is produced by the magnetic attraction between the rotor mounted permanent magnets and the stator teeth.

The reduction of cogging torque and torque ripple can also reduce the vibration and acoustic noise of motor.

The acoustic noise in PMSM is lower compared to the noise generated by SRM and IM.

Acoustic noise emitted by an electric machine is due to the electromagnetic forces and the stator frame interaction.

The acoustic noise generated by the mechanical and aerodynamic sources is mainly connected to the mechanical structure of the electric motor.

The load-induced sources of noise include:

- Noise due to coupling of the machine with a load, e.g., shaft misalignment, belt transmission, elevator sheave with ropes, tooth gears, coupling, and reciprocating compressor;
- Noise due to mounting the machine on foundation or other structure.

In many applications, PMSM has the quietest operation. SRM is the noisiest. PMSM has higher mode number, so less stator deformation as vibrational amplitude is inversely proportional to fourth power of mode order. PMSM exceeds the IM in power density and efficiency. The SRM provides a power density and efficiency comparable to the IM.

Figure 4.3 shows the noise sources and transmission in electrical motors, which has the following features:

- NVH sources generated by various components are transmitted through the medium (structure and air) to the recipient (human ear, sensors, and whole vehicle).
- All NVH sources are sensed by the human's ear and body.
- NVH causes both structure damages, vehicle handling and human discomfort.

The mechanical source consists of self, load-induced and auxiliaries which consist of bearing, balancing, and nonuniform air gap.

4.3 **Electromagnetic Forces and Torque Pulsation**

Electromagnetic forces contributing to the vibration and noise of an electric machine can be classified as the following categories [17–23].

- Forces acting on the surface of the stator teeth: These forces are normal to the surface of the iron and are responsible for the majority of the radial and tangential vibration in an electric machine.
- Forces acting on current-carrying stator windings: These forces can initiate vibration in the coils, which may lead to failure of insulation and short-circuit. This, in turn, can cause permanent damage into the stator laminations (due to large magnitude of the short circuit current). These forces are referred to as Laplace forces.

Magnetostrictive forces are another cause of vibration: these forces are caused in lieu of subjecting a ferromagnetic material to an external magnetic field. Heavily saturated stators subject to fast-changing magnetic fields can result in significantly large magnetostrictive forces.

Magnetic radial forces, cogging torque, and torque ripple as the main-electromagnetic sources of noise and vibration will be presented in this section.

4.3.1 **Electromagnetic Forces**

As an electrical machine operates, there will have forces and torques acting on its structure. The shaft torque is naturally the one to be maximized and all the other forces and torques are considered as parasitic in nature, i.e., they do not participate in the process of feeding active power to the desired process.

Three types of magnetic forces exist in an electrical machine [17– 23]:

1. The reluctance force, which acts on the boundaries of materials with different magnetic properties.
2. The Lorentz force, which acts on currents in the magnetic field.
3. The magnetostrictive force, which acts inside the iron core.

The air gap between the stator and rotor is the region where the reluctance force acts, and the tooth tips especially serve as an interface for the forces to deliver parasitic energy to the structure. The forces depend on the geometry of the slots, the length of the air gap, the winding arrangements, the degree of saturation, etc. The reluctance force is the most important excitation type, since its magnitude is usually the largest when compared with the other two forces.

The Lorentz force acts on the windings of the machine, but since the flux density in a slot is small, the force remains quite negligible.

Magnetostriction can play an important role in the excitation of structures [18, 20–23].

Since the reluctance force or stress is a 3-D vector and the air gap is cylindrical, a division into radial, tangential/circumferential, and axial components is reasonable. The radial components are usually the greatest in magnitude and they originate from the magnetic field trying to close the air gap. The magnitudes of the tangential components are usually smaller than those of the radial ones, but the static component which has a nonzero value when integrated around the rotor circumference produces the desired torque. Axial forces are usually negligible. To achieve the desired torque performance, additional radial and circumferential components are needed.

In the mechanical design of an electrical machine, a critical point is whether the structure can withstand all the forces acting on it.

One source of stresses is the magnetic forces acting on the machine, and it would be desirable to estimate their magnitudes. Sometimes, even a small magnetic force component, which is harmless to the structure, can be annoying to driver if it creates acoustic noise or vibration. The force component is dynamic in nature and has a certain spatial distribution around the air gap. The radial and tangential force components lead to the sound, and thus their properties are critical in magnetic noise generation.

For the event of axial variation of eccentricity in a tilted rotor, the magnetic force also adopts variation in the axial direction.

Attracting forces exist between the magnetic poles of the permanent magnets of the stator and the electromagnets of the rotor. The resulting force can be split into two components: One points tangentially to the arc of the rotor movement and the other points perpendicularly to the first one. The forces acting along the axis of the rotor coil have the same magnitude but point in opposite directions. The sum of those forces is zero. The remaining forces pointing tangentially to the rotor turn to rotate the armature anticlockwise.

The forces in motor cause vibrations in the structure of the electric motor and EV. The excitation forces can be divided into three kinds: electromagnetic force, mechanical force, and air friction force. Both the electromagnetic force and rotating torque which provide power to drive the electric machine are produced in the magnetic field in the air gap. The magnetic force causing vibrations of the motor is due to the flux in the air gap. Magnetic flux crossing an interface between an infinitely permeable and a vacuum medium only exists in a vertical direction to the surface. The fundamentals and theoretical formula of these forces are given as follows [17–23]. The fundamental magnetic equations of Maxwell are

$$\nabla \times E = -\frac{\partial B}{\partial t}$$

$$\nabla \cdot B = 0$$

$$\nabla \times H = J + \frac{\partial D}{\partial t} \tag{4.1}$$

$$\nabla \cdot D = \rho$$

where
   E is the electric field intensity
   B is the magnetic flux density
   H is the magnetic field intensity
   J is the surface current density
   D is the electric flux density

For stationary and quasi-stationary electromagnetic field distribution situations, the displacement current density is neglected, which gives the equation:

$$\nabla \times H = J \tag{4.2}$$

In three dimensions, when the divergence of the curl of any vector field is equal to zero, the following equation can be obtained:

$$\nabla \cdot J = \nabla \cdot (\nabla \times H) = 0 \tag{4.3}$$

The properties of macroscopic material are defined by the constitutive relations which are shown as below:

$$J = \sigma E$$

$$D = \varepsilon E = \varepsilon_0 \varepsilon_r E$$

$$B = \mu H = \mu_0 \mu_r H \tag{4.4}$$

Here, σ is the electric conductivity, ε is the electric permittivity which is given as $\varepsilon_0 * \varepsilon_r$, where ε_0 is the permittivity in free space and ε_r is the relative permittivity that determines the electric field solution in the insulators. It can be simple or anisotropic.

μ is the magnetic permeability, μ_0 is the permeability in free space, and μ_r is the relative permeability along with the magnetic coercivity determine the magnetic properties of the material. It can be simple (linear) or nonlinear (BH Curve) or anisotropic.

The solutions for transient state formulations are briefed as follows.

Consider that A is the magnetic vector potential that is defined as:

$$\nabla \times A = B \tag{4.5}$$

It is defined in the whole problem region and can be coupled with the electric scalar potential V which is defined by:

$$E = -\frac{\partial A}{\partial t} - \nabla V \tag{4.6}$$

For the electric scalar potential, one high value and low value of V are defined separately at the two ends of the conducting region. But V equals to zero at the nonconducting region. Therefore, combining the above equations, the magnetic flux density B and current density J can be calculated.

$$J = \sigma E = -\sigma \left(\frac{\partial A}{\partial t} + \nabla V \right) = \nabla \times H = \nabla \times \frac{B}{\mu} = \frac{1}{\mu} \nabla (\nabla \times A) \tag{4.7}$$

By using Coulomb Gauge,

$$\nabla \cdot A = 0 \tag{4.8}$$

The transient state formulations for electromagnetic field will get the solution.

$$\nabla \times \left(\frac{1}{\mu} \nabla \times A \right) = \nabla \times H_c - \sigma \left(\frac{\partial A}{\partial t} + \nabla V \right) + \frac{1}{\mu} \nabla (\nabla \cdot A) \tag{4.9}$$

where H_c is the coercive magnetic field intensity.

There are three types of magnetic force existed in an electrical machine. They are reluctance force that acts on the material boundaries with different magnetic properties,

Lorentz force which acts on windings in the magnetic field, and magnetostrictive force acting inside the iron core. The most important vibration excitation is the reluctance force produced in the air gap between the stator and rotor because of the larger magnitude and the less complexity compared with the other forces.

The reluctance force of electric motor is electromagnetic force which has two components, radial and tangential components. In a two-dimensional model, the material boundaries could be chosen as the edge of stator tip. The two components of electromagnetic force can be calculated according to the equation,

$$F_{rad} = \frac{L_{stk}}{2\mu_0} \oint_l \left(B_n^2 - B_t^2 \right) dl \tag{4.10a}$$

$$F_{tan} = \frac{L_{stk}}{2\mu_0} \oint_l B_n \cdot B_t \, dl \tag{4.10b}$$

where

B_n and B_t are the normal and tangential component of flux densities respectively

l is the length of the stator tip edge

L_{stk} is stack length of the machine

Machines having the diametrically asymmetric disposition of air gap is called eccentricity when the center of rotor and stator is not coincident. This often causes the undesired force waves and the electromagnetic force excitation will be unpredictable.

The radial force density distribution on the stator surface, which results from the air-gap magnetic field under no-load (open-circuit) and on-load conditions, is the main cause of electromagnetically induced noise and vibration and can be evaluated analytically by Maxwell's stress method discussed as above, $B_n = B_r(\theta_s,t)$, $B_t = B_\theta(\theta_s,t)$

$$f_{rad}(\theta_s,t) = \frac{1}{2\mu_0} \left[B_r^2(\theta_s,t) - B_\theta^2(\theta_s,t) \right] \tag{4.11}$$

where

f_{rad} is the radial component of force density

B_r and B_θ are the radial and tangential components of the magnetic flux density in the air-gap

μ_0 is the permeability of free space

θ_s is the angular position at the stator bore

t is the time

For PM machines, the air-gap flux density consists of the combination effect of the permanent magnet and the armature reaction fields

$$B_r(\theta_s,t) = B_{r,PM}(\theta_s,t) + B_{r,S}(\theta_s,t) \tag{4.12}$$

$$B_\theta(\theta_s,t) = B_{\theta,PM}(\theta_s,t) + B_{\theta,S}(\theta_s,t)$$

The index r and θ are used for the radial and tangential components, respectively.

Substituting the field flux densities into the equation of radial component of force density, the radial force density can be expressed as:

$$f_{rad}(\theta_s,t) = \frac{1}{2\mu_0} \left[B_{r,PM}(\theta_s,t) + B_{r,S}(\theta_s,t) \right]^2 \tag{4.13}$$

In the following analysis, the radial force density will be examined by considering the contribution due to the radial flux density components alone. This is done for the sake of simplicity; analogous equations like for the radial components hold true for the tangential components. Thus, from the above equation, we get the following:

4.3.1.1 **Air-Gap Flux Density Components:** Extended analytical expressions for the air-gap flux density components produced by the stator currents and permanent magnets can be obtained by taking into account stator and rotor MMF harmonics, rotor permeance effect, and the stator slotting. The resulting formulas for the main air-gap flux density components are presented in the following:

$$B_{\mathrm{PM}}\left(\phi_s,\theta\right) = B_{\mathrm{PM1}}\left(\phi_s,\theta\right) + B_{\mathrm{PM2}}\left(\phi_s,\theta\right)$$

$$B_S\left(\phi_s,\theta\right) = B_{S1}\left(\phi_s,\theta\right) + B_{S2}\left(\phi_s,\theta\right)$$

$$B_{\mathrm{PM1}}\left(\phi_s,\theta\right) = 2\left(\Lambda_{\min} + \Lambda_{so,0}\right)\sum_{\xi} \hat{\Theta}_{\mathrm{PM}}^{\xi} \cdot \cos\left(\xi p\left(\phi_s - \theta\right)\right)$$

$$B_{\mathrm{PM2}}\left(\phi_s,\theta\right) = \sum_{\xi}\sum_{k} \hat{\Theta}_{\mathrm{PM}}^{\xi}\Lambda_{so,k} \cos\left[\left(\xi \mp k\cdot Q_s\right)p\phi_s - \xi\cdot p\theta\right]$$

$$B_{S,I1}\left(\phi_s,\theta\right) = 2\Lambda_0 \sum_{v}\Theta_S^{v}\cos\left(p\theta - vp\phi_s + \delta\right)$$

$$B_{S,I2}\left(\phi_s,\theta\right) = \sum_{v}\sum_{j}\Theta_S^{v}\Lambda_j \cos\left[\left(1\pm 2j\right)p\theta - \left(v\pm 2j\right)p\phi_s + \delta\right] \tag{4.14}$$

in which

ξ and $\hat{\Theta}_{\mathrm{PM}}^{\xi}$ denote the space harmonics and corresponding amplitudes for the rotor MMF

k denotes the stator slotting harmonics

B_{PM1} is the PM flux density of the electric machine with uniform air-gap length

B_{PM2} represents the PM flux fluctuation as the results of the stator slotting

v and $\hat{\Theta}_S^{v}$ denote the space harmonics and corresponding amplitudes for the stator winding MMF

j denotes the rotor permeance harmonics

Λ_j is the flux-path permeance parameter

$B_{S,I1}$ is the current flux density of the electric machine with uniform air-gap length

$B_{S,I2}$ represents the air-gap flux density due to the interaction of the MMF stator harmonics and the magnetic reluctance path harmonics (rotor saliency)

The amplitude, harmonics, and corresponding angular frequencies for the flux density components due to magnets and stator currents for $\theta = wt$ are given in **Table 4.2**.

4.3.1.2 **Radial Force Density Components:** From the above equation, the components for the electromagnetic forces can be presented as sum of three components:

$$f_{r,\mathrm{PM}}\left(\theta_s,t\right) = \frac{1}{2\mu_0} B_{r,\mathrm{PM}}^2\left(\theta_s,t\right)$$

TABLE 4.2 Air-gap flux density components: amplitude, harmonics, and corresponding angular frequencies.

B-components		Amplitude	Frequency	Harmonics
$B_S(\phi_s, t)$	$B_{S,n}(\phi_s, t)$	$2\Lambda_0 \hat{\Theta}_S^{\nu}$	ω	νp
	$B_{S,r2}(\phi_s, t)$	$\Lambda_j \hat{\Theta}_S^{\nu}$	$(1 \pm 2j)\omega$	$(\nu \pm 2j)p$
$B_{PM}(\phi_s, t)$	$B_{PM,I}(\phi_s, t)$	$2(\Lambda_{min} + \Lambda_{so,0})^{\xi} \hat{\Theta}_{PM}$	$\xi\omega$	ξp
	$B_{PM,II}(\phi_s, t)$	$\xi\hat{\Theta}_{PM}\Lambda_{so,k}$	$\xi\omega$	$(\xi \mp k \cdot Q_s)p$

$$f_{r,S}\left(\theta_s, t\right) = \frac{1}{2\mu_0} B_{r,S}^2\left(\theta_s, t\right)$$

$$f_{r,\frac{PM}{S}}\left(\theta_s, t\right) = \frac{1}{2\mu_0} B_{r,PM}\left(\theta_s, t\right) \cdot B_{r,S}\left(\theta_s, t\right) \qquad (4.15)$$

where

$f_{r,PM}$ represents the radial forces due to rotor magnets

$f_{r,S}$ represents the radial forces due to stator reaction field

$f_{r,\frac{PM}{S}}$ represents the radial forces due to the interaction of the magnet field and the stator reaction field

1. *First Component:* From the above equations, the radial force component due to the rotor magnetic field can be presented as sum of three subcomponents

$$f_{r,PM}\left(\theta_s, t\right) = \frac{1}{2\mu_0} B_{PM1}^2\left(\theta_s, t\right) + \frac{1}{\mu_0} B_{PM1}\left(\theta_s, t\right) \cdot B_{PM2}\left(\theta_s, t\right) + \frac{1}{2\mu_0} B_{PM2}^2\left(\theta_s, t\right) \quad (4.16)$$

Using the above equations, the radial force components due to the rotor magnet fields are

$$f_{r,PM1}\left(\theta_s, t\right) = \frac{1}{2\mu_0} B_{PM1}^2\left(\theta_s, t\right)$$

$$= \frac{1}{2\mu_0}\left[2\left(\Lambda_{min} + \Lambda_{so,0}\right)\sum_{\xi} \hat{\Theta}_{PM}^{\xi} \cdot \cos\left(\xi p\left(\phi s - \theta\right)\right)\right]^2$$

$$= \frac{\left(\Lambda_{min} + \Lambda_{so,0}\right)^2}{\mu_0}\sum_{\xi_1}\sum_{\xi_2} \hat{\Theta}_{PM}^{\xi_1} \cdot \hat{\Theta}_{PM}^{\xi_2}$$

$$\cdot \left\{\cos\left[\left(\xi_1 - \xi_2\right)p\left(\phi_s - \theta\right)\right] + \cos\left[\left(\xi_1 + \xi_2\right)p\left(\phi_s - \theta\right)\right]\right\}$$

$$f_{r,\mathrm{PM2}}\left(\theta_s,t\right)=\frac{1}{\mu_0}B_{\mathrm{PM1}}\left(\theta_s,t\right)\cdot B_{\mathrm{PM2}}\left(\theta_s,t\right)$$

$$=\frac{1}{2\mu_0}2\left(\Lambda_{\min}+\Lambda_{so,0}\right)\sum_{\xi_1}\hat{\Theta}_{\mathrm{PM}}^{\xi_1}\cdot\cos\left(\xi_1 p\left(\phi s-\theta\right)\right)$$

$$\cdot\sum_{\xi_2}\sum_{k}\hat{\Theta}_{\mathrm{PM}}^{\xi_2}\Lambda_{so,k}\cos\left[\left(\xi_2\mp kQ_s\right)p\phi_S-\xi_2 p\theta\right]$$

$$=\frac{\left(\Lambda_{\min}+\Lambda_{so,0}\right)}{2\mu_0}\sum_{\xi_1}\sum_{\xi_2}\sum_{k}\Theta_{\mathrm{PM}}^{\xi_1}\Theta_{\mathrm{PM}}^{\xi_2}\Lambda_{so,k}$$

$$\cdot\left\{\cos\left[\left(\xi_1-\xi_2\pm kQ_s\right)\right]p\phi_S-\left(\xi_1-\xi_2\right)p\theta\right\}$$

$$+\cos\left\{\left[\xi_1+\xi_2\mp kQ_s\right]p\phi_S-\left(\xi_1+\xi_2\right)p\theta\right\}\right\}$$

$$f_{r,\mathrm{PM3}}\left(\theta_s,t\right)=\frac{1}{2\mu_0}B_{\mathrm{PM2}}^2\left(\theta_s,t\right)$$

$$=\frac{1}{2\mu_0}\left[\sum_{\xi_2}\sum_{k}\hat{\Theta}_{\mathrm{PM}}^{\xi}\Lambda_{so,k}\cos\left[\left(\xi_2\mp kQ_s\right)p\phi_S-\xi_2 p\theta\right]\right]^2$$

$$=\frac{1}{2\mu_0}\sum_{\xi_1}\sum_{\xi_2}\sum_{k}\hat{\Theta}_{\mathrm{PM}}^{\xi_1}\hat{\Theta}_{\mathrm{PM}}^{\xi_2}\Lambda_{so,k}^2\cdot\left\{\cos\left[\left(\xi_1-\xi_2\right)p\left(\phi s-\theta\right)\right]\right.$$

$$\left.+\cos\left[\left(\xi_1+\xi_2\mp kQ_s\right)p\phi_S-\left(\xi_1+\xi_2\right)p\theta\right]\right\}. \tag{4.17}$$

In the above equations, the $f_{r,\mathrm{PM1}}$ component presents the radial forces due to the rotor MMF harmonics alone; however, $f_{r,\mathrm{PM2}}$ and $f_{r,\mathrm{PM3}}$ represent the stator slotting effect on the radial forces.

2. *Second Component:* Analogous, from the above equations, the radial force component due to the stator fields is

$$f_{r,S}\left(\theta_s,t\right)=\frac{1}{2\mu_0}B_{S1}^2\left(\theta_s,t\right)+\frac{1}{\mu_0}B_{S1}\left(\theta_s,t\right)\cdot B_{S2}\left(\theta_s,t\right)+\frac{1}{2\mu_0}B_{S2}^2\left(\theta_s,t\right). \tag{4.18}$$

Using the above equations and for the case $v_1=v_2=v$, we have

$$f_{r,S1}\left(\theta_s,t\right)=\frac{1}{2\mu_0}B_{S1}^2\left(\theta_s,t\right)$$

$$=\frac{1}{2\mu_0}\left[2\Lambda_0\sum_{v}\hat{\Theta}_{S}^{v}\cdot\cos\left(p\theta-vp\phi_S+\delta\right)\right]^2$$

$$=\frac{\left(\Lambda_0\right)^2}{\mu_0}\sum_{v_1}\sum_{v_2}\hat{\Theta}_{S}^{v_1}\hat{\Theta}_{S}^{v_2}$$

$$\cdot\left\{\cos\left[\left(v_1-v_2\right)p\phi\right]+\cos\left[\left(v_1+v_2\right)p\phi+2p\theta+2\delta\right]\right\}$$

$$f_{r,S2}(\theta_s,t) = \frac{1}{\mu_0} B_{S1}(\theta_s,t) \cdot B_{S2}(\theta_s,t)$$

$$= \frac{1}{2\mu_0} 2\Lambda_0 \sum_{v_1} \hat{\Theta}_S^{v_1} \cdot \cos\left(p\theta - v_1 p\phi_s + \delta\right)$$

$$\cdot \sum_{v_2}\sum_{j} \hat{\Theta}_S^{v_2} \Lambda_j \cos\left[(1\pm 2j)p\theta - (v_2 \pm 2j)p\phi_s + \delta\right]$$

$$= \frac{\Lambda_0}{2\mu_0} \sum_{v_1}\sum_{v_2}\sum_{j} \hat{\Theta}_S^{v_1} \cdot \hat{\Theta}_S^{v_2} \cdot \Lambda_j$$

$$\cdot \left\{ \cos\left[((v_2 \pm 2j) - v_1) p\phi_s \mp 2j \cdot p\theta\right] \right.$$
$$\left. + \cos\left[((v_2 \pm 2j) + v_1) p\phi_s + 2(1\pm j)p\theta + 2\delta\right] \right\}$$

$$f_{r,S3}(\theta_s,t) = \frac{1}{2\mu_0} B_{S2}^2(\theta_s,t)$$

$$= \frac{1}{2\mu_0}\left[\sum_{v}\sum_{j} \hat{\Theta}_S^{v}\Lambda_j \cdot \cos\left[(1\pm 2j)p\theta - (v_2\pm 2j)p\phi_s + \delta\right]\right]^2$$

$$= \frac{(\Lambda_0)^2}{4\mu_0} \sum_{v_1}\sum_{v_2}\sum_{j}\left(\hat{\Theta}_S^{v_1}\cdot\hat{\Theta}_S^{v_2}\cdot\Lambda_j\right)^2 \cdot \left\{ \cos\left[(v_2 - v_1)p\phi\right] + \cos\left[(v_1 + v_2)p\phi + 2(1\pm 2j)p\theta + 2\delta\right]\right\}$$

$$(4.19)$$

In the above equation, $f_{r,S1}$ component presents the radial forces due to the stator winding MMF harmonics; $f_{r,S2}$ and $f_{r,S3}$ represent the stator slotting effects on the radial forces.

3. *Third Component:* The radial force component due to the interaction of the rotor and stator fields consists of four components.

From the above equation, for the radial force component due to interaction of the rotor and stator fields, we have.

$$f_{r,\frac{PM}{S}}(\theta_s,t) = \frac{1}{2\mu_0} \cdot \left\{ B_{PM1}(\theta_s,t)\cdot B_{S,I1}(\theta_s,t) + B_{PM1}(\theta_s,t)\cdot B_{S,I2}(\theta_s,t) \right.$$
$$\left. + B_{PM2}(\theta_s,t)\cdot B_{S,I1}(\theta_s,t) + B_{PM2}(\theta_s,t)\cdot B_{S,I2}(\theta_s,t) \right\} \qquad (4.20)$$

For the radial force component due to interaction of the rotor and stator fields, we have

$$f_{r,\frac{PM}{S1}}(\theta_s,t) = \frac{1}{2\mu_0} B_{PM1}(\theta_s,t)\cdot B_{S,I1}(\theta_s,t)$$

$$= \frac{1}{2\mu_0} 2\left(\Lambda_{\min} + \Lambda_{so,0}\right) \sum_{\xi} \hat{\Theta}_{PM}^{\xi} \cos\left(\xi p(\phi_s - \theta)\right) \cdot 2\Lambda_0 \sum_{v} \hat{\Theta}_S^{v} \cos\left(p\theta - vp\phi_s + \delta\right)$$

$$= \frac{\left(\Lambda_{\min} + \Lambda_{so,0}\right)\cdot\Lambda_0}{\mu_0} \sum_{\xi}\sum_{v} \hat{\Theta}_{PM}^{\xi} \cdot \hat{\Theta}_S^{v} \cdot \cos\left[(\xi + v)p\phi_s - (1+\xi)p\theta - \delta\right]$$
$$+ \cos\left[(\xi - v)p\phi_s + (1-\xi)p\theta + \delta\right]$$

$$f_{r,\frac{PM}{S2}}(\theta_s,t) = \frac{1}{2\mu_0}B_{PM1}(\theta_s,t) \cdot B_{S,I2}(\theta_s,t)$$

$$= \frac{1}{2\mu_0}2(\Lambda_{min}+\Lambda_{so,0})\sum_{\xi}\hat{\Theta}_{PM}^{\xi}\cdot\cos(\xi p(\phi_s-\theta))$$

$$\cdot\sum_{v}\sum_{j}\hat{\Theta}_S^v\Lambda_j\cos\left[(1\pm2j)p\theta-(v\pm2j)p\phi_s+\delta\right]$$

$$= \frac{(\Lambda_{min}+\Lambda_{so,0})}{2\mu_0}\sum_{\xi}\sum_{v}\sum_{j}\hat{\Theta}_{PM}^{\xi}\cdot\hat{\Theta}_S^v\cdot\Lambda_j\cdot\left\{\cos\left[(\xi+(v\pm2j))p\phi_s-((1\pm2j)+\xi)p\theta-\delta\right]\right.$$

$$\left.+\cos\left[(\xi-(v\pm2j))p\phi_s+((1\pm2j)-\xi)p\theta+\delta\right]\right\}$$

$$f_{r,\frac{PM}{S3}}(\theta_s,t) = \frac{1}{2\mu_0}B_{PM2}(\theta_s,t)\cdot B_{S,I1}(\theta_s,t)$$

$$= \frac{1}{2\mu_0}\sum_{\xi}\sum_{k}\hat{\Theta}_{PM}^{\xi}\Lambda_{so,k}\cos\left[(\xi\mp k\cdot Q_S)p\phi_s-\xi p\theta-\delta\right]\cdot2\Lambda_0\sum_{v}\hat{\Theta}_S^v\cos(p\theta-vp\phi_s+\delta)$$

$$= \frac{\Lambda_0}{2\mu_0}\sum_{\xi}\sum_{k}\sum_{v}\hat{\Theta}_{PM}^{\xi}\cdot\Lambda_{so,k}\cdot\hat{\Theta}_S^v\cdot\left\{\cos\left[((\xi\mp k\cdot Q_S)+v)p\phi_s-(1+\xi)p\theta-\delta\right]\right.$$

$$\left.+\cos\left[((\xi\mp k\cdot Q_S)-v)p\phi_s+(1-\xi)p\theta+\delta\right]\right\} \tag{4.21}$$

In the above equation, $f_{r,\frac{PM}{S1}}$ denotes the radial forces due to interaction of the rotor and stator MMF harmonics; $f_{r,\frac{PM}{S2}}$, $f_{r,\frac{PM}{S3}}$, and $f_{r,\frac{PM}{S4}}$ represent the effects of the stator slotting and rotor permeance on the radial forces.

In accordance with the above equation, the magnetic forces per unit area can be represented in general form as

$$f_{rad}(\phi_s,t) = \sum_{m=0}^{\infty}m\hat{F}_{rad}\cdot\cos(m\cdot\phi_s-\omega_m t-\varphi_m) \tag{4.22}$$

where $m\hat{F}_{rad}$ is the amplitude of the radial magnetic force (m-order), ω_m is angular frequency and $m = 0, 1, 2, 3, \ldots$ is corresponding order of radial magnetic forces.

Both the amplitude and the frequency of radial force are important in relation to acoustic behavior of the machines and they vary significantly from no-load to full-load conditions.

Tables 4.3–4.5 summarize all the resultant time-varying radial force components, in terms of their magnitude, frequency, and spatial order.

From Table 4.3, it can be seen that for the $f_{r,\,PM1}$ component, each rotor MMF harmonics leads to the mode 0, 2ξ, p and $(\xi_1 \pm \xi_2)$, p which rotate with the angular frequency 0, 2ξ, $p\omega$ and $(\xi_1 \pm \xi_2)$, $p\omega$, respectively. However, due to the stator slotting effect, additionally different modes which rotate with different speeds are induced on $f_{r,PM2}$ and $f_{r,\,PM}$ force density components. Therefore, there are many modes which are of

TABLE 4.3 First component: magnitude, frequency, and spatial order of time-varying components of radial force density.

	Flux density harmonics	Spatial order	Angular frequency	Magnitude
$f_{r,\,PM1}$	$\xi_1 = \xi_2 = \xi$	0 $2\xi \cdot p$	0 $2\xi \cdot p\omega$	$\dfrac{(\Lambda_{min} + \Lambda_{so,0})^2}{\mu_0}\sum_\xi (\hat{\Theta}_{PM}^\xi)^2$
	$\xi_1 \neq \xi_2$	$(\xi_1 \pm \xi_2)\cdot p$	$(\xi_1 \pm \xi_2)\cdot p\omega$	$\dfrac{(\Lambda_{min} + \Lambda_{so,0})^2}{\mu_0}\sum_{\xi_1}\sum_{\xi_2}\hat{\Theta}_{PM1}^{\xi_1}\cdot\hat{\Theta}_{PM1}^{\xi_2}$
$f_{r,\,PM2}$	$\xi_1 = \xi_2 = \xi$ $k = 1, 2, 3, \ldots$	$\pm kQ_S \cdot p$ $(2\xi \mp kQ_S)\cdot p$	0 $2\xi \cdot p\omega$	$\dfrac{(\Lambda_{min} + \Lambda_{so,0})^2}{2\mu_0}\sum_\xi\sum_k \left(\hat{\Theta}_{PM}^\xi\right)^2 \cdot \Lambda_{so,k}$
	$\xi_1 \neq \xi_2$ $k = 1, 2, 3, \ldots$	$(\xi_1 \mp \xi_2 \mp (\pm kQ_S))\cdot p\omega$	$(\xi_1 \mp \xi_2)\cdot p\omega$	$\dfrac{(\Lambda_{min} + \Lambda_{so,0})^2}{\mu_0}\sum_{\xi_1}\sum_{\xi_2}\sum_k \hat{\Theta}_{PM1}^{\xi_1}\hat{\Theta}_{PM1}^{\xi_2}\Lambda_{so,k}$
$f_{r,\,PM3}$	$\xi_1 = \xi_2 = \xi$ $k = 1, 2, 3, \ldots$	0 $(\xi \mp kQ_S)\cdot 2p$	0 $2\xi \cdot p\omega$	$\dfrac{1}{4\mu_0}\sum_\xi\sum_k \left(\hat{\Theta}_{PM}^\xi \cdot \Lambda_{so,k}\right)^2$
	$\xi_1 \neq \xi_2$ $k = 1, 2, 3, \ldots$	$(\xi_1 \mp \xi_2 \mp 2kQ_S)\cdot p$	0 $(\xi_1 + \xi_2)\cdot p\omega$	$\dfrac{1}{4\mu_0}\sum_{\xi_1}\sum_{\xi_2}\sum_k \hat{\Theta}_{PM1}^{\xi_1}\cdot\hat{\Theta}_{PM1}^{\xi_2}\cdot\Lambda_{so,k}^2$

TABLE 4.4 Second component: magnitude, frequency, and spatial order of time-varying components of radial force density.

	Flux density harmonics	Spatial order	Angular frequency	Magnitude
$f_{r,\,S1}$	$V_1 = V_2 = V$	0 $2v \cdot p$	0 $2p\omega$	$\dfrac{(\Lambda_0)^2}{\mu_0}\sum_V (\hat{\Theta}_S^V)^2$
	$V_1 \neq V_2$	$(V_1 - V_2)\cdot p$ $(V_1 + V_2)\cdot p$	0 $2p\omega$	$\dfrac{(\Lambda_0)^2}{\mu_0}\sum_{V_1}\sum_{V_2}\hat{\Theta}_S^{V_1}\cdot\hat{\Theta}_S^{V_2}$
$f_{r,\,S2}$	$V_1 = V_2 = V$ $j = 1, 2, 3, \ldots$	$\pm 2j \cdot p$ $(2V \pm 2j)\cdot p$	$\pm 2j \cdot p\omega$ $2(1 \pm j)\cdot p\omega$	$\dfrac{(\Lambda_0)^2}{2\mu_0}\sum_V\sum_j \left(\hat{\Theta}_S^V\right)^2 \cdot \Lambda_j$
	$V_1 \neq V_2$ $j = 1, 2, 3, \ldots$	$((V_2 \pm 2j) - V_1)\cdot p$ $((V_2 \pm 2j) + V_1)\cdot p$	$\mp 2j \cdot p\omega$ $2(1 \pm j)\cdot p\omega$	$\dfrac{\Lambda_0}{2\mu_0}\sum_{V_1}\sum_{V_2}\sum_j \hat{\Theta}_S^{V_1}\hat{\Theta}_S^{V_2}\Lambda_j$
$f_{r,\,S3}$	$V_1 = V_2 = V$ $j = 1, 2, 3, \ldots$	0 $2V \cdot p$	0 $2(1 \pm 2j)$	$\dfrac{(\Lambda_0)^2}{4\mu_0}\sum_V\sum_j \left(\hat{\Theta}_S^V \cdot \Lambda_j\right)^2$
	$V_1 \neq V_2$ $j = 1, 2, 3, \ldots$	$(V_1 - V_2)\cdot p$ $(V_1 + V_2)\cdot p$	0 $2(1 \pm 2j)$	$\dfrac{(\Lambda_0)^2}{4\mu_0}\sum_{V_1}\sum_{V_2}\sum_k \left(\hat{\Theta}_S^{V_1} \cdot \hat{\Theta}_S^{V_2} \cdot \Lambda_j\right)^2$

the same order, however, with different speed, e.g., for $f_{r,\,PM1}$ and $\xi_1 = \xi_2$, we have mode 0 with the angular frequency 0, and for $f_{r,\,PM2}$ and $\xi_1 = 1$, $\xi_1 = 11$, $k = 1$, and $Q_S = 12$, we have another mode 0 with the angular frequency $12P\omega$ and so on. The same is true also for other force components. Thus, as the results of the stator slotting and the rotor permeance effect, additional modes of the same order, however, with different speed are induced in the magnetic force.

TABLE 4.5 Third component: magnitude, frequency, and spatial order of time-varying components of radial force density.

	Flux density harmonics	Spatial order	Angular frequency	Magnitude
$f_{r,\,PM1/S1}$	$\xi = v$	0 $2\xi\cdot p$	$(1-\xi)\cdot p\omega$ $(1+\xi)\cdot p\omega$	$\dfrac{(\Lambda_{min}+\Lambda_{so,0})\cdot\Lambda_0}{\mu_0}\sum_\xi\sum_v\hat\Theta_{PM}^\xi\cdot\hat\Theta_S^v$
	$\xi \neq v$	$(\xi\pm v)\cdot p$	$(1+\xi)\cdot p\omega$	$\dfrac{(\Lambda_{min}+\Lambda_{so,0})\cdot\Lambda_0}{\mu_0}\sum_\xi\sum_v\hat\Theta_{PM}^\xi\cdot\hat\Theta_S^v$
$f_{r,\,PM2/S2}$	$\xi = v$ $j = 1, 2, 3, \ldots$	$(\xi\pm j)\cdot 2p$ $\mp 2j\cdot p$	$((1\pm 2j)+\xi)\cdot p\omega$ $((1\pm 2j)-\xi)\cdot p\omega$	$\dfrac{(\Lambda_{min}+\Lambda_{so,0})}{2\mu_0}\sum_\xi\sum_v\sum_j\hat\Theta_{PM}^\xi\,\hat\Theta_S^v\cdot\Lambda_j$
	$\xi \neq v$ $j = 1, 2, 3, \ldots$	$(\xi\pm(v\pm 2j))\cdot p$	$((1\pm 2j)\pm\xi)\cdot p\omega$	$\dfrac{(\Lambda_{min}+\Lambda_{so,0})}{2\mu_0}\sum_\xi\sum_v\sum_j\hat\Theta_{PM}^\xi\,\hat\Theta_S^v\cdot\Lambda_j$
$f_{r,\,PM3/S3}$	$\xi = v$ $k = 1, 2, 3, \ldots$	$(2\xi\mp k\cdot Q_s)\cdot p$ $\mp k\cdot Q_s\cdot p$	$(1+\xi)\cdot p\omega$ $(1-\xi)\cdot p\omega$	$\dfrac{\Lambda_0}{2\mu_0}\sum_\xi\sum_v\sum_j\hat\Theta_{PM}^\xi\cdot\Lambda_{so,k}\cdot\hat\Theta_S^v$
	$\xi \neq v$ $k = 1, 2, 3, \ldots$	$((\xi\mp k\cdot Q_s)\pm v)\cdot p$	$(1+\xi)\cdot p\omega$	$\dfrac{\Lambda_0}{2\mu_0}\sum_\xi\sum_v\sum_j\hat\Theta_{PM}^\xi\cdot\Lambda_{so,k}\cdot\hat\Theta_S^v$
$f_{r,\,PM3/S3}$	$\xi = v$ $k = 1, 2, 3\ldots$ $j = 1, 2, 3\ldots$	$(2\xi\mp k\cdot Q_s\pm 2j)\cdot p$ $(\mp k\cdot Q_s\pm 2j)\cdot p$	$(1\pm 2j+\xi)\cdot p\omega$ $(1\pm 2j-\xi)\cdot p\omega$	$\dfrac{1}{4\mu_0}\sum_\xi\sum_k\sum_v\sum_j\hat\Theta_{PM}^\xi\cdot\Lambda_{so,k}\cdot\hat\Theta_S^v\,\Lambda_j$
	$\xi \neq v$ $k = 1, 2, 3, \ldots$ $j = 1, 2, 3, \ldots$	$((\xi\mp k\cdot Q_s)\pm(v\pm 2j))\cdot p$	$(1\pm 2j\pm\xi)\cdot p\omega$	$\dfrac{1}{4\mu_0}\sum_\xi\sum_k\sum_v\sum_j\hat\Theta_{PM}^\xi\cdot\Lambda_{so,k}\cdot\hat\Theta_S^v\,\Lambda_j$

In the practical analysis, finite-element method has been used to calculate the magnetic flux scattering of the electric machine; then, the harmonic electromagnetic forces can be calculated by commercial software such as ANSYS Maxwell with stress analysis and Fourier analysis. These forces are subsequently used as the applied forces in the simulated stator model to study the vibrations of different models. There are many Maxwell transient solvers, such as ANSYS Maxwell which is a commercial interactive software using the finite-element methods to solve electric and magnetic problems. It determines the electromagnetic field by solving Maxwell's equations within a finite region of space with user-defined initial conditions and appropriate boundary conditions.

An appropriate set of equations and their terms can be solved using varied solvers, such as Maxwell static solver, eddy current and transient solver. Electric and magnetic transient solver solves different transient fields. Magnetic transient solver computes time-varying magnetic fields, which is a time-domain solver solving instantaneous magnetic fields at each time step. The sources of the time-varying magnetic field can be moving sources, moving permanent magnets, and time-varying voltages or current sources in linear of nonlinear materials.

Furthermore, the induced fields such as proximity and skin effects can also be considered.

Electric transient solver computes time-varying electric fields caused by time-varying voltages, charge distributions, or applied current excitations in inhomogeneous materials [36–38].

FIGURE 4.4   Magnetic radial and tangential forces of two motor designs [38].

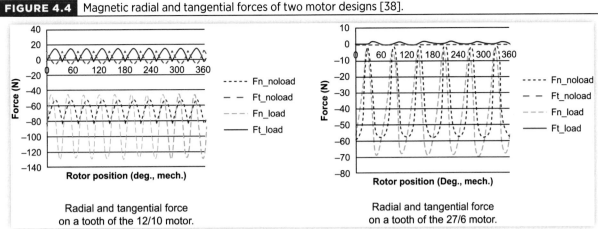

Radial and tangential force
on a tooth of the 12/10 motor.

Radial and tangential force
on a tooth of the 27/6 motor.

As an example, several case studies of surface-mounted PMSM slot/pole configurations, such as the 12/10, 9/6, 12/8, 15/10, and 27/6 configurations are given in **Figure 4.4** which shows the magnetic radial and tangential forces of two motor designs [38].

Resonant vibration occurs when the frequency of radial vibration force coincides with the specific natural frequency of the stator mechanical structure having the same order of vibration mode.

When the vibration mode order associated with dominant radial exciting force is high, noise and vibration are lower due to higher mechanical stiffness.

Electromagnetic sources of noise and vibrations in electric motors can be summarized as follows [18, 23–40]:

- Phase unbalanced
- Slot opening
- Input current waveform distortion
- Magnetic saturation
- Magnetostrictive expansion of core laminations
- Unbalanced magnetic pull static and dynamic eccentricity error produced unbalanced magnetic pull, resulting in the bending of shaft
- Reduced mechanical stiffness of the shaft and first critical speed of the rotor due to increased magnetic pull
- Unbalanced force due to the diametrically asymmetric disposition of the stator slots and coils—Proper selection of slot/pole combination no. will help in reducing the unbalanced magnetic pull
- Torque pulsation to be presented in next section.

It is important to determine magnetic fields and the reluctance forces they produce in the air gap of an electrical machine, as the reluctance forces are the main origin of magnetic noise.

A magnetic field in the air gap of an electrical machine is needed to produce required torque. The main quantity for the magnetic field is the flux density. In an ideal electrical motor, there exists the fundamental flux density distribution or fundamental wave BS1 with a frequency equal to the supply frequency S1 and the wave number r is equal to the machine pole-pair number p. However, since a real electrical motor has slots and distributed windings, is eccentric, and is magnetically saturated to some extent, the flux density is added with undesired harmonics. Therefore, the total flux density distribution in the air gap is a sum of rotating flux density waves with various frequencies.

Consider a simple case: in the air gap, if the stator has a magnetic flux density wave $B_{m1} \cdot \cos(\omega_1 t + k\alpha + \Phi_1)$, the rotor has a magnetic flux density wave $B_{m2} \cdot \cos(\omega_2 t + l\alpha + \Phi_2)$, their product has the total wave of magnetic flux density,

$$P_{mr} = 0.5 B_{m1} \cdot B_{m2} \cdot \cos[(\omega_1 + \omega_2)t + (k+l)\alpha + (\Phi_1 + \Phi_2)]$$
$$+ 0.5 B_{m1} \cdot B_{m2} \cdot \cos[(\omega_1 - \omega_2)t + (k-l)\alpha + (\Phi_1 - \Phi_2)] \qquad (4.23)$$

where

B_{m1} is amplitude of magnetic flux density wave in stator
B_{m2} is amplitude of magnetic flux density wave in rotor
ω_1, ω_2 are angular frequency of magnetic fields of stator and rotor
k, l are integer variables (1, 2, 3, …)

The magnetic stress or pressure waves act in radical direction on the stator and rotor active surfaces, which cause the deformation and subsequently cause the vibration and noise.

The mixed product of stator and rotor winding space harmonic create forces at frequencies

$$f_r = f_1\left[\frac{n \cdot Z_r}{p} \cdot (1-s) + 2\right], f_1 \cdot \left[\frac{n \cdot Z_r}{p} \cdot (1-s)\right] \qquad (4.24)$$

where

f_1 is supply frequency [Hz]
n is integer value $n = 0, \pm 1, \pm 2, …$
p is number of pole pairs
N_{rs} is number of rotor slots
s is slip

The mixed product of stator winding and rotor eccentricity space harmonics create forces with frequencies

$$f_r = f_1\left[\frac{n \cdot N_{rs}}{p} \cdot (1-s) + 2\right], \quad f_1\left[\frac{n \cdot N_{rs}}{p} \cdot (1-s)\right],$$
$$f_1\left[\frac{n \cdot N_{rs}}{p} \cdot (1-s) + \frac{1-s}{p}\right], \quad f_1\left[\frac{n \cdot N_{rs}}{p} \cdot (1-s) + 2 + \frac{1-s}{p}\right] \qquad (4.25)$$

The mixed product of stator winding and rotor saturation harmonics create forces at frequencies,

$$f_r = f_1\left[\frac{n \cdot N_{rs}}{p} \cdot (1-s) + 4\right], \quad f_1\left[\frac{n \cdot N_{rs}}{p} \cdot (1-s) + 2\right] \qquad (4.26)$$

The air gap width depends only on position in the static eccentricity. The magnetic field in the air gap rotates as synchronous speed. That is given by the mains frequency and with the number of pole pair's induction motor. Modulation of magnetic field in one period is function, which is represented by a variable air gap, i.e., a function of its conductivity. Static eccentricity is defined as the rotor axis offset from the axis of the stator. The air gap has a variable character. There is a stronger interaction of stator and rotor magnetic field at the point where the gap is smaller. Influence of the static eccentricity manifests as the emergence of side frequency bands, which are shifted from the mains frequency f_1 of the synchronous frequency f. For static eccentricity is the angular frequency $\Omega\varepsilon = 0$.

Static eccentricity is straight-line. The frequency for static eccentricity is twice power frequency, $f_{stat} = 2f_1$.

The relative eccentricity ε is defined as

$$\varepsilon = \frac{e}{g} = \frac{e}{R-r},$$

where

R is Inner stator core radius

r is outer rotor radius

e is rotor eccentricity

g is ideal uniform air-gap for $e = 0$

Dynamic eccentricity occurs when the rotor fails or its affiliates. Ratios are complicated by the fact that the width of air gap is a function of position and time. The variable air gap varies at the rotation of the rotor. There are sidebands that appear in the frequency range of vibrations of electric motor. The frequency of dynamic eccentricity is

$$\Omega_\varepsilon = \Omega \cdot (1-s) = \frac{\omega}{p} \cdot (1-s) = 2 \cdot \pi \cdot \frac{f}{p} \cdot (1-s) \tag{4.27}$$

The frequency generated by the dynamic eccentricity is

$$f_{DYN} = f_1 \pm (1-s) \cdot f_{SO} \tag{4.28}$$

The frequency generated by eccentricity is also given by

$$f_{exc}\left[\left(n_{rt} \cdot R \pm n_d \right) \cdot \frac{1-s}{p} \cdot n_{\omega s} \right] \cdot f \tag{4.29}$$

where

R is the number of grooves engine

s is the chute

p is the number of pole pair

Electromagnetic sources of noise and vibrations in an electrical vehicle also include torque pulsations which are listed as follows.

- Cogging torque
- Distortion of the sinusoidal or trapezoidal distribution of the magnetic flux density in the airgap, or commutation torque ripple

- The difference between the airgap permeances in the magnetic axis, or reluctance torque.
- Current ripple resulting from the causes like PWM
- Phase current commutation

The electromagnetic torque pulsations are presented in next section.

4.3.2 Electromagnetic Torque Pulsation

4.3.2.1 Cogging Torque: Based on the tangential component of the electromagnetic force, electromagnetic torque of a PMSM motor can be calculated according to Maxwell's stress tensor method as:

$$T = \frac{L}{\mu_0} \oint_s rB_n B_t dl \tag{4.30}$$

The torque can be better calculated by computing the average value of the electromagnetic torque over the entire air-gap surface

$$T = \frac{L}{\mu_0} \oint_s rB_n B_t ds \tag{4.31}$$

where

S is the air gap surface constituted by the layers between stator and rotor
g is the air gap length

The electromagnetic torque can be calculated by the derivative of the magnetic co-energy with respect to rotor angle at constant current.

Cogging torque is the circumferential component of the magnetic force that drives to align of the stator teeth with the pole magnet of the rotor [39]. The peak cogging torque occurs when the interpolar axis aligns with the edge of the stator due to the variation of magnetic energy Wm with rotor position θ being the highest at this position. The cogging torque is given by

$$T_{\text{cog}} = \sum_{i=1,2,3...}^{\infty} K_{sk} T_{i\_pk} \sin(iN_c\theta) \tag{4.32}$$

where

K_{sk} is the skew factor and equal to unit when there is no skewing
N_c is the least common multiple (LCM) of slot number Qs and pole number P
T_{i_pk} is the magnitude of the ith harmonic component of the cogging torque
θ is the rotor position in mechanical degree

The number of time periods of a cogging torque during one slot-pitch rotation is given as

$$N_p = \frac{P}{HCF\{Q,P\}} \tag{4.33}$$

in which *HCF* is the highest common factor of Q and P. The cogging torque of above equation can be expressed in terms of N_p using the relation between *LCM* and *HCF* of Q and P. It becomes

$$T_{\text{cog}} = \sum_{i=1,2,3\ldots}^{\infty} K_{sk} T_{i\_pk} \sin\left(iN_p Q\theta\right) \tag{4.34}$$

A low N_p means that peaks of all the elementary torques due to the interaction of slot opening and the edge of the magnet pole occur at the same rotor position and add up to result in a significant peak cogging torque. A high N_p means that the peaks of elementary torques are distributed along the slot pitch resulting in lower peak cogging torque but with higher fundamental frequency. The period of a cogging torque can be expressed as

$$T_p = \frac{360°}{LCM\{Q,P\}} \tag{4.35}$$

Therefore, a high *LCM* value or low *HCF* value of Q and P minimizes cogging torque. Since the position of the pole magnet edge with respect to the slot opening is closely related to the peak cogging torque, the geometrical parameters such as the length of the pole magnet arc and slot openings are crucial in minimizing cogging torque. The skewing of pole magnet and stator teeth can minimize cogging torque by creating an offset between the slot opening and the pole magnet edge.

Cogging torque is a pulsating torque with zero average value due to the tendency of the PM to align with the stator iron. This occurs whenever magnet flux travels through a varying reluctance. A simplified expression of cogging torque is given by

$$T_{\text{cogg}} = -\frac{1}{2}\phi_g^2 \frac{d\mathfrak{R}}{d\theta} \tag{4.36}$$

where
  ϕ_g is the magnet flux crossing the air gap
  $\mathfrak{R}$ is the total reluctance through which the flux passes

If the reluctance $\mathfrak{R}$ does not vary as the rotor rotates, the derivative in equation is zero and the cogging torque is zero. Based on the equation, cogging torque is independent of flux direction as the magnet flux φg is squared. The reluctance of the PM and the iron core is negligible compared to air gap reluctance and, thus, the reluctance refers to the air gap reluctance.

Cogging torque is due to the interaction between the permanent magnets and the stator slots. The most common technique for cogging torque reduction is to have stator slot skewing. The idea is to reduce the change of reluctance with position and thus to reduce the cogging torque. The change in reluctance can be minimized if the slot openings are spread over the surface area of the magnet. The slots are skewed so that each magnet sees a net reluctance that stays nearly the same as slots pass by. In this way, changes along the axial dimension reduce the effect of changes along the circumferential dimension. As a result, the $d\mathfrak{R}/d\theta$ experienced by the entire magnet decreases, and consequently, the cogging torque decreases. The same effect can also be obtained by shaping the stator slots using any of the following techniques:

(a) Bifurcated slots

(b) Empty or dummy slots

(c) Closed slots

(d) Teeth with different width of the active surface (teeth pairing)

The same results can be achieved by skewing the PMs where each magnet can be represented by a few straight rotor bar segments offset from each other by a fixed angle. This type of magnet skew is called step skew. There is another type of skew called continuous skew where each magnet pole is a skewed rotor bar. The cogging torque is also a function of stator slots and rotor poles. The closer the number of slots to the poles, the higher the cogging torque period and the lower its amplitude. The proper selection of slot-pole combination can reduce cogging torque significantly. The use of slotless stator is another way to eliminate cogging torque. Since the PM field and stator teeth produce cogging torque, a slotless motor can totally eliminate the cogging torque. A slotless structure requires an increased air gap, which in turn reduces the PM excitation field. To keep the same air gap magnetic flux density, the height of PMs must be increased. Slotless PMSMs use more PM material than slotted motors.

The existing research demonstrated that the use of teeth pairing with two different tooth widths has significant improvement in the cogging torque. The airgap permeance function is modified as the period is doubled with teeth pairing compared to single tooth situation. The fundamental component of air gap permeance $nNLG$ as well as the cogging torque can be eliminated with appropriate combinations of two different tooth widths a and b. Without teeth pairing, the period of airgap permeance is $2\pi/Ns$ and with teeth pairing, the period is $4\pi/Ns$. According to virtual work method, electromagnetic torque is equal to the derivative of the magnetic co-energy with respect to rotor angle at constant current. The cogging torque and permeance are given by

$$T_{\text{cogg}}(\theta) = -\frac{\partial W_g \theta}{\partial \theta} \propto \int_0^{2\pi} G^2(\alpha) B^2(\alpha,\theta) d\alpha \propto \sum_{n=0}^{\infty} nN_L G_{nN_L} \sin(nN_L\theta)$$

$$G_{nN_L} = \frac{N_s}{\pi} \frac{1}{nN_L} 2\left[\sin nN_L \frac{a}{2} + \sin nN_L \frac{b}{2} \right] \qquad (4.37)$$

where

N_L is the *LCM* of N_s and N_p

W_g and B are the air gap energy function and air gap flux density function, respectively

4.3.2.2 Torque Ripple:
The spatial harmonics of the stator magnetomotive force (MMF) that rotates asynchronously with the rotor cause variation of flux across the flux barriers of the rotor. At the same time, there are high order harmonics in the airgap flux density of the rotor due to its rectangular distribution.

Interaction of these unwanted harmonics produces torque ripple in an internal permanent magnet (IPM) motor. A typical analytical expression of torque ripple regarding rotor and stator MMFs is given as follows,

$$T_{\text{ripple}} = -\frac{P}{2}\frac{\mu_0}{g}r_g l\pi \sum_{\substack{n=6m\mp 1 \\ m=1,2,3\ldots}} \left(nf_{s,n}f_{r,n}\sin(n\pm 1)\theta_r \pm \gamma_d\right)$$ (4.38)

where

P is the number of poles

r_g is the air gap radius

l is the stack-length

g is the air gap length

n is the harmonic number

θ_r is the rotor angular position (electrical)

γ_d is the current angle measured from d-axis

$f_{s,n}$ is the nth order stator MMF harmonic

$f_{r,n}$ is the nth order rotor MMF harmonic

The above expression is valid of an FSCW IPM machine with the exception that the main torque producing harmonic in such a machine is not the fundamental, but its order is equal to pole-pair number.

A winding function specific to an FSCW stator could be used to derive stator MMF f_s. However, the two observations about the amplification of torque ripple by the harmonic order $n = 6m \mp 1$ and null contribution from the even order harmonics are valid in an FSCW IPM machine. Thus, by reducing the discontinuity effect of the stator slots and rotor flux barriers in the MMF waves, the higher order harmonics and the amplitude of the torque ripple can be minimized.

Torque ripple can cause undesirable vibrations in system. Torque ripple consists of two components: electromagnetic torque fluctuation and cogging torque. Electromagnetic torque ripple is caused by the harmonic interaction between the BEMF and the phase currents associated with the motor electrical dynamics.

Torque ripples are produced due to the interaction of stator currents with flux linkages or BEMFs of different frequencies. The frequencies of the torque ripple are significantly higher than the fundamental frequency and their mean value is zero.

The interaction of the stator current with the magnetic flux linkage harmonics produces a constant torque if the harmonics are of the same number and there is no torque pulsation.

In general, the instantaneous torque has two components: constant with average T_0, and periodic component $T_r(\theta_e)$, which is a function of time or electrical angle θ_e superimposed on the constant component.

The periodic component causes the torque pulsation called the torque ripple. The torque ripple can be defined as a percentage of the average or RMS torque in the following way:

$$t_r = \frac{T_{r,p-p}}{T_{av}}\times 100\% = \frac{T_{\text{max}} - T_{\text{min}}}{T_{av}}\times 100\%$$ (4.39)

The three main contributions of torque ripple in PMSM are given by: (i) reluctance torque; (ii) cogging torque; and (iii) mutual torque. The contribution from reluctance torque in surface-mounted PMSMs can be ignored as the difference in the *d*- and *q*-axes reluctances are negligible. The mutual torque produces torque ripple due to the distortion of the magnetic flux density distribution in the air gap or the distortion in stator current. The interaction between the rotor magnetic flux and variable permeance of the air gap due to the stator slot opening causes the cogging torque. Researchers have focused on reducing both the mutual torque and the cogging torque to minimize the torque ripple in SPM machines. Some researchers have considered torque ripple problem primarily from a design point of view, while others emphasized on the control aspects [40–50].

According to existing research, the reduction of cogging torque and torque ripple can significantly reduce the vibration and acoustic noise of system.

Cogging torque of electrical motors is the torque due to the interaction between the permanent magnets of the rotor and the stator slots of a PMSM motor. It is also known as detent or "no-current" torque. This torque is position-dependent and its periodicity per revolution depends on the number of magnetic poles and the number of teeth on the stator. Cogging torque is an undesirable component for the operation of a motor. It is prominent at lower speeds, with the symptom of jerkiness. Cogging torque results in both torque and speed ripple. However, at high speed, the motor moment of inertia filters out the effect of cogging torque [51–58].

Most of the techniques used to reduce cogging torque also reduce the motor counter-electromotive force and the resultant running torque.

Lower torque ripple in a PMSM motor ensures the smooth running of the motor and does not guarantee less vibration or noise. The electromagnetic radial forces, when they act on a stator, cause the deformation of the stator core leading to vibrations in a PMSM motor.

A summary of techniques used for reducing cogging torque is given as follows:

- Slotless winding
- No slot and teeth on the stator (no change in the airgap permeance then no cogging torque)
- Increase effective airgap of the motor—more PM material to achieve the same airgap flux density as of the slotted machine
- Skewing stator slots/stack or magnets/magnetization profile
- Reduction in the effective back-emf of the motor
- Less effective in case of the rotor eccentricity
- Manufacturing difficulty with the windings when automated winding machines are used
- Shaping of the stator slots/optimizing the magnet pole arc or width
- Selection of slot/pole combination, subject to the consideration of lamination dimensions and operating speed range
- Bifurcated slots/using fractional slots per pole
- Empty (dummy) slots
- Closed slots
- Teeth with different width of active surface
- Modulating drive current waveform

Figure 4.5 shows several shape designs of stator slots for cogging torque reduction.

FIGURE 4.5 Different shapes of stator slots for cogging torque reduction.

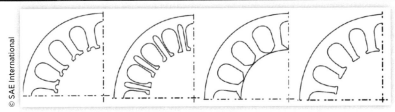

FIGURE 4.6 Different shapes of permanent magnet designed for cogging torque reduction and the performance comparison [18].

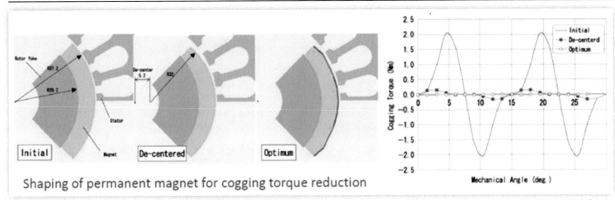

Shaping of permanent magnet for cogging torque reduction

Reduction of cogging torque can be attained by changing the shape of magnet. Magnet thickness lower at the edges compared to central part can help reduce both cogging torque and torque ripple. Decentered PMSMs with bifurcated stator slot can help in reducing the cogging torque.

Figure 4.6 shows different shapes of permanent magnet designed for cogging torque reduction and the performance comparison [18].

4.4 **Motor System NVH**

Electric motor noise and vibrations are predominantly related to electromagnetic forces in the airgap. Time-varying magnetic field produces radial and tangential forces. The magnetic field is impacted by the nonuniform air gap and line current, leading to many harmonics in forces [62].

The total sound power emission from an electrical motor can be considered as a combination of three uncorrelated noise sources working together [48]. These sources are magnetic, cooling/aerodynamic noise, and mechanical/rotational noise. A block diagram of the acoustical model for electrical motor is given in **Figure 4.7**.

FIGURE 4.7 Block diagram of acoustical model of electrical motor.

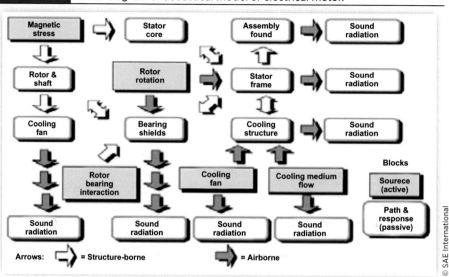

Magnetic noise is due to the temporal and spatial variations of the magnetic force distribution in the air gap.

The cooling noise is due to the operation of a cooling fan. Mechanical noise results from: (1) an unsmooth body (rotor) rotates in a cavity that has obstacles and discontinuities; (2) the shaft and the bearings interact. The magnitude of each source depends heavily on the motor type. The following details the major issues affecting each of the sources in an electrical motor.

- Magnetic noise P_{magn}: It is affected by shaft load, voltage, current, frequency, supply type, winding parameters, slot geometry, saturation, eccentricity, etc.
- Cooling noise P_{cool}: It is affected by fan type, axial-, radial-, or mixed flow, rotational speed and fan diameter, airflow velocity, cooling method-closed/open, and water/air.
- Mechanical or rotational noise P_{rot}: It is affected by type of cooling-closed or open, type of bearings, and speed.

The total sound power level L_{Wtot} of an electrical machine in decibels can be expressed as

$$L_{Wtot} = 10\log_{10}\left(\frac{P_{magn} + P_{cool} + P_{rot}}{P_{ref}}\right) \qquad (4.40)$$

Here P_{ref} = 1 pW is the reference sound power. The above equation shows that the total sound power level of an electrical machine is the summative result of all of the sources. The equation is useful in considering the reduction of the total sound power of a machine.

The rotating force components tend to deflect the structure they act on, and they make the machine stator, frame, and foundation vibrate. The level of vibration depends on the factors in three aspects:

1. The frequency, wave number, and amplitude of the force wave.
2. The existence of any proper or suitable structural mode at the exciting frequency.
3. The structural damping of the stator, frame, and foundation.

The conditions for high-vibration level and thus high-noise radiation are that the frequency of the force wave matches any of the structural natural frequency. In this case, only the damping properties of the structure limit the vibration level. The higher the damping, the lower the vibration levels, and vice versa. In welded steel frame machines, the side plates of the frame are usually the main radiators of magnetic noise. The bearing shields or machine ends rarely radiate remarkable levels of sound power.

The most effective way to reduce radiated sound is to reduce the normal vibration of the surface. Because the vibration velocity is directly proportional to the radial tooth force amplitude, which in turn is proportional to the square of flux density B, the following holds [51]:

$$L_{\text{Wmagn}} \triangleq 10\log_{10}\left(B^4\right) \quad \text{or} \quad L_{\text{Wmagn}} \triangleq 10\log_{10}\left(U^4\right) \tag{4.41}$$

The symbol U represents the supply voltage. Based on Equation (4.41), if the terminal voltage of the machine is reduced by 50%, the sound power level will be reduced by 12 dB.

4.4.1 The Electromagnetic Noise and Vibrations

The factors influencing electromagnetic noise and vibrations are briefed as follows.

1. The slots, distribution of windings in slots, input current waveform distortion, air gap permeance fluctuations, rotor eccentricity, and phase unbalance influence mechanical deformations and vibration. The parasitic higher harmonic forces and torques are due to the magnetomotive force space/time harmonics, slot harmonics, eccentricity harmonics, and saturation harmonics. Especially, radial force waves in AC motors, which act both on the stator and rotor, produce deformation of the magnetic circuit.
2. The stator-frame (or stator-enclosure) structure is the primary radiator of the motor noise. If the frequency of the radial force is close to or equal to any of the natural frequencies of the stator-frame system, resonance occurs, leading to big stator system vibration and noise.
3. Magnetostrictive noise is due to the periodic deformation of the core material. This noise of electrical motors in most cases can be neglected due to low frequency $2f$ and higher order $r = 2p$ of radial forces, where f is the fundamental frequency and p is the number of pole pairs. However, radial forces due to the magnetostriction can reach about 50% of radial forces produced by the air gap magnetic field. It could be significant in motors/generators with high-power transmission as in EVs and HEVs.

Magnetostrictive offset has been identified with even harmonics of the steady-state induced components resulting from the periodic deformation of the core material. Magnetostrictive noise is identified with random fluctuations of the magnetostrictive offset caused by frictional forces exceeding the magnetostrictive stress when the core material is near zero deformation.

4. Parasitic oscillation torques are produced due to higher time harmonics in the stator winding currents in inverter-fed motors. These parasitic torques are generally greater than oscillating torques produced by space harmonics. Moreover, the voltage ripple of the rectifier is transmitted through the intermediate circuit to the inverter and produces other kind of oscillating torque.

The audible noise of rotating electrical motor can be classified as ventilation, bearing, and electromagnetically caused noise.

The magnetically excited vibrations are the most important ones. Key for the analysis of acoustic noise of electric machines is the identification of excitation of the stator vibrations and accurate analysis of electromagnetic forces acting on the machine's body and stator body's natural modes to avoid resonances.

To avoid resonance excited by harmonic magnetic radical forces, the analysis and design of the stator to separate the natural modes are critical.

Electromagnetic noise is generated when the natural frequencies of vibration of induction motors match or are close to the excitation frequencies of the electromagnetic force. To avoid noise and vibration, it is necessary to estimate the amplitude of the radial electromagnetic forces as well as the natural frequencies of the structure.

The input current interacts with the magnetic field producing high-frequency forces that act on the inner stator core surface. These forces excite the stator core and frame in the corresponding frequency range and generate mechanical vibration and noise.

Stator system analysis can be conducted by using a circular cylindrical shell as a simplified model of the stator. Stator is considered as a cylindrical radiator with sound power and sound pressure generated at a given position. The simplified models can give a rapid estimation of the vibration and sound. **Figure 4.8** shows simplified model of stator system.

FIGURE 4.8 Simplified model of stator system

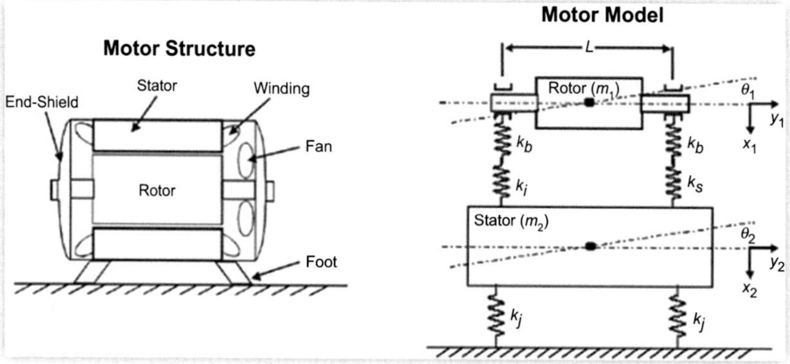

In reality, due to the curvature of the cylindrical shell of a stator, the vibrations of three orthogonal directions, radial, axial, and tangential, are coupled to each other. Any excitation in one direction could cause vibrations in all three directions. The electromagnetic force has radial and tangential and axial components. In principle, all components may excite the vibration in radial direction. But since the radial component is approximately an order of magnitude larger than the tangential component, it might be reasonable to neglect the contribution from the tangential component in the electromagnetic force, which means that the coupling in the vibration between the radial and tangential directions could be neglected. The advanced analysis can be implemented by using FEM. The simplified estimation of natural frequency of the stator system is given as

$$f_m = \frac{1}{2\pi} \sqrt{\frac{K_m}{M_m}} \tag{4.42}$$

where

K_m is the lumped stiffness (N/m)
M_m is the lumped mass (kg) of the stator system at mode no. m.

The circumferential vibration mode $m = 0$, the natural frequency is given as

$$f_0 = \frac{1}{\pi D_c} \sqrt{\frac{E_c}{\rho_c k_i k_{md}}} \tag{4.43}$$

where

E_c is Young's elasticity modulus
D_c is the mean diameter of the stator core
ρ_c is the mass density of the stator core
K_i is the stacking factor (=0.96)
k_{md} is the mass addition for displacement and is defined as

$$K_{md} = 1 + \frac{M_t + M_w + M_i}{M_c} \tag{4.44}$$

where

M_t is the mass of all stator teeth
M_w is the mass of stator windings
M_i is the mass of insulation
M_c is the mass of the stator core cylinder (yoke)

The relatively sophisticated formula is presented as follows

$$f_m \approx \frac{1}{2\pi} \sqrt{\frac{K_m^{(c)} + K_{mn}^{(f)} + K_m^{(w)}}{M_c + M_f + M_w}} \tag{4.45}$$

The natural frequencies of the stator system can be approximately evaluated as

$$f_m \approx \frac{1}{2\pi} \sqrt{\frac{K_m^{(c)} + K_{mn}^{(f)} + K_m^{(w)}}{M_c + M_f + M_w}} \tag{4.46}$$

where

$K_m^{(c)}$ is the lumped stiffness of the stator core for the mth circumferential vibrational mode

$K_{mn}^{(f)}$ is the lumped stiffness of the frame for the mth circumferential and nth axial vibrational mode

$K_m^{(w)}$ is the lumped stiffness of the stator winding for the mth circumferential vibrational mode

M_c is the lumped mass of the stator core

M_f is the lumped mass of the frame

M_w is the lumped mass of the stator winding

It has been assumed the frame is a circular cylindrical shell with both ends constrained mechanically by end bells.

These values can be corrected with the aid of correction factors obtained from the FEM numerical analysis or measurements. Then, using the damping coefficient ζ_m, the amplitudes of radial velocities for the mth circumferential mode are calculated, i.e.,

$$A_{mr} = \frac{F_{mr} / \left[(2\pi f_m)^2 M \right]}{\sqrt{\left(\left[1 - \left(\frac{f_r}{f_m} \right)^2 \right]^2 + \left[2\zeta_m \left(\frac{f_r}{f_m} \right) \right]^2 \right)}} \tag{4.47}$$

The damping factor ζ_m affects the accuracy of computation. Detailed research has shown that the damping factor is a nonlinear function of natural frequencies f_m.

For mode number larger than one, different natural frequency equation is required. The stator system is a complex structure which consists of the laminated stack with yoke and teeth, winding distributed in slots, encapsulation (potting), and frame (enclosure or casing). The above equation can only be used to estimate the natural frequency of the stator core alone. The structure of a motor and the exciting forces are rather complex in comparing with plates or cylinders. It will require a very sophisticated process and analysis using FEM [66, 67]. **Figure 4.9** shows an example of various natural vibration modes of a stator with winding.

FIGURE 4.9 Various natural vibration modes of a stator with winding.

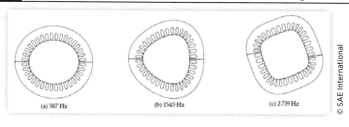

4.4.2 Mechanical Noise and Vibrations

The mechanical sources of noise and vibrations in electrical motors include the following aspects.

Mechanical noise is due to many factors, such as bearings defects, ovality, sliding contacts, bent shaft, rotor unbalance, shaft misalignment, couplings, U-joints, gears, etc. In principle, the mechanical source of noise has a mixed character. The noise caused by unbalance of rotating parts and noise of bearing can be substantially reduced if the machine is designed and fabricated very well. Dynamic balancing in production serves to reducing the noise of mechanical source. Compliance with the manufacturing tolerances and technological processes is the best solution to reduce the noise of mechanical source. Usually, specific change in noise features from this source can mean failure of the mechanical parts inside the motor. For example, the bearing's failure such as damaged ball exhibits changes in the noise spectrum. There are specific frequencies corresponding specific damage.

Motor bearings can be either a sliding or rolling bearings. Rolling bearings can create multiple vibration frequencies, which have their origin mainly in the uneven parts or rolling themselves paths to the bearing rings. If bearing has mechanical damage, it leads to uneven movement of the whole system and thus increases the vibration and noise of the electric machine.

The factors impacting noise and vibrations in motor also include the following factors: shaft misalignment, coupling misalignment, rotor unbalance, inaccurately machined parts, bearings, and their defects such as lubrications/contacts defects, bent shaft, system resonance, universal-joint and gears effects.

Angular and parallel shaft misalignment: shaft misalignment produces a mechanical vibration with the frequency; $f_s = 2n_m$, n_m is the shaft speed in rev/s.

Dynamically unbalanced rotor: unbalanced rotor, bent shaft, eccentricity, rubbing parts, etc. It produces a vibrating frequency of once per revolution or multiple of the number of revolution per cycle. $f_s = kn_m$, n_m is the shaft speed in rev/s, $k = 1, 2, 3, \ldots$.

Loose stator stack: loose stator lamination results in the following vibration frequency with the frequency sidebands of 1000 Hz. $f_{lam} = 2f$, f is the line frequency. Stator stack related noise can be reduced by encapsulation of the stator stack.

Rolling bearing generates mechanical impulses when the rolling element passes the defective groove, causing small radial movement of the rotor. The frequency at which the defect in the outer race causes an impulse when the ball or the roller passes the defective is of the race

$$f_{or} = \frac{N_b}{2} n_m \left(1 - \frac{d_b}{D} \cos \alpha \right) \tag{4.48}$$

or

$$f_{or} = 0.4 N_b n_m$$

in which
n_m is the shaft speed in rev/s
D the is pitch diameter
N_b is the number of rolling elements
d_b is the diameter of rolling element
α is the contact angle of rolling element

The frequency at which the defect in the inner race causes an impulse when the ball or the roller passes the defective is of the race

$$f_{ir} = \frac{N_b}{2} n_m \left(1 + \frac{d_b}{D} \cos \alpha \right) \quad \text{or} \quad f_{ir} = 0.6 N_b n_m \qquad (4.49)$$

Moreover, the ball spin frequency

$$f_{bs} = \frac{D}{2 d_b} n_m \left(1 - \left(\frac{d_b}{D} \right)^2 \cos^2 \alpha \right) \qquad (4.50)$$

The cage fault frequency

$$f_{cf} = \frac{1}{2} n_m \left(1 + \frac{d_b}{D} \cos \alpha \right) \qquad (4.51)$$

The irregularities in the ball cage yield specific frequency

$$f_{bc} = n_m \left(\frac{d_i}{d_i + d_o} \right) \qquad (4.52)$$

where d_i and d_o is the diameter of inner and outer contact surface.

The irregularities in the shape of ball yield specific frequency

$$f_{re} = n_m \frac{d_i d_o}{d_r \left(d_i + d_o \right)} \qquad (4.53)$$

The variation of bearing stiffness results in the specific frequency

$$f_{st} = N_b k n_m \frac{d_i}{d_i + d_o}, \quad k = 1, 2, 3, \dots \qquad (4.54)$$

Since the rotor unbalance causes vibration and noise emission from the stator, rotor, and rotor support structures, the rotor should be precisely balanced.

The noise due to rolling bearings depends on the accuracy of bearing parts, mechanical resonance frequency of the outer ring, running speed, lubrication conditions, tolerances, alignment, load, temperature, and presence of foreign materials.

The noise of sliding bearings is lower than that of rolling bearings. The vibration and noise produced by sliding bearing depends on the roughness of sliding surfaces, lubrication, stability and whirling of the oil film in the bearing, manufacture process, quality, and installation.

4.4.3 Aerodynamic Noises of Motors

The primary aerodynamic noise source of motor is from the cooling fan. In addition to the fan structure, any obstacle in the airstream produces noises. In nonsealed motor, the noise of the cooling fan is emitted by the vent holes. In a totally sealed motor, the noise of external fan predominates. The spectral distribution of fan noise includes both broadband noise ranging from 100 to 10,000 Hz and tonal noises (siren noise) which are generated at blade passage frequency (BPF) and its harmonics: BPF = Number of blade × RPM/60 (Hz).

Aerodynamic noise arises most often around the fan. Noise can also be created on the neck's stator slot windings or rotor. The aerodynamic noise sources can also include the noise produced by airflow inside and outside the electrical motors.

The ventilation noise is characterized in a broad-band frequency band of about 500…1000 Hz. The main root cause of the fan noise is the formation of turbulent airflow around the blades. This noise is characterized by the spectrum of a wide range, which has continuous character.

Tonal noise can be eliminated by increasing the distance between the impeller and the stationary obstacle. The fan noise can be approximated as

$$L_A = 60 \log U_2 + 10 \log D_2 \cdot b_2 + \sum k_1 \qquad (4.55)$$

where

U_2 is outer speed of fan on the circuit (m/s)
D_2 is outer diameter of the fan (m)
B_2 is fan width (m)
k_1 is constants for the correction

The vortex frequency is expressed by

$$f_v = 0.185 \cdot \frac{v}{D_2} \qquad (4.56)$$

The frequency of the pure tone due to the fan blades is given by

$$f_f = N_b \cdot \frac{N}{60} \qquad (4.57)$$

where

N is the speed (rev/min)
N_b is the number of fan blades

The sound power level of aerodynamic noise is

$$L_w = 67 + 10 \log_{10}^{(P_{out})} + 10 \log_{10}^{(p)}$$

$$L_w = 40 + 10 \log_{10}^{(Q)} + 20 \log_{10}^{(p)} \qquad (4.58)$$

where

P_{out} is motor rated power (kW)
p is fan static pressure (Pa)
Q is flow rate [m³/s]

Reducing aerodynamic noise in electrical motors can be done by the following approaches:

- Reducing the required amount of coolant used for ventilation of electrical motors.
- Optimal design of fan, especially the number and shape of the fan blades.
- To minimize the noise by preventing vibration machine parts from contacting cooling medium.

Electromagnetic noise occurs mainly due to the air gap between the rotor and the stator. Whining noise as a major annoyance source in electric motors is electromagnetic noise. Active noise control using countershafts is the best approach to eliminate this noise source. Various other treatments such as inequality slots and dummy slots have been used to reduce this kind of noise but at the expense of its efficiency.

The mechanical noise can be produced due to bearing wear and failure. As a result of bearing wear, air gap eccentricity can increase, and this can generate serious stator core damage and even destroy the winding of the stator. High mechanical unbalance in the rotor increases centrifugal forces on the rotor. Looseness or decreased stiffness in the bearing pedestals can increase the forces on the rotor. Critical speed shaft resonance increases forces and vibration on the rotor core.

Aerodynamic noise occurs primarily due to the cooling fan attached to the motor. The cooling fan creates a typical tonal noise which is subjectively annoying. With the increasing demand for high-performance lightweight electric vehicles, the electric motor has become more and more compact. As such, heating is the major issue of an electric motor. For this purpose, it is necessary to provide a special cooling arrangement like an axial fan. Hence, it becomes necessary to study the acoustic behavior of the motor noise due to this airborne source. Aerodynamic noise arises most often around the fan, or in the vicinity of the machine that behaves like a fan. This noise is predominant at higher motor speed and also in electric vehicle due to higher speed fluctuation. The approaches of reducing this noise source include using optimized profile blades instead of straight blades, uniform distribution of fan blades, and even number of fan blades. The fan noise is generally associated with blade passing frequency which is the discrete frequency noise or the tonal noise. The blade-passing tone results from the air impulses which occur once a blade passes a specific point. Each time when the blade passes a point in space, an impulsive force fluctuation is experienced. The number of impulses experienced per second is known as blade passing frequency [18, 59–65].

All the electrified vehicles include an electrified powertrain made of bearings, gears, cooling system, and an electric motor. This system can generate significant acoustic noise and this noise can be separated in terms of:

- Aerodynamic noise (e.g., wind, fans)
- Mechanical noise (e.g., tire/road, bearings, gear mesh)
- Electromagnetic noise (e.g., magnet/slot interactions, PWM)

To quantify and relate the psychoacoustic effect of electromagnetically excited noise to the design parameters of the electrical motor and drive, measurements are usually carried on a full electric car with maximum acceleration from 0 to 110 km/h.

As an example of this kind of test, microphones are placed at driver's ear and 40 cm from electric motor under car hood. A 3D accelerometer is mounted on the motor housing. The electric motor is a wound rotor synchronous machine with $Z_s = 48$ stator slots and $p = 2$ pole pairs. The electric drivetrain includes a reducer and the motor is air-cooled so gear noise and aerodynamic noise are present inside measurements.

Figure 4.10 shows the acoustic noise spectrogram measured during run-up. Aerodynamic and mechanical noise remain below 1000 Hz at max speed, while electromagnetic noise results in tonalities. One can distinguish two types of electromagnetic noise. The first one is due to excitations which are proportional to speed; they are due to interactions between pole field harmonics (rotor), stator slot harmonics, and armature field harmonics (stator). The second one creates excitations in "V-shape" centered around multiples of 10 kHz: these are created mainly by PWM current harmonics combined with fundamental rotor and stator fields.

Theory shows that the first type of excitations occurs in open circuit (very light torque level) at multiples of the stator slot number (H48, H96, etc.—H denoting the mechanical orders, not the electrical orders), while it occurs at multiples of H24 (H24, H48, H72, etc.) at partial load. These electromagnetic excitations are pulsating, which means they excite in phase all the stator teeth, both in radial and tangential directions. It is known that the vibroacoustic behavior of most electric motors used in automotive traction is dominated by pulsating forces [6]. In this case, one can deduce from the spectrogram that the breathing mode of the lamination has a natural frequency lying in the 4900–5200 Hz range. It was also shown that the main PWM excitations are actually pulsating [2]. The excitation frequencies are given by harmonic orders $fswi\pm3fs$, $fswi\pm5fs$, etc. and $2fswi$, $2fswi\pm2fs$, $fswi\pm4fs$, etc. These excitations do not meet any structural resonance. The main PWM lines are far from the breathing mode of the stator close to 5 kHz; hence, PWM noise comes from a forced excitation of the motor.

Figure 4.11 shows the numerically separated sound sources using software MANATEE and LEA.

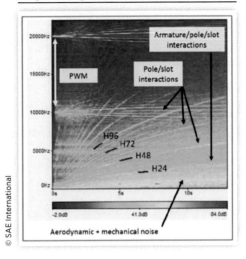

FIGURE 4.10 Spectrogram of noise measured close to electric motor at max torque during runup.

© SAE International

FIGURE 4.11 Separation of motor sound sources.

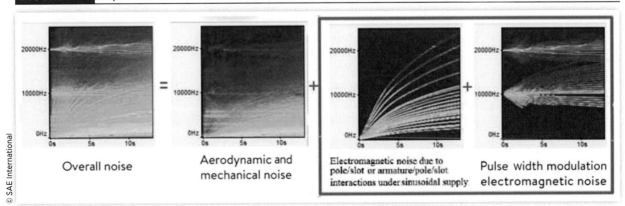

© SAE International

It can be seen that the following noise sources are separated: (1) mechanical and aerodynamic noise; (2) electromagnetic noise due to pole/slot or armature/pole/slot interactions under sinusoidal supply; (c) electromagnetic noise due to PWM. Using order extraction feature and without the need of a tachometric measurement, both armature/pole/slot interaction and PWM noise are isolated from aerodynamic and mechanical noise. This makes it possible to preliminarily identify the acoustical impact of each noise component by listening to each of them independently, before going more into details on psychoacoustics' metrics calculation.

References

1. Valavi, M., Nysveen, A., and Nilssen, R., "Effects of Loading and Slot Harmonic on Radial Magnetic Forces in Low-Speed Permanent Magnet Machine with Concentrated Windings," *IEEE Transactions on Magnetics* 51, no. 6: 1–10, 2015, doi:10.1109/TMAG.2014.2377014.

2. Magnequench, N., "Noise and Vibration in PM Motors – Sources and Remedies," Magnequench Technology Centre, Singapore, 2011.

3. Akiyama, Y., Nakamura, A., Takaku, K., and Sugiura, O., "Slot Ripple of Induction Motor and FEM Simulation on Magnetic Noise," *IAS '96.Conference Record of the 1996 IEEE Industry Applications Conference (IAS), Annual Meeting*, San Diego, CA, 1996, vol. 1, 644-651, doi:10.1109/IAS.1996.557104.

4. Arenas, J.P. and Crocker, M.J., "Properties of the Resistance Matrix and Applications in Noise Control," *Proceedings of the 8th International Congress of Sound and Vibration (ICSV 8)*, Hong Kong, 2001, 2599-2606.

5. Arkkio, A., "Analysis of Induction Motors Based on the Numerical Solution of the Magnetic Field and Circuit Equations," Ph.D. thesis, Department of Electrical and Communication Engineering, Helsinki University of Technology (HUT), Espoo, 1989.

6. Arnold, R.N. and Warburton, G.B., "The Flexural Vibrations of Thin Cylinders," *Proceedings of the Institution of Mechanical Engineers* 167: 62-80, 1953, doi:10.1243/PIME_PROC_1953_167_014_02.

7. Belahcen, A. et al., "Radial Forces Calculation in a Synchronous Generator for Noise Analysis," *Proceedings of the 3rd Chinese International Conference on Electrical Machines*, Xi'an, China, 1999, vol. 1, 119–122.

8. Inigo, G.D.M., "Analysis of Force and Torque Harmonic Spectrum in an Induction Machine for Automotive NVH Purposes," Master's thesis, Energy and Environment Department, Chalmers University of Technology Gothenburg, Sweden, 2016.

9. Ming, C., Le, S., Giuseppe, B., and Lihua, S., "Advanced Electrical Machines and Machine-Based Systems for Electric and Hybrid Vehicles," *Energies* 8, no. 9: 9541-9564, 2015, doi:10.3390/en8099541.

10. SAE International, "Motor Trend 2004 Car of the Year Winner: Toyota Prius," 2004, https://www.motortrend.com/cars/toyota/prius/2004/04-coy-win/.

11. Chan, C.C. and Chau, K.T., *Modern Electric Vehicle Technology*, 1st edn. (Oxford: Oxford University Press, 2001), 16-28, ISBN-13:978-0198504160.

12. Cheng, M. and Chan, C.C. "General Requirement of Traction Motor Drives," *Encyclopedia of Automotive Engineering*, vol. 3 (Hoboken: Wiley, 2015), 1261–1278, doi.org/10.1002/9781118354179.auto041.

13. Chan, C.C. and Chau, K.T., "An Overview of Power Electronics in Electric Vehicles," *IEEE Transactions Industrial Electronics* 44: 3–13, 1997, doi:10.1109/41.557493.

14. Zhu, Z.Q. and Howe, D., "Electrical Machines and Drives for Electric, Hybrid, and Fuel Cell Vehicles," *Proceedings of the IEEE* 95, no. 4: 746–765, 2007, doi:10.1109/JPROC.2006.892482.

15. Dorrell, D.G., Knight, A.M., Popescu, M., Evans, L. et al., "Comparison of Different Motor Design Drives for Hybrid Electric Vehicles," *IEEE Energy Conversion Congress and Exposition*, Atlanta, GA, September 12–16, 2010, 3352-3359, doi:10.1109/ECCE.2010.5618318.

16. Lakshmikanth, S., Natraj, K.R., and Rekha, K.R., "Noise and Vibration Reduction in Permanent Magnet Synchronous Motors – A Review," *International Journal of Electrical and Computer Engineering (IJECE)* 2, no. 3: 405-416, 2012, doi:10.11591/ijece.v2i3.322.

17. Xin, G., "Simulation of Vibrations in Electrical Machines for Hybrid-Electric Vehicles," Master thesis, Department of Applied Mechanics, Chalmers University of Technology, Sweden, 2014.

18. Janne, R., "Unit-Wave Response-Base Modeling of Electromechanical Noise and Vibration of Electrical Machines," Ph.D. thesis, Department of Electric Engineering, Helsinki University of Technology, Helsinki. 2009.

19. Anwar, M.N. and Husain, I., "Radial Force Calculation and Acoustic Noise Prediction in Switched Reluctance Machines. Industry Applications," *IEEE Industry on Industry Applications* 36, no. 6: 1589-1597, 2000, doi:10.1109/IAS.1999.799157.

20. Banharn, S., "Prediction of Torque and Radial Forces in Permanent Magnet Synchronous Machines Using Field Reconstruction Method," Master thesis, Department of Electrical Engineering, University of Texas at Arlington, 2010.

21. Bashir, M.E., "Dynamic Eccentricity Fault Diagnosis in Round Rotor Synchronous Motors," *Energy Conversion and Management* 52, no. 5: 2092-2097, 2011, doi:10.1016/j.enconman.2010.12.017.

22. Hong-Seok, K. and Kwang-Joon, K., "Characterization of Noise and Vibration Sources in Interior Permanent-Magnet Brushless DC Motors," *IEEE Transactions on Magnetics* 40, no. 6: 3482-3489, 2004.

23. Islam, R., Husain, I., Fawdoun, A., and Mclaughlin, K., "Permanent Magnet Synchronous Motor Magnet Designs With Skewing for Torque Ripple and Cogging Torque Reductin," *IEEE Transactions on Industry Application* 45, no. 1: 152-160, 2009.

24. Marcel, J., Ondrej, V., and Vitezslav, H., "Chapter 8: Noise of Induction Machines," in: *Induction Motors-Modelling and Control*, (London, UK: Intech Open) 2012, doi:10.5772/38152.

25. Sang-Ho, L. and Jung-Pyo, H., "A Study on the Acoustic Noise Reduction of Interior Permanent Magnet Motor with Concentrated Winding," in *IEEE Industry Applications Society Annual Meeting*, Vancouver, BC, Canada, 2008, 1-5, doi:10.1109/08IAS.2008.38.

26. Minh-Khai, N., Young-Gook, J., and Young-Cheol, L., "Acoustic Noise Reduction in Single Phase SRM Drives by Random Switching Technique," *International Journal of Electronics and Communication Engineering* 3, no. 2: 292-296, 2009.

27. Hashemi, N.N. and Lisner, R.P., "A New Strategy for Active control of Acoustic Noise in Converter-Fed induction Motors," *Proceedings AUPEC02*, Melbourne, Australia, Sep.-Oct. 2002, 1-6, ISBN:0-7326-2206-9 .

28. Ahmad, A.K. and Osama, M., "Wavelet Filtering for Position Estimation of Permanent Magnet Machine in Carrier Signal Injection Based Sensorless Control," in *North American Power Symposium (NAPS)*, Tempe AZ, 2009, 1-5, doi:10.1109/NAPS.2009.5484072.

29. Torregrossa, D., Peyraut, F., Cirrincione, M., Espanet, C. et al., "A New Passive Methodology for Reducing the Noise in Electrical Machines: Impact of Some Parameters on the Modal Analysis," *IEEE Transactions on Industry Applications* 46, no. 6: 1899–1907, 2010, doi:10.1109/TIA.2010.2057491.

30. Sutthiphornsombat, B., Khoobroo, A., and Fahimi, B., "Mitigation of Acoustic Noise and Vibration in Permanent Magnet Synchronous Machines Drive Using Field Reconstruction Method," in *IEEE Vehicle Power and Propulsion Conference (VPPC)*, Lille, France, Sep.-3, 2010, 1–5, doi:10.1109/VPPC.2010.5728997.

31. Nicolas, B., Dan, I., Frederic, G., Michel, H.et al., "Design of Permanent Magnet-Synchronous Machine in Order to Reduce Noise Under Multi-Physic Constraints," in *IEEE International Electric Machines & Drives Conference (IEMDC)*, Niagara Falls, 2011, 29-34, doi:10.1109/IEMDC.2011.5994863.

32. Yongxiang, X., Yanmei, Y., Qingbing, Y., Jibin, Z. et al., "Reduction of the Acoustic Noise in PMSM Drives by the Periodic Frequency Modulation," *International Conference on Electrical Machines and Systems (ICEMS)*, Beijing, China, Aug. 2011, 20-23, doi:10.1109/ICEMS.2011.6073927.

33. Nejadpak, A., Mohamed, A., Mohammed, O.A., and Khan, A.A., "Online Gain Scheduling of Multi-Resolution Wavelet-Based Controller for Acoustic Noise and Vibration Reduction in Sensorless Control of PM-Synchronous Motor at Low Speed," *IEEE Power and Energy Society General Meeting*, Detroit, MI, Jul. 24–29, 2011, 1-6.

34. Asmo, T., "Electromagnetic Forces Acting between the Stator and Eccentric Cage Rotor," Ph.D. thesis, Helsinki University of Technology (Espoo, Finland), Helsinki, 2003, ISBN:951-22-6682-2.

35. Gurakuq, D. and Dieter, G., "The Influence of Permeance Effect on the Magnetic Radial Forces of Permanent Magnet Synchronous Machines," *IEEE Transactions on Magnetics* 49, no. 6: 2953-2966, 2013, doi:10.1109/TMAG.2013.2241073.

36. Marcel, J., Ondrej, V., and Vitezslav, H., *Noise of Induction Machines* London, UK: Intech Open, (2012), doi:10.5772/38152.

37. Mohammed, R.I., "Cogging Torque, Torque Ripple and Radial Force Analysis of Permanent Magnet Synchronous Machines," Ph.D. thesis, Electrical Engineering Department, University of Akron, May 2009.

38. Mohammad, S.I., Rakib, I., and Tomy, S., "Noise and Vibration Characteristics of Permanent-Magnet Synchronous Motors Using Electromagnetic and Structural Analyses," *IEEE Transactions on Industry Applications* 50, no. 5 (2014): 3214-3222, doi:10.1109/TIA.2014.2305767.

39. Rukmi, D., Kazi, A., and Faz, R., "Cogging Torque and Torque Ripple in a Direct-Drive Interior Permanent Magnet Generator," *Progress in Electromagnetics Research B* 70, no. 1: 73–85, 2016, doi:10.2528/PIERB16072001.

40. Hwang, S.M. and Lieu, D.K., "Reduction of Torque Ripple in Brushless DC Motors," *IEEE Transactions on Magnetics* 31, pt. 2, no. 6: 3737–3739, 1995.

41. Wang, J., Xia, Z.P., and Howe, D., "Comparison of Vibration Characteristics of 3-Phase Permanent Magnet Machines with Concentrated, Distributed and Modular Windings," *3rd IET International Conference on Power Electronics, Machines and Drives-PEMD*, Dublin, 2006,doi:10.1049/cp:20060157.

42. Asano, Y., Honda, Y., Murakami, H., Takeda, Y. et al., "Novel Noise Improvement Techniques for a PMSM with Concentrated Winding," *Proceedings of the Power Conversion Conference*, Osaka, Japan, 2002, vol. 2, 460–465, doi:10.1109/PCC.2002.997562.

43. Ko, H.-S. and Kim, K.-J., "Characterization of Noise and Vibration Sources in Interior Permanent Magnet Brushless DC Motors," *IEEE Transactions on Magnetics* 40, no. 6: 3482–3489, 2004, doi:10.1109/TMAG.2004.832991.

44. Gieras, J.F., Wang, C., Lai, J.C. S., and Ertugrul, N., "Analytical Prediction of Noise of Magnetic Origin Produced by Permanent Magnet Brushless Motors," *Proceedings IEEE International Conference Electrics Machines & Drives Conference*, Antalya, Turkey, May 2007, vol. 1, 148–152, doi:10.1109/IEMDC.2007.383568.

45. Islam, R. and Husain, I., "Analytical Model for Predicting Noise and Vibration in Permanent Magnet Synchronous Motors," *IEEE Transactions Industry Applications* 46, no. 6: 2346–2354, 2010, doi:10.1109/TIA.2010.2070473.

46. Zhu, Z.Q., Ishak, D., Howe, D., and Chen, J.T., "Unbalanced Magnetic Forces in Permanent Magnet Brushless Machines with Diametrically Asymmetric Phase Windings," *IEEE Transactions Industry Applications* 43, no. 6: 1544–1553, 2007, doi:10.1109/TIA.2007.908158.

47. Magnussen, F. and Lendenmann, H., "Parasitic Effects in PM Machines with Concentrated Windings," *IEEE Transactions Industry Applications* 43, no. 5: 1223–1232, 2007, doi:10.1109/TIA.2007.904400.

48. Timar, P.L., Fazekas, A., Kiss, J., Miklos, A. et al., *Noise and Vibration of Electrical Machines* (Amsterdam: Elsevier, 1989), ISBN:0444988963.

49. Pollock, C. and Wu, C.Y., "Analysis and Reduction of Vibration and Acoustic Noise in the Switched Reluctance Drive," *IEEE Transactions Industry Applications* 31, no. 1: 91–98, 1995, doi:10.1109/IAS.1993.298911.

50. Gieras, J.F., Wang, C., and Li, J.C., *Noise of Polyphase Electric Motors*, 1st edn. (Boca Raton, FL: CRC Press, 2005), ISBN:9780824723811.

51. Zhu, Z.Q. and Howe, D., "Electromagnetic Noise Radiated by Brushless Permanent Magnet DC Drives," *Proceedings of the International Conference on Electrical Machines and Drives*, Durham, UK, Sep. 8–10, 1993.

52. Zhu, Z.Q., Ishak, D., Howe, D., and Chen, J.T., "Unbalanced Magnetic Forces in Permanent Magnet Brushless Machines with Diametrically Asymmetric Disposition of Phase Windings," *IEEE Transactions on Industry Applications* 43, no. 6: 1544-1553, 2007, doi:10.1109/TIA.2007.908158

53. Zhu, Z.Q., "Noise and Vibration in Fractional-Slot Permanent Magnet Brushless Machines," *UK Magnetic Society Seminar on Noise in Electrical Machines*, Cardiff, Feb. 2009.

54. Islam, M.S., and Islam, R.. "Cogging torque minimization in PM motors using robust design approach," *IEEE Transactions on Industry Applications* 47: 1661–1669, 2011.

55. Wang, J., Xia, Z.P., and Howe, D., "Comparative Study of 3-Phase Permanent Magnet Brushless Machines with Concentrated, Distributed and Modular Windings," *Proceedings of the IET International Conference on Power Electronics, Machines and Drives 2006*, Ireland, Apr. 4–6, 2006.

56. Miller, T.J.E., "Faults and Unbalance Forces in the Switched Reluctance Machine," *IEEE Trans. Ind. Appl.* 31, no. 2: 319–328, 1995, doi:10.1109/28.370280.

57. Verma, S.P. and Balan, A., "Measurements Techniques for Vibrations and Acoustic Noise of Electrical Machines," *Proceedings of 6th International Conference Electrical Machines and Drives*, London, UK, Sep. 8-10, 1993

58. Shenbo, Y., Qing, Z., Xiulian, W., Jiang, D. et al., "Analysis of Noise and Vibration from Permanent Magnet Synchronous Machine," *Proceedings of the ICEMS 2003*, Beijing, China, Nov. 9–11, 2003.

59. Jung, J., Kim, D., Hong, J., Lee, G. et al., "Experimental Verification and Effects of Step Skewed Rotor Type IPMSM on Vibration and Noise," *IEEE Transactions on Magnetics* 47, no. 10: 3661–3664, 2011, doi:10.1109/TMAG.2011.2150739.

60. Vijayraghavan, P. and Krishnan, R., "Noise in Electric Machines: A Review," *IEEE Trans. Ind. Appl.* 35, no. 5: 1007–1013, 1999, doi:10.1109/IAS.1998.732298.

61. Shin-Eisu chemical Co. Ltd, "Magnetic Circuit Analysis/SPM Motor," http://www.shinetsu-rare-earth-magnet.jp/e/circuit, accessed 2019.

62. Hameyer, K., Henrotte, F., and Delaere, K., "Electromagnetically Excited Audible Noise in Electrical Machines," *J. KSNVE* 13: 109–118, 2003.

63. Le Besnerais, J., "Fast Prediction of Variable-Speed Acoustic Noise due to Magnetic Forces in Electrical Machines,"*2016 XXII International Conference on Electrical Machines (ICEM)*, Lausanne, Switzerland, 2016, 2259–2265, doi:10.1109/ICELMACH.2016.7732836

64. MANATEE Software, "Magnetic Acoustic Noise Analysis Tool for Electrical Engineering," Computer Software (version 1.05), 2016.

65. LEA Software, "GENESIS' Brochure," 2019.

NVH of BEV

The key system of a BEV consists of the following components. An auxiliary battery provides electricity to power the vehicle accessories. A charge port allows the vehicle to connect to an external power supply in order to charge the traction battery pack. A DC/DC converter converts higher voltage DC power from the traction battery pack to the lower voltage DC power needed to run the vehicle accessories and recharge the auxiliary battery. An electric traction motor uses power from the traction battery pack; this motor drives the vehicle's wheels. Some vehicles use motor generators that perform both the drive and regeneration functions. An onboard charger takes the incoming AC electricity supplied via the charge port and converts it to DC power for charging the traction battery. The charger monitors battery characteristics such as voltage, current, temperature, and state of charge while charging the pack. A power electronics controller manages the flow of electrical energy delivered by the traction battery, controlling the speed of the electric traction motor and the torque it produces. A thermal cooling system maintains a proper operating temperature range of the engine, electric motor, power electronics, and other components. A traction battery pack stores electricity for use by the electric traction motor. A transmission transfers mechanical power from the electric traction motor to drive the wheels [1–4].

5.1 Metrics to Assess BEV NVH

The major NVH challenges in electric vehicles (EVs) have been widely realized and addressed, which should be solved if EVs are to be accepted by the buying public. EVs have significantly lower forcing functions because the internal combustion engine (ICE) is replaced by a motor. Moreover, EVs use lightweight materials to achieve the highest levels of efficiency and range, which exhibits higher vibration and noise for a given set of sources. Eliminating the IC engine from an EV completely changes the source side of the "source-path-receiver" equation conventionally used in NVH engineering. An EV does not require an engine, multispeed transmission, transfer case, fuel tank, air intake or exhaust systems, long driveshafts, or center bearings. This leaves much more space to accommodate the electric motor(s) and battery pack, with additional room for creative interior occupant positioning. More importantly, the sources of noise and vibration (NV) in EVs have been reduced in quantity and are potentially located in different, often more beneficial, areas of the vehicle compared to an IC engine car. The sources are also greatly simplified, which also reduces the complexity of the various noise paths and the probability of cross-coupling and system interaction.

However, there is still a tremendous challenge for EV noise control. The elimination of the IC engine will reduce masking background noise and expose many of the noise behaviors of a vehicle that were previously acceptable but will now seem plainly audible and perhaps annoying. As such, the main efforts for EV NVH are different from that for ICE vehicles. Some estimation of how the noise reduction effort was ranked in terms of the basic overall effort is (1) motor/gearbox noise—15%, (2) ancillary device noise—15%, (3) tire/road noise—40%, and (4) wind noise—30%. In addition, other conventional issues, such as buzz, squeak, and rattle, remain as additional issues. By comparison, for a conventional IC engine vehicle, it is estimated that the amount of engineering effort spent on powertrain NVH is up to 40–50% of the total NVH effort on a vehicle. This is because the vibration and noise levels generated by an IC engine and its drivetrain are highly complex, crossing many subsystems of the vehicle. Moreover, the energy levels are very high due to the combustion and reciprocating forces generated by an IC engine [5–14].

From the physical point of view, EVs show a lower overall sound pressure level in the vehicle interior than combustion-driven vehicles. This reduction in sound pressure level usually results in an increased perception of comfort, thus supporting the most important dimension of vehicle sound quality perception.

Moreover, sound quality has also been critical to the NVH of EV, which has many unique properties that need to be addressed. In reality, many different sound concepts are designed and developed, which vary in spectral content and sound characteristics, such as total loudness, modulation, loudness adaption level, and so on, under varied vehicle operational conditions as described in Ref. [15].

To simulate the NHV of EVs for movement of varied situations, a pattern of speed similar for all the sounds has been used. This included several phases of driving, such as idling, acceleration, constant speed (typically 50 km/h), deceleration, and idling, as shown in **Figure 5.1**.

Mapping the speed of the car with the parameters of the sound (frequencies and amplitudes of the components) makes the sounds evolve and "simulates" the movement of the car.

Analogous to an IC engine, for this mapping, the rule used is that the frequencies of the components increase when the speed increases. This pattern creates realistic conditions to facilitate the perception of speed, acceleration, or deceleration.

FIGURE 5.1   Speed pattern of cars for sound evaluation [18].

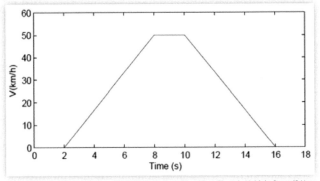

Opposite to this positive effect, several facts result in reduced sound quality of EV interior sounds. First, masking is reduced so that component sounds become audible, which are masked in combustion-driven vehicles. Since lightweight materials have to be used to reduce the overall vehicle weight, the component sounds might even be worse than in combustion-driven vehicles.

Furthermore, the electric drive train usually generate high-frequency tonal components, yielding to uncomfortable and annoying perception. Since the combustion engine sound component is missing, the overall spectral balance is broken and a spectral gap is introduced between the low-frequency tire noise and the high-frequency wind noise. These issues can only be solved by tedious, cost, and weight-increasing countermeasures. Other issues are related to the driver. For example, the electric engine does not give enough feedback about the current status of operation to the driver, the sound does not address his or her emotion, and a transportation of brand attributes is hindered.

All of the above-listed issues can be addressed by a proper generation of interior sounds for EVs. The necessity to perform this sound generation is supported by legislation, which prescribes the generation of sound for the exterior. Exterior sound generation thus becomes mandatory, resulting in the fact that sound design has to be performed and can no longer be avoided. Furthermore, portions of the exterior sounds usually are also audible in the vehicle interior and can result in undesired sound perception, which can be compensated by an aligned interior sound generation. Nevertheless, sound design for EVs is a complex task, especially since experience in the field of sound generation is missing and the customer acceptance is not well understood yet. Furthermore, comprehensive tools have to be developed, allowing a flexible generation of a variety of different sound characters and an easy and time-efficient vehicle adaptation.

The results on sound quality evaluations for ICE vehicles and EV vehicles are quite different. The detailed quantitative indexes of sound quality will be presented in Chapter 8.

Following is an example given by SIEMENS. **Figure 5.2** shows the comparison of overall sound pressure levels and a specific sound quality index for ICE vehicles and EVs, from which we can see that even overall sound pressure levels of EVs are much

FIGURE 5.2 Comparison of overall sound pressure levels and specific sound quality index for an ICE vehicle and an EV [19].

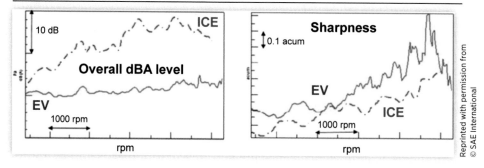

TABLE 5.1 Sound quality index for EV noise [17].

Articulation Index	ICE 56%
	EV 73%
Prominence Ratio	EV order *4×np*
	9.27 dB
	> 9 dB threshold
Tone to Noise Ratio	EV order *4×np*
	11.03 dB
	> 8 dB threshold

Case: WorldAutoSteel FSV concept study.

Benchmark Vehicle EV PWT – Acceleration WOT.

Reprinted by permission from Springer Nature, Fiebig A., Schulte-Fortkamp B. (2019) Acceptance of synthetic driving sounds in the interior of electric vehicles. In: Siebenpfeiffer W. (eds) Automotive Acoustics Conference 2015. Proceedings. Springer Vieweg, Wiesbaden. https://doi.org/10.1007/978-3-658-27648-5_2

lower than that of ICE vehicles; the specific sound quality index for the EV is much higher than that of the ICE vehicle. **Table 5.1** shows a certain sound quality index for EV noise.

Moreover, the characteristic of a noise phenomenon can be described using simple subjective and onomatopoeic terms. This is especially helpful for test drivers or engineers that do not have a deep background in NVH and are unfamiliar with psychoacoustic metrics, such as tonality, sharpness, or modulation. The characterization is especially easy when using opposing property pairs like high/low, dark/bright, soft/rough, and so on. Based on the conventional metrics used, the following selection was proposed, which includes the following basic properties: (1) auditory/haptic (hear/feel), (2) frequency location (bass/mid/treble), (3) frequency range (broad/narrow), (4) course (steady/modulated, rising/falling/wobbling), (5) duration (short/long, abrupt/fading), (6) impulsiveness (click/damped, hit/brush), (7) tonality (tonal, noisy), and (8) definition (defined/stochastic). Comparative properties have also been used, such as humming/buzzing, rumble, howling, whistling, squealing, tacking, clacking, rattle, hissing, droning, scrubbing, and so on. In practice, it is also possible to use example sounds to give the user an idea of what each character attribute means.

5.2 NVH Development of BEV, NVH Targets, and Target Decomposition/Cascading

In the NVH development of BEV, much effort is dedicated to the reduction of motor powertrain-related NVH, in addition to using conventional approaches to deal with wind and road/tire NVH. For example, the multistage system level method is used to design, optimize, and enhance electric motor NVH performance of General Motors' Chevrolet Bolt BEV. Detailed examples are given as follows. First, the rotor EM (electromagnetic) design optimizes magnet placement between adjacent poles asymmetrically, along with a pair of small slots stamped near the rotor outer surface to lower torque ripple and radial force. The size and placement of stator slot openings under each pole are optimized to lower torque ripple and radial force. Next, motor stator level FE (finite element) analysis and modal test correlation are performed to benchmark the orthotropic stator material properties and accurately predict modal results within 7% error below 2 kHz. Furthermore, tangential and radial EM forces are applied to the motor-in-fixture subsystem FE model, which predicts surface vibration and pseudo-sound power on the motor housing. Analysis results are validated by test data and are used to benchmark electric motor as the BEV noise source. Analysis also helps to identify key motor orders and RPM for NVH optimization. Lastly, optimized EM and motor mechanical designs are modeled in the drive unit (DU) for transmission level NVH analysis. The multistage system level model is used to study key design parameters, such as EM force coupling with structural modes, motor mounting design, DU ribbing, and stiffness optimization. Key design concepts and parameters that have the most influence on radiation sound power from DU are identified, and subsequently optimized for improved noise performance of the Bolt EV [18–26].

1. Electromagnetic design to reduce torque ripple and radial force (noise source control)

2. Motor component-level FE analysis and modal test correlation to ensure a high-fidelity mechanical model for the electric motor (motor structural path design)

3. Motor-on-fixture subsystem-level FE analysis and vibration test correlation to validate EM force and stator coupling (motor subsystem model)

4. Motor-in-DU transmission-level FE analysis to predict pseudo sound power response and perform parametric/topology optimization to reduce motor noise at critical motor orders and RPMs (structural path and radiation control)

In the NVH development of BEVs, in addition to dealing with motor NVH, as well as wind and road/tire MVH, it is also important to deal with the NVH of gearbox components and housing, mounts, driveshafts, and power electronics housings. This would provide a far deeper understanding of the full system response and avoid the problems arising from the individual simulation of separate subsystems. For system modeling, a proprietary design and simulation package for drivelines were used to model the shafts and bearings of the gearbox and motor, gears, mounts, driveshafts, stator, and the housing of the gearbox, motor, and power electronics. Like many FEA-based approaches, the method involves multiple coupled shaft systems, with a linear model used to derive mode shapes forced in the frequency domain. The housing is included by taking an FEA model of it, and performing a dynamic condensation using Craig-Bampton modal reduction. Component mode synthesis is used to link the reduced dynamic model of the housing to the dynamic model of the internal components.

Figure 5.3 shows the most important components within NVH development of EVs.

Despite the lower sound levels of an EV and the missing masking effect of the ICE, several other components, such as tires, gear, wind, and the Heating, Ventilation, Air Conditioning (HVAC) system, are detectable in the sound signature of the car and may affect the interior sound quality. In addition, new sound sources appear, for example, from motor or battery cooling fans or ducts. Besides these new sources, new driving modes are also introduced in EVs, including regenerative braking, which is typically characterized in the noise footprint by a whining noise.

FIGURE 5.3   Overview of the most important components within NVH development [16].

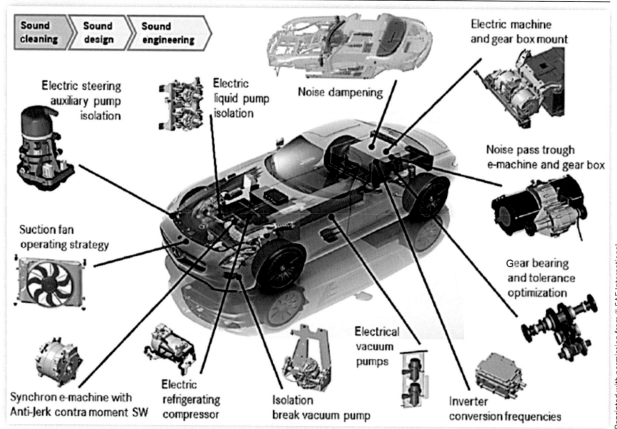

Overall, the EV NVH efforts are summarized as follows [16, 27, 28]:

Vehicle layout

- Different weight distribution
- Different powertrain component structural stiffness

Body

- Reduction of body weight by utilizing more composites
- Different connection points
- Transparency for mid-high-frequency noise
- Interior sound characterization, design, and control (e.g., active noise control)

Subframes

- Important changes to known configurations

Accessories

- Less masking (fan, power-steering system, HVAC, hydraulic pumps, etc.)

Electrical motor and power electronics

- High torsional load at low RPM
- Electro-magnetic excitation of motor-housing
- Operation as motor or generator
- Good balancing
- Off-zero modulation (power electronics)

Batteries

- Battery cooling
- NVH additional accessories (pump)

Road noise
Wind noise
Pedestrian safety

- Warning sounds

5.3 **Motor NVH, Motor Whine**

The noise radiation of electric motors is caused by the radial force waves created by the motor stator currents. The origin of the noise emission is the phase current in the stator coils, which generates time-varying radial forces in the air gap between the stator and rotor poles. These radial forces induce vibrations in the stator, which are propagated through the mechanical structure. The highest vibration levels occur when the natural modes of the stator are excited by the pulsating radial forces. Finally, the deformations of the machine, the stator and the attached components, cause pressure fluctuations perceived as sound by the human ear.

Inverter-induced noise is another NVH challenge. Moreover, additional noise components from the rest of the motor can make the noise spectrum even more complex. Examples are unbalance, misalignment, looseness and distortion, defective bearings, gearing and coupling inaccuracies, rotor/stator misalignments, bent rotor shafts, and so on. Also, the way the electric powertrain is mounted inside the vehicle is very important because it affects the transmission of noise to the driver compartment [29, 30].

The EM disturbances, such as radial electromagnetic forces and torque ripple, can radiate directly from the e-machine housing (airborne component) and be transmitted through the structural attachments, like stator bolts, the stator ring, powertrain mounts, and so on (structure-borne component). In the e-machine driven by the PWM switching inverter, current is not perfectly sinusoidal but contains a different level of harmonics. Current harmonics impact torque ripple, which in turn would translate into undesirable NVH. The e-machine whine noise is tonal in nature, typically in the 400–2000-Hz range, and can be annoying to the customer. The tonal noise issues can play a larger role on hybrid vehicles and BEVs due to the lack of sufficient "normal" masking noise from the ICE under motoring and brake regeneration modes of operation. While modal decoupling and acoustic cover can be employed to mitigate the whine noise, a source-level improvement is the most cost-effective and robust approach during the development phase of an EV.

To calculate radial electromagnetic forces, torque ripple during the preliminary design phases of e-machines assumes ideal sinusoidal air gap flux and stator current distributions. In reality though, the air gap flux and current contain different levels of harmonics that render significant distortion of these waveform distributions. Current harmonics in e-machines are induced by various factors, including non-ideal spatial distribution of stator windings, rotor geometry, PWM switching of inverter drives, dead time during switching, control bandwidth limitations, and manufacturing and assembly issues (e.g., rotor eccentricity). The current harmonics result in unpredictable and abrupt changes to the magnitude and phase angle of torque ripple at different time instances, and torque and speed conditions during the operation of the e-machine.

For example, the Nissan Leaf is a zero-emission car that has been designed for the mass market with improvement in motor NVH performance [31]. The electrical powertrain (e-PT) component of the Nissan Leaf is an integrated unit with the electric motor, inverter, and the charging system, in which the power delivery module (PDM) is the combined unit and consists of the charger, DC/DC converter, and the junction box.

Figure 5.4 is the radiated noise influence rate of PT noise radiation measurement, in which the integration exerted a bad influence on the e-PT noise radiation level at two places: the low-frequency area (at 1.7 kHz) and the high-frequency area (around 6 kHz). The analysis showed that the main cause for the noise radiation generated at 1.7 kHz was the PDM (structural component), and the noise radiation generated at 6 kHz was the motor.

The main specific acoustic signature of the electric powertrain comprises orders (frequency components related to the rotational speed) due to primarily EM forces, tones, and/or orders due to DC/AC converter pulse-width modulation (PWM) and finally gear mesh order(s) from the reducer. All those phenomena yield tonal and high-frequency sound character, which can be perceived as annoying when prominent.

FIGURE 5.4 Radiated noise influence rate of PT noise radiation measurement [32].

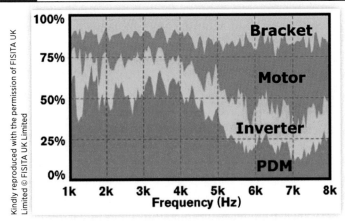

For quantification of whining noise, tone-to-noise ratio (TNR) and prominence ratio (PR) are two established metrics, standardized in ECMA-74 [4], which relates the tonal energy to adjacent broadband noise. They are preferably described using psycho-acoustic metrics in contrast to sound pressure levels, which will be presented in Chapter 8. TNR and PR have been employed to quantify levels of electric motor noise. In a recent study, the relationship between perceived annoyance and TNR/PR levels was investigated based upon synthetic tones added to the broadband masking noise recordings from one electric car [33].

Depending on the design of the motor, the EM force pulses and corresponding torque pulses from the motor can be very strong. These can be forces radiated as noise directly from the motor housing and can also be transmitted structurally to the support structure through the motor mounts.

The EM forces are generally lower than the combustion and reciprocating mass forces of an IC engine, and significantly, they are at a much higher frequency. As a result, the rubber isolation systems used to mount the electric motor to the body can be tuned more efficiently and achieve a much higher level of isolation than with an IC engine. Also, the noise radiated directly from the motor is generally quite high in frequency (>1000 Hz), which means that conventional acoustical materials are highly effective at blocking and absorbing this airborne noise energy. So, while it is true that electric motors present a lesser challenge for noise control engineers compared to IC engines, strategies for mechanical and acoustic isolation of the motor must still be effectively executed, which is not a trivial task.

In addition to the motor, other parts of the electromotive system create noise, particularly the gearbox and the power electronics unit 28, 34–36. With respect to the gearbox, most EVs use direct-drive, single-speed gear sets. The standard knowledge base on gear design (e.g., helical vs. straight cut gears), and more importantly, gear finishing, applies to EV gearboxes. The goal should be to create a system that generates high-frequency gear whine no greater than the noise generated by the motor. The power electronics unit provides the high voltage, high-current energy to the motor.

This is a sophisticated computer-controlled switch that has variable or fixed switching frequencies almost always above 10,000 Hz. This high-energy switching can generate quite a bit of radiated noise at the base switching frequency and at higher harmonics.

Table 5.2 shows the root causes of electric motor NVH issues.

The PWM technology is widely used in inverter-fed motor control, including electrified vehicles. PWM will cause high-frequency harmonics due to its amplitude switching. Harmonics of electromagnetically induced forces will excite the motor structure and cause mechanical resonances and noise radiation.

The motor magnetic noise is decreasing when PWM switching frequency is increasing. A noise and efficiency tradeoff is necessary before increasing the switching frequency to reduce the noise. Certain optimum switching frequency could be identified.

5.4 Gearbox NVH, Gear Mesh Whine, and Rattle, Driveline NVH

The sources of NVs in an electric drivetrain can be grouped into two categories: (1) the EM noise related to the motor topology, stator and rotor structure, power supply, and so on; and (2) the pure mechanical noise related to the full assembly [37].

Conventionally, gear whine noise mainly affects vehicle ride comfort through the middle and high-frequency noise, and the frequency range of gear whine noise is 500–5000 Hz, which increases with the input speed, showing the narrowband order excitation. It is often accompanied by harmonic components, mostly in the first-order and second-order harmonic. Its high-frequency characteristics will make people feel irritable and is difficult to accept. Gear whine noise of the automobile transmission is mainly caused by forced vibration of the transmission case under the excitation of load gears. Transmission error (TE) caused by the time-varying meshing stiffness excitation, processing and assembly error excitation, and others makes the actual meshing line deviate from the theoretical meshing line, resulting in tooth interference, collision, and exciting force causing the vibration of transmission mechanism and housing.

NVH problems in gearing are mainly concerned with the smoothness of the drive [38, 39]. The parameter that is employed to measure smoothness is the TE. This parameter can be expressed as a linear displacement at a base-circle radius defined by the difference of the output gear's position from where it would be if the gear teeth were perfect and infinitely stiff. Many references have attested to the fact that a major goal in reducing gear noise is to reduce the TE of a gear set. Experiments show that decreasing TE by 10 dB (approximately three times less) results in decreasing transmission sound level by 7 dB. TE results not only from manufacturing inaccuracies, such as profile errors, tooth pitch errors, and run-out, but from poor design.

The pure tooth involute deflects under load due to the finite mesh stiffness caused by tooth deflection. A gear-case-and-shaft system also deflects due to load. While running under load, tooth contact stiffness varies, which excites the parametric vibration and, consequently, noise.

TABLE 5.2 Electric motor NVH diagnostic chart [28].

Cause	Frequency of vibration	Phase angle	Amplitude response	Power cut	Comments
Misalignment: ① Bearing	Primarily 2 x Some 1 x Radial High at drive end and Axial	Phase angle can be erratic.	Steady.	Drops slowly with speed.	① 2 x can dominate during cost-down. ② 2 x is more prevalent with higher misalignment.
Misalignment: ② Coupling	Primarily 1 x Some 2 x Radial High at DE and Axial	Drive 180° out Phase with non drive end.	Steady.	Level drops slowly with speed.	① Parallel causes radial forces and angular causes axial. ② Load dependent.
Rub- ① Seal/or bearing	1/4x, 1/3x, 1/2x or 10-20xcan be seen Primarily 2 x Some 1 x. Radial.	Erratic.	Erratic depending upon severity.	Disappears suddenly at some lower speed.	① Full rubs tend to be 10 to 20x higher. ② Bearing misalignment can give rub symptoms.
② Rotor	1/4x, 1/3x, 1/2x, & 1x with slip freq. side bands. Radial.	Erratic.	High.		① Severe pounding.
Looseness: ① Bearing (non-rotating)	2 x 3 x may be seen Radial	Steady.	Fluctuates	Disappear at Some lower speed	① Bearing seat looseness. ② Looseness at bearing split.
② Rotor Core (rotating)	1-10x with 1, 2, & 3 predominant. Radial	① Can exist relative to type of looseness ② General core loose gives erratic symptom.	Erratic, high amplitude	① Drops with speed. ② Can disappear suddenly.	① End plates loose. ② Core ID loose.
③ Pedestals (non-rotating)	1-10x with 2 & 3 predominant Radial & Axial	Steady.	Fluctuates.	Disappears at some lower speed.	
④ External Fans	1&3 x Radial & Axial - OE (fan end)	N/A	Fluctuates.	① Drops with speed. ② Can disappear suddenly.	
Unbalance Rotor	1x rotor speed. Radial	① NDE & DE in phase. ② Couple gives out of phase condition	Steady.	Level drops slowly.	Rotor has unbalance - can be due to thermal problems.
Unbalance of External Fan	① 1X Radial high at NDE (fan end). ② 1X Axial with high at fan end.	① Couple DE 180° out of phase with EO.	Steady.	Level drops slowly	

TABLE 5.2 (*Continued*) Electric motor NVH diagnostic chart [28].

Cause	Frequency of vibration	Phase angle	Amplitude response	Power cut	Comments
Coupling Unbalance	1 x Radial & higher on drive end		Steady	Level drops slowly	Unbalance due to coupling or key
Bent Shaft Extension	2 x Primarily 1 x may be seen Axial	EO 180° out of phase with DIE.	Steady.	Level drops slowly.	DE runout should give higher 2x axial at that end Normal runout on core - 1-2 mil.
Eccentric Air Gap	Strong 120 Hz Radial	N/A	Steady	Immediately drops	Difference between max. and min. air gap divided by ave. should be less than 10%.
Soft Foot Eccentric rotor	1x Primarily Some 60 & 120 Hz Radial	Unsteady.	Modulates in amplitude with slip	Immediately drops	① Eccentricity limit 1-2 mil. ② Slip beat changes with speed/load.
Loose stator core.	120 Hz. Axial & radial	Frame & bearing brackets in phase at 120 Hz	Steady	Immediately drops	① Look for relative motion of core with respect to housing.
Rotor Bow (Thermal Bow)	1x Primarily Some 120 Hz may be seen	Unsteady	① Changes with temperature. ② Time or load related. ③ Varies at Freq. slip x poles	Some drop but high level would come down with speed.	① Heat related. ② Examine rotor stack for uneven stack tightness or looseness. ③ Shorted Rotor Iron ④ Check bar looseness.
	May have Modulators on 1X & 2X vib. - Radial				
Broken rotor bars	1x and modulates at slip x # poles	Dependent upon where broken bars are located.	STRONG BEAT POSSIBLE. - Varies @ Freq. Slip x poles	Immediately drops	① Sparking in the air gap may be seen. ② Long term variation in stator slot frequencies can be indicator of bar problems.
	May have high stator slot frequencies On slower speed				
	Motors		- Amplitude increased with load		③ Broken bars cause holes in magnetic field. ④ Large current fluctuations. ⑤ Current analysis shows slip frequency side bands.

TABLE 5.2 (*Continued*) Electric motor NVH diagnostic chart [28].

Cause	Frequency of vibration	Phase angle	Amplitude response	Power cut	Comments
Loose bars.	① 1 x Possible balance effect with thermal sensitivity Radial ② Stator slot freq. plus sidebands @ ±(# Poles'Slip)	1.1 x vibration will be steady 2. Stator slot freq. with modulate causing a fluctuation in phase angle on overall vibration	Steady	① Stator slot freq. will immediately disappear. ② Imbalance effect can suddenly disappear at some lower speed.	Excessive looseness can cause balance problems in high speed motors.
Interphase fault	60 & 120 Hz Radial	N/A	Steady and possible beat.	Immediately disappears.	
Ground fault	60 Hz & 120 Hz slot freq. - Radial	N/A	Steady and possible beat.	Immediately disappears.	
Unbalanced Line Voltages	120 Hz Radial	N/A	Steady 120 Hz & Possible beat.	Immediately disappears.	
Electrical Noise Vibration	(RPM x # of Rotor slots)/60 +/-120, 240, etc. - Radial	Due to modulation overall vibration will fluctuate	Steady	Immediately disappears.	Increases with increasing load.
System Resonance	1 x RPM or other forcing frequency One plane - usually Horizontal	Varies with load and Speed	Varies	Disappears rapidly	Foundation may need stiffening - may involve other factors
Strain	1 x RPM		Steady		Caused by casing or foundation distortion from attached structure (piping).
Poorly shaped Journal	2x Rotational Usual	Erratic	Steady	May disappear at lower speed	May act like a rub.
Oil Film Instability (Oil Whirl)	Approx. (43-.48)*rotational	Unstable	Steady		
Anti-Friction Bearing Problems	Various Frequencies dependent on bearing design	Unstable	Steady		Four basic frequencies.
Resonant Parts Top Cover Fit	At forcing Frequency or Multiples	N/A	Steady♦	Drops rapidly	May be adjacent parts
Approx.	120 Hz. Radial	N/A	Steady	Disappears immediately.	① Magnification of 120 Hz electrical ② Top cover rests on basic core support.

The meshing noise is caused by the action of gears that not conjugate (i.e., the profiles of the teeth do not conjugate perfectly) and it results in dynamic forces on the teeth, which can generate vibrations of the wheels and shaft.

The dynamic actions during the meshing are transmitted through the bearings to the walls of the shell, and from there, through structural vibrations or sound waves radiated to the rest of the vehicle. It should be noted that the walls of the gear box usually radiate noises. **Figure 5.5** outlines the gear whine noise generation process.

Excitations generated at the gear mesh, such as TE, friction, and axial shuttling forces [2], induce torsional and bending vibrations in the shafting. The bearing forces resulting from the shaft vibrations excite the housing, which then radiates noise.

The gear rattle is associated with the characteristic noise that unselected impacting gears radiate to the environment [40]. It is induced by force-order vibration in the presence of backlash in the unengaged gear pairs, resulting in oscillatory response within their backlash range. A tribo-dynamic model of a front wheel drive manual transmission has been developed to study idle rattle, considering the hydrodynamic contact film reaction and flank friction. The model includes the torsional motions of the idle gears and the lateral motions of the supporting output shafts.

A power supply of sinusoidal voltage that varies from positive to negative peak voltage in each cycle produces an EM attracting force between the stator and rotor, which is at a maximum when the magnetizing current flowing in the stator is at a maximum either positive or negative at that instant in time [28]. As a result, there will be two peak forces during each cycle of the voltage or current wave reducing to zero at the point in time when the current and fundamental flux wave pass through zero, as demonstrated in **Figure 5.6**. This is with an ideal current that does not consider the harmonics introduced by inverter switching. This will result in a frequency of vibration equal to two times the frequency of the power source (twice line frequency vibration). This particular vibration is extremely sensitive to the motor's foot flatness, frame and base stiffness, and how consistent the air gap is between the stator and rotor around the stator bore. It is also influenced by the eccentricity of the rotor.

The simplified load-related magnetic force frequencies and mode shapes are given as follows. The frequencies of the load-related magnetic forces applied to the stator teeth and core equal the passing frequency of the rotor bars plus side bands at + or − $2f$, $4f$, $6f$, and $8f$ Hz, where f is the line frequency. A magnetic force is generated at the passing frequency of the rotor slot (FQR), which is motor speed in revolutions per second multiplied by the number of rotor slots as

$$FQR = RPM * Nr / 60 \, Hz, \tag{5.1}$$

where Nr is the number of rotor slots. As an example, a typical two-pole 3570 RPM motor with 45 rotor slots has an FQR = 2680 Hz. The side bands are created when the amplitude of this force is

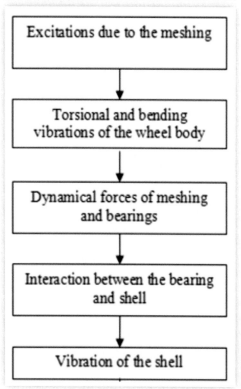

FIGURE 5.5 Gear whine noise generation process.

Excitations due to the meshing

↓

Torsional and bending vibrations of the wheel body

↓

Dynamical forces of meshing and bearings

↓

Interaction between the bearing and shell

↓

Vibration of the shell

© SAE International

FIGURE 5.6 One period flux wave and magnetic force wave.

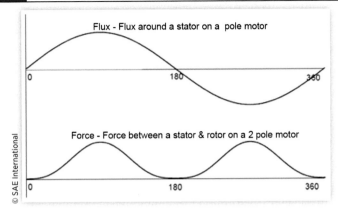

modulated at two times the frequency of the power source. On a 60-Hz system, the 120-Hz modulation produces the side bands, giving excitation frequencies of FQR, FQR + 120, FQR—120, FQR + 240, FQR—240 Hz, and so on.

The forces applied to the stator teeth are not evenly distributed to every tooth at any instant in time; they are applied with different magnitudes at different teeth, depending upon the relative rotor- and stator-tooth location. This results in force waves over the stator circumference. The mode shape of these magnetic force waves is a result of the difference between the number of rotor and stator slots as

$$M = (Ns - Nr) \pm KP, \tag{5.2}$$

where Ns is the number of stator slots, Nr is the number of rotor slots, P is the number of poles, and K is the all integers 0, 1, 2, 3, and so on.

In the e-machine driven by a PWM switching inverter, the current is not perfectly sinusoidal but contains different levels of harmonics [25]. Current harmonics impact torque ripple, which in turn would translate into undesirable NV. There is very limited literature referencing the influence of current harmonics on torque ripple and e-machine NVH. This chapter specifically addresses the impact of current harmonics on the 6th and 12th electrical orders of the torque ripple and relevant e-machine whine noise and e-machine disturbance, including torque ripple and radial electro-magnetic forces. These disturbances can radiate directly from the e-machine housing (airborne component) and also can be transmitted through the structural attachments like stator bolts, the stator ring, powertrain mounts, and so on (structure-borne component).

In practical analysis, the EM forces have a beam modeled using FEM model. An induction motor is considered as an example for the electric system. The model of the e-motor has been created using FEM software. The magnetic field has been solved in the time domain, considering the electric circuit and the voltage with a three-phase electricity source [41].

The current of the three-phase source is shown in **Figure 5.7a**. The fluctuation in the current is also indirectly another source of the NVH issue, which is affected also by the DC–AC converter, which will be detailed in following section.

FIGURE 5.7 (a) Currents of the three-phase electrical source, (b) ripple torque diagrams (for varied effect of skew angle and eccentricity), and (c) total radial forces (with and without eccentricity) [42].

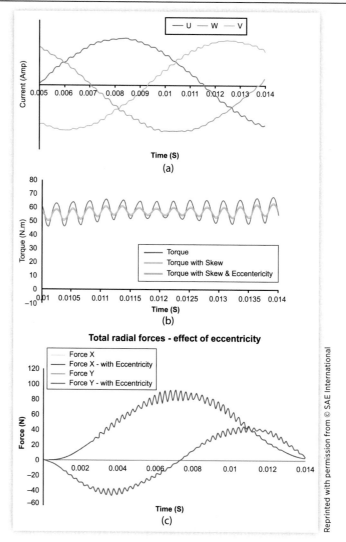

Reprinted with permission from © SAE International

Figure 5.7b shows the ripple torque diagrams as the result of the EM FEM model. The expected harmonics are approximated in the following equations:

$$m_s = \frac{MN_s}{P} \pm 1$$

$$m_r = \frac{MN_r}{P} \pm 1$$

(5.3)

where m_s is the order of the stator harmonics, m_r is the order of the rotor harmonics, M is an integer, N_s is the number of stator slots, N_r is the number of rotor slots, and P is the number of pole pairs.

Figure 5.7b shows the ripple torque diagrams for varied effects of skew angle and eccentricity. By adding a skew angle to the rotor, the torque fluctuation reduces considerably. The rotor of an e-motor always has some eccentricity due to manufacturing errors or by the dynamics of the rotor and the forces applied to the shaft and bearings. A summary of the different eccentricity types in e-motors is eccentricity of rotation axis (tilt angle, offset) and part eccentricity (tilt angle and offset), for example, the axis is at the center, but the rotor part is assembled eccentric. The eccentricity, in many cases, does not change the output torque of the e-motors significantly, as shown in Figure 5.7b.

The effects of eccentricity on the bearing forces of the rotor in this case for an axis eccentricity of 0.05 mm in the X-direction are shown in **Figure 5.7c.**

The EV powertrain is much different from a traditional vehicle in the following ways. The power source is a motor instead of an engine. There is no clutch in the powertrain. There is no torsional shock absorber in the powertrain. A secondary gear reducer replaces the speed transmission. Therefore, the EV powertrain presents new characteristics: the powertrain is a weak damping system; the motor dynamic response is fast, so impact vibration problems stand out remarkably; motor torque exciting characteristics are much different from an engine.

A simplified gear reducer and differential assembly model is given as follows by considering gear backlash. The reducer is a secondary gear reducer in this book and simplified as a gear pair. It is also assumed that the gear shaft and support are rigid, then the theory of the vibration and impact analysis is used to study torsional vibration problem with gear backlash.

Figure 5.8 shows a gear backlash model and gear torsional vibration model. Ignoring static transfer error in this chapter and the dynamic equation of gear reducer and differential assembly model is

$$J_{g1}L^2\left(\ddot{\theta}_{g1}\right)+R_{g1}c_m\left(R_{g1}L\left(\dot{\theta}_{g1}\right)-R_{g2}L\left(\dot{\theta}_{g2}\right)\right)+R_{g1}k_mf\left(x\right)=T_{g1}$$

$$J_{g2}L^2\left(\ddot{\theta}_{g2}\right)+R_{g2}c_m\left(R_{g1}L\left(\dot{\theta}_{g1}\right)-R_{g2}L\left(\dot{\theta}_{g2}\right)\right)=R_{g2}k_mf\left(x\right)=T_{g2}$$

$$x=R_{g1}\theta_{g1}-R_{g2}\theta_{g2}$$
$$f\left(x\right)=x-b\left(\text{when } x\geq b\right)\text{or}$$
$$f\left(x\right)=0\left(\text{when}-b\right)<x<b)\text{or}$$
$$f\left(x\right)=x+b\left(\text{when } x\leq-b\right).$$

(5.4)

FIGURE 5.8 Electric vehicle powertrain model diagram and gear torsional vibration model.

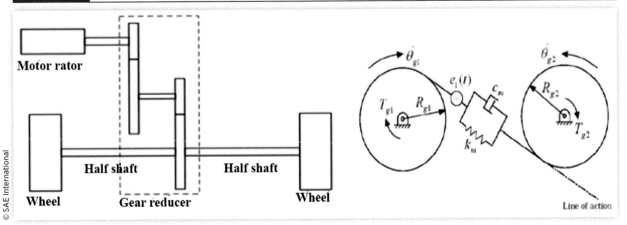

where J_{g1} is the rotation inertia of the driving gear of the gear reducer; J_{g2} is the equivalent output rotation inertia of all driven gears; θ_{gi}, $L(\theta_{gi})$, and $L_2(\theta_{gi})$ ($i = 1, 2$) are angle, angular speed, and angular acceleration of the driving gear and driven gear, respectively; ig is the gear ratio of gear reducer and differential assembly; R_{g1} is the equivalent radius of the driven gear; K_m is meshing stiffness of the gear pair; Cm is the meshing damping of the gear pair; x is the relative displacement between gears; b is gear backlash; T_{g1} is the driving torque of motor; and T_{g2} is the sum of load torque of two half shafts.

The comprehensive model needs to integrate the effects of the shaft, wheel, and the motor rotor.

An electric driveline consists of several subsystems and functions: control, battery, inverter, gearbox, and electric motor. The cooling system is significant because of its great impact on the performances of the e-machine and inverter. Housing, inverter, and coupled excitations play a critical role in electric driveline vibrations and must be accurately represented to attain realistic simulation results. A few insights to the integration of these subsystems are discussed in the following paragraphs [43].

Inverter integration: The inverter should be integrated in the simulations for two reasons. First, current harmonics can be supplied to the e-machine. Second, noise radiation from the inverter housing can be considered. The effect of the inverter on electric current harmonics was shown in a Campbell diagram. The difference between the ideal and the actual case is not negligible. The inverter switching changes the amplitudes of some e-machine orders. The inverter also introduces extra harmonics, such as the *fPWM − 2 × fPhase* frequency.

Housing and mounting brackets integration: As stated previously, a considerable difference in the e-machine and the full assembly eigenmodes is usually expected when comparing the e-machine only. The e-machine noise can be radiated from a gearbox or vice versa. Vibration from the e-machine or gear train has often been coupled through mounting bracket/arms leading to tonal interior noise problems via this structure noise path. The eigenmodes of the brackets or mounts can couple with stator modes and become excessively excited. Analysis of the complete system is imperative for preventing tonal noises from gear whine, the e-machine, and the inverter.

As an example, innovation targets at the project outset included [44] use of a high speed switched reluctance motor up to 23,000 RPM, a high degree of integration between motor and gearbox, a single lubrication/cooling system for the entire drivetrain, and a high efficiency gearbox optimized for NVH. Romax set about applying NVH simulation as early as possible in the design process with a goal to properly achieve a design for low noise. While this was applied to an SRM, it has subsequently been applied to PMSM and induction motors. For the first design loop, the first stage was to identify the most promising of a large number of proposed basic concepts based on key targets, such as cost and dynamic performance. Software was used to rapidly iterate through all of the proposed layouts to narrow down the field and benchmark concept layouts to identify those with the best chance of good noise performance. At this stage, unit excitations were used to enable understanding of the system's dynamic response. This included torque ripple, radial forces, and imbalance from the motor and TE from the gears selected. The decision remained as to how the driveline and power electronics should be assembled into the vehicle. Two driveline layout options were identified, referred to as T and L layouts. Of course, it was not possible to fully design a housing for both layouts: project timing required that the choice had to be made without either design being fully modeled. As a result, a simple representative housing was modeled for both structures and the system simulation was carried out. This time, the simulation was more involved. Driveline-mount stiffnesses were included and the assessment was made based on total structure-borne vibration, measured at the mounts, and by summing the housing kinetic energy, indicative of the total radiated noise. The same excitations were used as in the initial simulation. It was possible to identify which concept had the best fundamental dynamic behavior. For the selected concept, it was possible to compare which noise mechanism (torque ripple, radial forces, etc.) was most significant at each speed and identify the problematic modes of vibration associated with the peaks.

A motor gearbox assembly driven by a PWM inverter is one of the key components of an EV [45]. However, the motor gearbox assembly is one of the dominant noise sources of EVs. In the process of DC–AC conversion, the PWM inverter generates a high-frequency current distortion, and as a result, the EM force along the air gap between the rotor and stator induces the high-frequency vibration of the motor. Also, in the process of mechanical energy transmission by gears, vibration due to gear meshing force is generated. Furthermore, the reaction force from the mount that supports the assembly causes vibration. To optimize the design of the motor gearbox assembly, the trade-off between the main functions and the NV performance must be addressed.

The research first introduced the formulation of the structural model of the gearbox assembly and the excitation force model, which includes EM force, gear meshing force, and mount reaction force, as shown in **Figure 5.9**.

A simulation model was proposed that can predict the transient vibration of a motor gearbox assembly driven by a PWM inverter up to 10 kHz. The model considers excitation by gear meshing, reaction force from mount, and EM forces including the carrier frequency component of the inverter. By utilizing the novel techniques of structural

FIGURE 5.9 Schematic of the motor gearbox assembly. Three types of excitation force are examined: (1) EM force, (2) gear meshing force, and (3) mount reaction force [45].

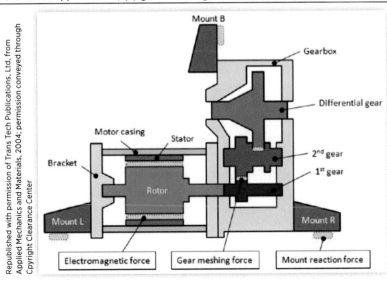

FIGURE 5.10   System model of an EV to simultaneously predict main functions and NVH performance. Models on the left are behavioral models of EV and hatched models are the vibration simulation models of the assembly [45].

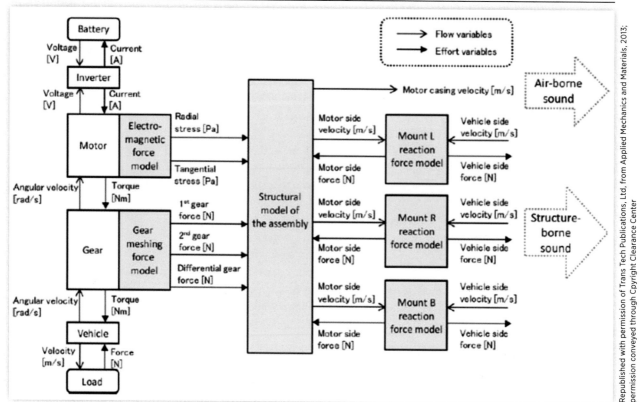

model reduction and state space modeling from the energy point of view, the model is able to predict the vibration of the gearbox assembly in the operating condition to simulate with a system level EV simulator. Verification testing revealed that the model can predict the trend of the dominant vibration component caused by the EM force of the motor, including the carrier frequency of the inverter.

Figure 5.10 shows the system model of an EV to simultaneously predict main functions and NVH performance.

To solve the problem, a structural vibration model is coupled with the EM force model. **Figure 5.11** shows the results of the transient simulation of the waterfall of acceleration of the gearbox.

The measured gearbox vibration in the out-of-plane direction is shown in **Figure 5.12**.

Transmission whine is primarily generated by TE, which is a deviation of meshing gears from a perfectly conjugate (smooth) motion due to manufacturing tolerances, tooth corrections, and elastic deflection due to transmitted torque. The TE excites the geared shaft and bearing system leading to dynamic forces at the bearings, which in turn excite the transmission casing, causing it to radiate noise processes for the generation of gear whine noise. The process starts with excitations at the gear mesh and ends with sound being radiated from the gearbox surfaces. Possible excitations include TE, friction, and axial shuttling forces in the gear mesh. These excitations induce torsional

FIGURE 5.11   Simulated gearbox vibration in STFT waterfall [45]: (a) the inverter carrier frequency, (b) the motor time harmonic order, and (c) the first gear meshing for model validation.

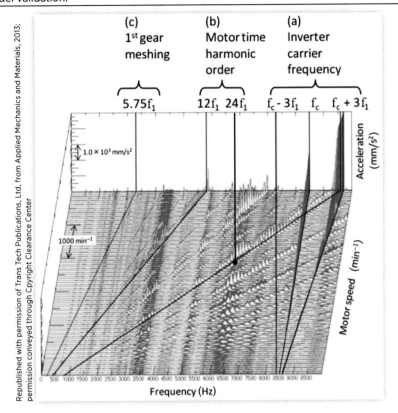

and bending vibrations in the shafting. The resulting bearing forces excite the housing, which then radiates noise.

A typical example of electrical drive unit (EDU) [46–48] consists of a battery-powered electric motor that transmits torque through an input shaft, then two gear reductions via an intermediate shaft to an open differential. The differential then transmits the torque to the rear wheels of the vehicle. The entire assembly, as described, is considered to be the EDU. The EDU is soft-mounted to the vehicle via a subframe. In this design, a four-mount architecture is adopted. The motor is located on the right with two gear reductions to the left, and a differential unit and outputs located on the bottom-left. The major noise concern comes from EDU gear mesh. The inevitable gear TE is the source, which consists of errors from gear tooth surface imperfection, as well as errors coming from loaded gear deflections. The dynamics of EDU internal shafts, supporting bearings, and EDU housing determines the system vibration response to the gear mesh due to TE. To create an optimized EDU for NVH performance, four distinct areas were evaluated: EDU housing and bearing stiffness and dynamics need to be optimized for proper support of gears under load. Gear geometry optimization needs to be done with the given working environment and predicted gear deflection. Gear tooth profile modifications are needed to compensate gear deflection as a function of loads, targeting to minimize the loaded TE. EDU shafting torsional dynamics, which are defined at the gear mesh force level and dynamics, need optimized. Through

FIGURE 5.12 Measured gearbox vibration in STFT waterfall. The order components picked are caused by (a) the inverter carrier frequency, (b) the motor time harmonic order, and (c) the first gear meshing for model validation [45].

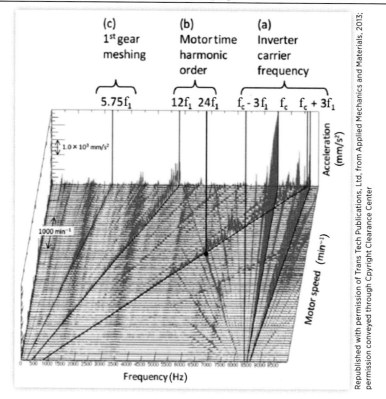

Republished with permission of Trans Tech Publications, Ltd, from Applied Mechanics and Materials, 2013; permission conveyed through Copyright Clearance Center

analytical design studies and optimization, the gear mesh dynamic force due to a given level of gear TE can be minimized. EDU housing radiation efficiency needs optimized. As the EDU gear mesh noise is in the mid-to-high-frequency range (400–2500 Hz), and the EDU rigid body modes controlled by the mount bushings are much lower, the major concern for gear mesh noise is airborne. In other words, the noise received in the vehicle is expected to mainly come from noise radiation directly from EDU housing. Therefore, optimization of the housing structure relative to its dynamics and radiation efficiency becomes crucial.

Bearing selection is a key component of any EDU NVH development for EDU housing and bearing dynamics. Careful consideration must be made to ensure that micro and macro geometry, as well as material characteristics, are compatible with both the supported and supporting elements around them. An optimal bearing was to be judged by the following criteria: consistent dynamics within prescribed bearing preloads, consistent dynamics under thermal loading, and minimized gear deflection under load.

For the gear train in the EDU, there were two sets of meshing gears to consider for gear geometry optimization. The first reduction was a set of helical gears that transfers torque from the motor input shaft to the intermediate shaft. The second reduction then transmits the torque from the intermediate shaft to the differential, also via helical gears.

Evaluation of gear geometry was performed using derived gear-loaded deflections from nonlinear FEA model analyses on the full EDU assembly. Gear design analytical tools were then used for gear tooth geometry optimization. The tools used include internally developed tools, as well as commercial FEA and other numerical calculation tools.

After the bearing design and gear mesh transmission error (MTE) were optimized, EDU shafting torsional dynamics optimization was conducted. A more robust FE model was developed to determine optimal torsional dynamics for the EDU shafting. The target is to minimize the dynamic gear mesh force generated by a given level of gear mesh MTE. With a design worst-case dynamic TE assumed, optimization on shaft architecture and geometry was conducted, focusing on minimizing the mesh force response levels to meet the developed EDU housing vibration target.

The primary focus of the study was the input shaft that connected the electric motor to the first gear reduction, as this was to have the greatest frequency content throughout the motor operating range. A secondary consideration was made to the intermediate shaft but was found to have limited design space to achieve the needed effect on overall performance and thus will not be discussed in further detail.

A final consideration was given to the overall housing design of the EDU for optimizing EDU housing noise radiation efficiency. As discussed previously, an airborne noise target was created to evaluate the efficiency of the EDU. With the assumption that the structural vibration transmissibility from gear mesh force to EDU housing has been optimized through previous engineering steps, the focus here was shifted to minimize the surface radiation of the EDU housing.

5.5 Power Electronics NVH, Inverter High-Frequency Sideband/Tonal Noise

An inverter is usually built with power semiconductors of silicon carbide (SiC) metal-oxide-semiconductor field-effect transistors (MOSFETs) or conventional silicon (Si) insulated gate bipolar transistors (IGBTs) [49–56].

For EVs, the conversion of DC-to-AC voltage takes place in the inverter through PWM. The inverter consists of power electronic transistor, such as IGBTs or MOSFETs, which work as on/off switches. By varying the width of the voltage pulses, the voltage and the speed of the motor is electronically controlled. There are several modulation strategies and the widely used carrier-based techniques entail tonal components that can be perceived as annoying. Those components are in contrast to the main magnetic noise orders previously described as speed-independent, but also speed-dependent sideband harmonics (also referred to as off-zero harmonics) can be prominent. The harmonics are due to the inverter output current waveforms being distorted and not purely sinusoidal. **Figure 5.13** shows the structure of a typical inverter [59, 60].

The audible noise in switched-mode power converters/inverters comes from two main sources: the cooling system (liquid-cooled or air-cooled) and magnetic components, such as transformers, input filter inductors, and power-factor-correction chokes. The noise from magnetic components consists mainly of two parts, depending on the excitation mechanisms. The first and most dominant part of the noise is caused by magnetization of the core. As is well known, one property of ferromagnetic materials is called magnetostriction, which causes

FIGURE 5.13 (a) Structure of a typical inverter and (b) packaging in Prius inverter/converter.

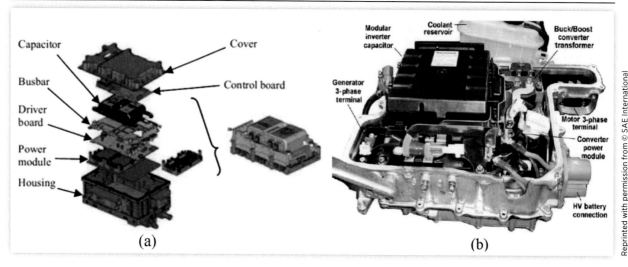

(a)   (b)

the magnetic core to change shape or dimension during the process of magnetization. Therefore, magnetostriction can cause a mechanical interaction between the core and the windings that leads to vibration. The second part of the noise from magnetic components is caused by EM forces created by the magnetic field of the currents in the components' windings.

The mechanical vibration of a magnetic component generates audible noise in power electronic devices. Such noise is undesirable in EV applications because these power devices are normally placed closer to the user and the noise characterization is obviously different from a traditional vehicle. Therefore, it is necessary to study such switching-mode power electronic devices in order that appropriate measures may be taken to reduce the audible noise when designing an EV.

Typically, inverters convert a DC voltage to a variable current or voltage at the fundamental frequency, which is applied to the motor [58]. The fundamental frequency generated by the drive causes the rotational speed of the motor. The inverter hereto uses a PWM approach with a switching or carrier signal in a frequency range between 6 and 20 kHz. Voltage and current PWM harmonics caused by the inverter provoke losses in the rotor and stator. Accordingly, the switching frequency is a very important parameter and has an extensive impact on the voltage, current, and sound spectra of the powertrain. Each type of PWM control scheme causes different harmonics. By means of these schemes, an independent control of the current and the frequency is possible in order to adjust the torque produced by the motor, as well as the speed at which it operates. The various harmonics will transmit to structural components, which will radiate them as noise. Hence, it is required to include an analysis of this phenomenon in NVH investigations on electric powertrains.

A decomposition of the different harmonic components in the noise signature of an electric powertrain leads to speed-dependent orders, speed-independent resonances, and speed-related off-zero harmonics.

Off-zero harmonics of fixed switching applications are grouped with several sidebands around a basis frequency in a rotational speed-frequency diagram.

If the fixed switching frequency f_s component is of order zero, the most prominent sideband orders for vibration and noise are $\pm mp$ and the most prominent orders for phase current are ±np with respect to mechanical frequency. p is the pole-pair number, m is an odd integer, and n an even integer. Alternatively, the corresponding frequencies to the prominent orders with respect to force density, and consequently also to vibration and acoustic signals, can be expressed as

$$f = f_s \pm mp * f_0, \tag{5.5}$$

where f_0 is the fundamental electrical frequency. The relationships between current, flux density, force density, surface vibrations, and radiated noise are complex.

In the following example [59, 60], **Figure 5.14** shows a typical noise source spectrum above the invertor for a fixed switching frequency of 10 kHz and a speed varying between 500 and 8500 RPM. The PWM modulation patterns are clearly noticeable.

The magnetic noise can in turn be categorized into two groups that are easily distinguishable from time/frequency analyses of EV motor vibration data. The first group comprises tonal components that are related to the physical design of the electric motor and increase in frequency linearly with rotor speed. The second group of tonal components is located in the region of the switching frequency of the DC/AC inverter, often further up in the frequency range, and is primarily related to the PWM process. Figure 5.14 shows surface vibration in the housing of an electric motor during a full-load acceleration [59].

The PMW components comprise a speed-independent tone at 10 kHz (equal to the switching frequency) and adjacent speed-dependent sidebands. The main magnetic motor components mainly caused by the radial forces in the motor are also marked out.

Electrified vehicles use power electronics to control the rotational speed of the electric motor. Semiconductor-based components are implemented to transform a DC battery voltage into a controllable AC voltage with a base frequency proportional to the rotational speed. This is achieved by an on-off switching of the electronic transistors or switches. The control strategy of these switches is called modulation. One of the most common modulation methods in electric powertrains is the PWM strategy. A drawback of this particular technique is that it generates undesired higher frequency and off-zero

FIGURE 5.14 Surface vibration on the housing of an electric motor during a full-load acceleration.

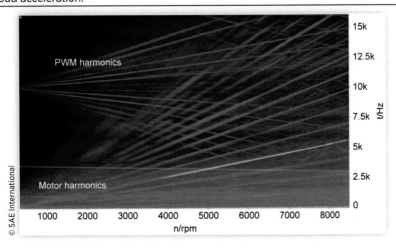

harmonics in the voltage and current signals. These off-zero harmonics will consequently also appear in the magnetic forces, the stator vibrations, and the radiated noise. The harmonic structure of the PWM components depends on four parameters: the applied control strategy, the switching frequency, the inverter bridge configuration, and the motor configuration. A method to avoid noise problems is to increase the switching frequency above the human audible threshold of 20 kHz. However, the switching losses will increase and impact the efficiency of the whole driveline.

Electrified vehicles feature new source components, such as the electric motor(s), the control unit(s), and the large battery, with very different frequency characteristics and working conditions than those in ICE cars. Typical are the tonal components caused by the magnetic fields during electric driving and regenerative braking. In addition, the PWM mechanism of the variable speed drive generates high-frequency, off-zero harmonics in the noise spectra. These harmonics are perceived as unpleasant whining or whistling noise in the vehicle interior.

Noise generation due to inverter switching has the following features and counter-measures [61–63]:

- The switching frequency ranges from 6 to 20 kHz, it can be significant to become perceivable and annoying to the customer.
- The countermeasures are to increase its frequency to be high enough that the customer will likely not be able to detect it, incorporate the inverter into the motor isolation system to reduce its transmitted vibration, or wrap the inverter in absorptive/barrier layers to block the airborne noise transfer path.

It should be noted that under very light loads and no loads, increased switching frequency will result in a significant increase in switching losses. Consequently, in these situations, the switching frequency for noise reduction should just decrease to the value at which the power and efficiency requirements are still met. Thus, mechanical methods should be applied to eliminate the audible noise under very light loads, as well as under no-load operation.

The following is an example [64]. The changes in the appearance of the motor and inverter from the Nissan 2011 model to the 2013 model are shown in **Figure 5.15a**. As illustrated in the figure, integrating the motor and inverter eliminated the three-phase high-voltage wiring harnesses and connectors previously used to connect them. For connecting the three-phase wiring between the motor and inverter, a terminal block is positioned on the motor, as will be described later, and three-phase busbars from the inverter are connected directly to the terminal block. Although not shown in the figure, the inverter members previously used to connect the inverter to the vehicle frames were also eliminated. This elimination of three-phase wiring harnesses, inverter members, and other parts by integrating the motor and inverter and the downsizing of each component reduced the weight and volume of the motor-inverter alone by 11.7 kg and 5.1 L, respectively, compared to the 2011 model. The reduction of NV issues was successfully resolved by the development work carried out to integrate the motor and inverter in addition to attain size and weight reductions, and reduction of heat transfer.

Figure 5.15b compares the motor vibration measured for the 2011 and 2013 models. The motor produces EM excitation forces when it generates torque. These EM forces are transmitted through the structures of the connected parts and resonate with

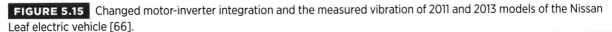

FIGURE 5.15 Changed motor-inverter integration and the measured vibration of 2011 and 2013 models of the Nissan Leaf electric vehicle [66].

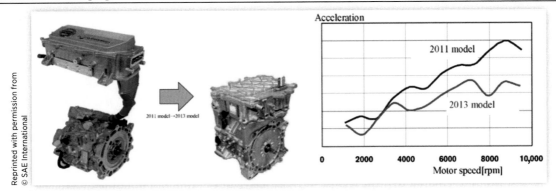

the natural frequencies of the parts they pass through to produce noise. In order to address this issue, the EM circuit of the motor was optimized so as to reduce EM forces, and the shapes of the parts the forces pass through were also optimized, thereby reducing NV.

5.6 NVH of Specific Components/ Accessories: Cooling Fans, Power Steering Pumps, Air-Conditioning Systems, and Others

In the development of EVs, EV-specific components and some accessories similar to those used in conventional vehicle can cause unique challenges for NVH. These systems can range from the EV specific inverters to more common parts, such as cooling fans, power steering pumps, air-conditioning (A/C) systems, and hydraulic pump NVH. The challenges for these components can be related to inherent issues from new EV parts and/or the mounting structure of the more common parts that can no longer be mounted to the ICE, as well as reduced masking levels from the absence of the engine.

The NVH response and perceptions of EVs are totally different with ICE vehicles. Although the disturbing NVH of the different auxiliary components is largely masked in a vehicle with a combustion engine, they are very present in a vehicle with an electric motor and a considerably lower SPL [65–73].

After attaining the perfect balance between road, wind, and motor noise, there is still more work to do for vehicle NVH. An all-EV must still have an A/C and heating system, properly functioning brakes, ABS/ESP systems, and other mechanical systems. Without the engine to drive these systems, a whole host of electrically driven ancillary devices will be needed to provide this functionality. The vibration and acoustic challenges presented by these ancillary devices are by themselves relatively minor, but as a whole could become significant. These include a vacuum pump to the power brake booster, heat exchanger cooling fan(s) for motor and/or battery cooling, fluid pumps for EV system cooling (if using water-to-air heat exchangers), A/C compressor for cabin cooling, ABS module/pump, and an electric steering rack. Some EVs will have more and some will have fewer of the systems, depending on their design. Nonetheless, all EVs will most certainly have some kind of ancillary system noise to overcome. Due to acoustic masking, most of these sounds will drop into background noise once the vehicle is moving at sufficient speed. However, at full stop (say at a traffic signal) where the background noise levels are very low (often in the 35–40-dBA range), one can imagine all of these systems buzzing, gurgling, and howling away and drawing the ire of vehicle occupants. Worse, the sounds that these ancillary devices make will be mostly independent of vehicle operating conditions, such as vehicle speed. Imagine, for example, the sound of an automotive repair shop air compressor cycling on and off as the demands of the shop deplete air supply. While not inherently annoying, it is the kind of acoustic event that if it were in an EV, it might seem disconnected from the operation of the vehicle as it randomly cycles on and off. As such, it may be perceived as an annoying condition by the vehicle occupants.

In all of these ancillary device cases, the strategy must be to make them nearly imperceptible to occupants. This is potentially a tall order, given the number of such devices and their vibration and acoustic output characteristics, and the inherently low ambient noise of the vehicle interior, especially at a full stop.

For example, the noise origin of ventilation is crucial to observe, especially in machines with high rotational speed. Detailed analysis of the fan noise shows that the main source in this case is the fan nearest to its surroundings. The device often exceeds other sources of noise, such as rotor wings, radial or axial cooling channels in the machine, input and output caps, and the like.

Frequency analysis of noise ventilation origin shows that the spectrum has a broadband character, either discrete or vice versa. In the first case, the aerodynamic noise is created from turbulent airflow near the fan blade and near the entrance, but also the output edges of blades. These pulsations are uneven, both in space and in time, so the frequency spectrum created from wind noise is broadband and contains all components of the audible band. In contrast, the discrete nature of the spectrum, sometimes the siren phenomenon, can arise. This phenomenon arises if the fan or obstacles behind it (such as a blade with these obstacles) are not the profile of velocity uniform air flow around the wheel circumference, leading to periodic pulsation of pressure. Then, the siren noise is produced naturally. The noise of mechanical origin is primarily inflicted on roller bearings and the unbalance of rotating machine parts. Rolling bearings can create multiple frequency components, which have their origin mainly in the inequality as part of rolling paths of the bearing rings. In principle, the noise of mechanical origin has a mixed character.

5.6.1 Cooling Fans NVH

Noise generated by cooling fans has become a major concern for vehicle manufacturers [74–79]. The low-frequency tonal noise (typically below 1 kHz frequency range) can be transmitted inside the passenger compartment and can become a source of discomfort for passengers.

Low-speed axial flow fan noise can basically be decomposed in broadband and discrete tone components. Broadband noise is due to unsteady rotating excitations and tonal noise at multiples of blade passing frequencies (BPFs) is due to periodic pressure loads on fan blades by chopping the incoming non-uniform stationary flow. Whether using the vehicle-cooling fan to provide additional cooling to the EV components or having dedicated fans for specific components, it is important to make sure fan-radiated noise levels do not exceed masking levels. Some EVs have water pumps for cooling the electrical components, especially for high-power output systems. It has better cooling efficiency, but the leakage and freezing protection are the disadvantages compared to air cooling. Cooling fan noise needs to be especially controlled at low speed and stop conditions, which could be the only noise perceivable by the customer.

Three basic noise sources of the fan include the following:

1. Broadband aerodynamic noise generated by the turbulent flow.

2. Discrete tones at the BPF Fp (Hz) given by
   Fp = (rotation in RPM × number of blades/60) and the harmonics ($2Fp$, $3Fp$, etc.).

3. Mechanical noise due to mounting, bearing, balancing, and so on.

The cooling fan usually requires free air access and inlet/outlet passages. Noise attenuators, such as silencers, are much less effective if it is not possible to make a closed chamber around it, like for many other engines. Due to the turbulent nature of the air flow in an air-cooling fan, a sound attenuating device will not be as effective as in uniform flow. As such, it is critical to reduce NVH by working on the fan noise sources.

Generally, the principal noise sources of low tip speed, axial flow fans can be separated into the categories of non-rotational and rotational. The non-rotational includes blade interactions with inflow distortion and turbulence, and with nearby fixtures, while the rotational noise includes laminar boundary layer vortex shedding, blade interactions with the tip clearance vortex, and blade stall. The above mechanisms will generate aero-acoustic dipole noise. Monopole noise related to the blade thickness can be neglected for subsonic fans. Finally, quadrupole noise due to the unsteady momentum transport in turbulent flow is also negligible for subsonic fans. Therefore, as proposed by Neise, the dominant aero-acoustic noise source for subsonic fans comes from dipole sources. The dipole sources can be divided into two types, due to the steady rotating forces and unsteady rotating forces. For a uniform inflow without disturbances, periodic pressure fluctuations at discrete harmonics exist and the corresponding tonal noise is related to BPF. **Figure 5.16** shows the overview of the aero-acoustic sound generation mechanisms for fans [78].

The characteristics of fan noises are quantified as follows. The rotor self-noise is the turbulent and laminar vortex shedding at the blade rear sections and at the blade tip. The noise of the ingestion of turbulence in the main air flow could be generated by the heat exchanger, fan supports, or other upstream obstructions. The turbulence leads to random variations in angles of incidence at blade leading edges, causing fluctuating blade loads and surface pressures over a broad range of frequencies. Besides the broadband noise levels, discrete peaks of sound pressure associated with the BPF. This frequency is the product of the fan rotation speed (RPM) and the number of blades. The noise is caused by the pressure pulsation, which is generated when a fan blade is passing a sharp and close disturbance, such as a support beam. Consider that the rotor blade surfaces are considered as pressure loads applied to the fluid domain. The geometry of the system is put to polar coordinates with r_1, ϕ_1 of a point on the fan surface; ϕ, θ, an r are the spherical coordinates of a point in the acoustic domain; and Ω is the rotational velocity of the fan. Only axial (z) components of the fan forces are considered.

FIGURE 5.16 Overview of the aero-acoustic sound generation mechanisms for fans [75].

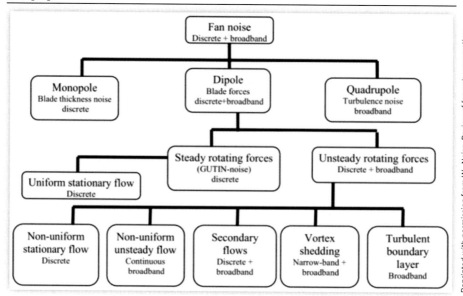

Reprinted with permission from W. Neise, Review of fan noise generation mechanisms and control methods, Fan Noise Symposium, CETIM, France, (1992)

If the upstream flow is uniform, forces acting on a blade are independent of the circumferential coordinate φ_1. In a fixed reference frame, the forces are periodic in time with an angular frequency $\omega = B\Omega$ (where B is the number of blades). The axial component of the pressure acting on the blade can be written as

$$f_z\left(t; r_1,\varphi_1\right) = \sum_{-\infty}^{+\infty} A_s\left(r_1\right)e^{-is\omega_1\left(t-\frac{\varphi_1}{\Omega}\right)} = f_z^0\left(r_1\right)\sum_{-\infty}^{+\infty}\alpha_s\left(r_1\right)e^{isB\varphi_1}e^{-is\omega_1 t} \qquad (5.6)$$

where $i = \sqrt{-1}$ and $\delta_s(r_1) = A_s(r_1)/f_z^0(r_1)$ is a time Fourier transform coefficient representing the complex source strength at the sth harmonic of the BPF and f_z^0 is the time average value of the axial pressure. If the upstream flow is non-uniform, but stationary, due to obstructions, the amplitude factor f_z^0 is now a function of φ_1. It is possible to expand f_z^0 into circumferential harmonics using a spatial Fourier series

$$f_z^0\left(r_1,\varphi_1\right) = f_z^0\left(r_1\right)\sum_{q=-\infty}^{+\infty}\beta_q\left(r_1\right)e^{iq\varphi_1} \qquad (5.7)$$

Inserting Eq. 5.6 into Eq. 5.7 gives the expression of the fluctuating axial pressure

$$f_z\left(t; r_1,\varphi_1\right) = f_z^0\left(r_1\right)\sum_{S=-\infty}^{+\infty}\sum_{q=-\infty}^{+\infty}\alpha_s\left(r_1\right)\beta_q\left(r_1\right)e^{i(sB+q)\varphi_1}e^{-is\omega_1 t} \qquad (5.8)$$

In Eq. 5.8, β_q is a spatial Fourier coefficient that accounts for the non-uniformity with respect to φ_1. Even if β_q is small, non-uniform upstream flow generally leads to considerably larger radiated sound at low Mach number. The axial fluctuating forces appear as dipole terms in the Helmholtz integral, so the acoustic pressure expressed in spherical coordinates can be expressed as

$$p(t;r,\varphi,\vartheta) = \iint_{r_1\,\varphi_1} f_z\left(t;r_1,\varphi_1\right)g_{1z}\left(t;r_1,\varphi_1,r,\vartheta,\varphi\right)r_1 dr_1 d\varphi_1 \tag{5.9}$$

Where the integral is taken over the fan area and g_{1z} is the dipolar Green's function,

$$g_{1z} = -ik\cos\vartheta\frac{e^{ikr}}{4\pi r}\sum_{m=-\yen}^{+\yen}i^m J_m\left(kr_1\sin J\right)e^{im(\varphi-\varphi_1)}e^{-i\omega t} \tag{5.10}$$

where J_m is the cylindrical Bessel function of order m. In order to express the resulting far field radiation at the propeller blade passage frequency, it is necessary to set $\omega = s\omega_1$ and $k = s\omega_1/c = sk^1$. When solving the Helmholtz integral, it can be shown that the integration over ϕ_1 is zero when m = sB + q and is equal to 2π when m = sB + q.

Consequently, the acoustic pressure due to the pressure distribution over the blades is

$$p(t;r,\varphi,\vartheta) = \frac{ik_1\cos\vartheta}{4\pi r}\sum_s\sum_q i^{sB+q}e^{isk_1 r}e^{i(sB+q)\varphi-is\omega_1 t} \times \int_0^a sf_z^0 \alpha_x\left(r_1\right)\beta_q\left(r_1\right)J_{sB+q}\left(sk_1 r_1 \sin\vartheta\right)2\pi r_1 dr_1 \tag{5.11}$$

In Eq. 5.11, the s and q indexes represent the time Fourier harmonics and spatial (circumferential) Fourier harmonics of the fluctuating pressure distribution on the fan area, respectively. Therefore, Eq. 5.11 provides a direct aero-acoustic model of the axial fan radiation. **Figure 5.17** shows the sound power spectrum generated from a cooling fan.

The sound power generated from a cooling fan can be estimated as follows [76]. The sound power level (Lw) generated by fans (without the drive motor) can be easily predicted for each of the octave bands from 63 to 8000 Hz.

$$L_w = K + 10\log_{10}\overline{Q} + 20\log_{10}P_a + C dB \tag{5.12}$$

where Q is the flow rate (m³/s), Pa is the static pressure (kPa), K is the specific sound power level for each of the octave bands based on a volume flow rate of 1 m³/s and a total pressure of 1 kPa, and C is a constant to be added only at the octave band containing the BPF.

For the design consideration of a cooling fan, from the previous formula, the most critical design parameter will be the fan tip speed, Utip, and characteristic factor, C, without affecting the pressure drop, flow rate, or fan efficiency. However, they interact with each other, such as changing the tip speed will affect the pressure drop and flow rate if the same shape of the impeller blade is used. By the rule of thumb, wider blades and more blades will reduce the noise level at the same pressure drop and volumetric flow rate. Increasing the number of blades will increase vortex generation, which needs to be considered by

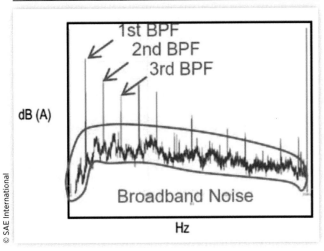

FIGURE 5.17 Sound power spectrum generated from a cooling fan.

optimizing the design blade shape for a given load and condition, such as using a forward sweeping blade.

An application of a cooling fan is in an automotive air conditioning system, which is typically, compact [79]. The fan or blower provides air movement and circulation. It is possible to change the flow rate to achieve an appropriate heat exchange balance and, thereby, a pleasant climate condition inside the cabin. As to the noise generated, the main constraint is the packaging space limitation, which results in highly complex flow ducts between the fan and the outlet vents.

There is not enough duct length for the flow to develop and expand through plenums and ducts. The sudden changes in duct size and direction of the airflow generate flow separation, vortex shedding, turbulence, and unwanted flow circulation inside the air ducts. These features contribute to increasing the level of noise generation. Additionally, due to the packaging constraints and close proximity of all possible noise sources, such as the fan, evaporator core, flow ducts, and louvers, the identification of dominant noise sources is not a simple task in most cases.

Fans or blowers of automotive heating, ventilation, and air conditioning (HVAC) systems are usually of the centrifugal type. Although this type of fan is used to produce low-pressure heads, it often produces a very bothersome rumble noise at low frequencies.

An automotive HVAC unit consists of two primary subsystem assemblies: the heating and cooling subassembly and the blower unit, which consists of a blower wheel and blower scroll. The primary function of the centrifugal blower unit is to draw the air (from inside or outside the vehicle) and blow it across the heating and cooling section based on how the customer chooses to operate the control head. The blower unit assembly consists of the fresh and recirculation air inlet ducts with an optional air filter. The blower wheel in a scroll draws air axially and blows it radially through the evaporator or heater core. The blower motor and wheel assembly, in general, generates broad-band air-rush noise induced by the turbulent kinetic energy and some rotational order-related tones. The blower motor brush commutator interface generates slot passing a frequency that readily propagates structurally. The rotating blower wheel with symmetrically located multiple blades often generates impulses as they pass the cutoff edge of the scroll producing airborne tone harmonics. In some designs, this is more sensitive than others.

Blade passing frequency (BPF) tone is perceived as the pure tone produced when the blades of the wheel rotate past the housing cutoff enclosed in a scroll case. Hence, the BPF is calculated by multiplying the number of blades times the rotating speed in revolutions per second. In the HVAC assembly, this cutoff or nose shape is necessary to straighten and accelerate airflow in the enclosed scroll casing.

Potential countermeasure items of consideration are listed below.

1) Modify the cutoff nose shape where BPF tone is theoretically generated. Several modified cutoff shapes and designs were evaluated mainly for their impact on the audibility of the tone. However, most of the options tried did not impact the BPF tone and some even made it sound louder.

    This confirmed that the original cutoff shape was already optimized for airflow, efficiency, and overall air-rush noise levels.

2) Evaluate two alternative fan designs with thicker/thinner blades. The result was that both the blower wheels exhibited similar levels of the BPF tones.

3) Apply passive and dissipative noise suppression in the panel ducts. Since this issue was observed in the panel mode and the path was airborne via the ducts, mufflers with lined sound absorption were tried on the HVAC subsystem with little impact on suppression of the BPF tone. However, the use of mufflers was ruled out based on cost and packaging space constraints under the instrument panel. Foam absorption in ducts improved noise slightly but was deemed insufficient. In addition, there was an additional risk of peel off in real-world applications.

4) Improve noise performance by control strategy. Since the major problem of BPF tone was observed in vehicles with automatic temperature control systems, there was some potential of implementing "no-fly" blower speed zones so that it would not operate at the BPF tone RPM range. However, the needed width of the BPF tonal range was too large for the expected usage and this was also rejected as an option.

5) Reduce pressure drop of ducts or cowl inlet. If the standard variant panel ducts could be used on the luxury version, the BPF tone could be mitigated to acceptable levels. However, this was also ruled as a non-feasible option due to packaging constraints.

6) Use a thicker air filter and/or a honeycomb pattern inlet to straighten the flow of air before entering the fan. This option helped suppress the BPF tone, but there was no packaging space and the large side effect of reduced airflow volume was not acceptable.

7) Modify the bell-mouth inlet ring to guide and straighten the flow of air entering the blower inlet for better uniformity.

This last countermeasure offered a significant noise improvement with potentially minimal reduction in airflow volume. Several iterations and locations of the inlet ring guide rib flushed to the bell-mouth inlet were evaluated to choose the ideal option with minimal circumference and height of ring, which would theoretically have the least reduction to airflow volume. The ideal countermeasure was initially chosen as a half-moon ring along the bell-mouth inner edge.

5.6.2 Power Steering Pump Noise

The power steering system is operated by a hydraulic pump that pumps highly pressurized oil to the steering system with minimum effort from the driver. Typical hydraulic power for the steering is provided by a rotary-vane pump schematically shown in **Figure 5.18**. The pump contains a set of retractable vanes that spin inside an oval chamber.

There are three types of steering system noises, as shown in **Figure 5.19**.

Fluid-borne noises are typically generated in the fluid reservoir or in the power steering hydraulic lines and could propagate through the power steering mounting bracket. Hiss and clunk are basically the fluid-borne noises but the transmission of these noises to the driver is a system-dependent phenomenon.

FIGURE 5.18 Power steering pump and system operation.

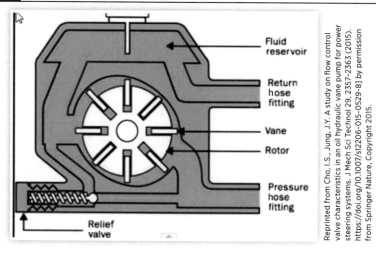

Fluid reservoir

Return hose fitting

Vane

Rotor

Pressure hose fitting

Relief valve

Reprinted from Cho, I.S., Jung, J.Y. A study on flow control valve characteristics in an oil hydraulic vane pump for power steering systems. J Mech Sci Technol 29, 2357-2363 (2015). https://doi.org/10.1007/s12206-015-0529-8] by permission from Springer Nature, Copyright 2015.

FIGURE 5.19 Root cause of power steering noise.

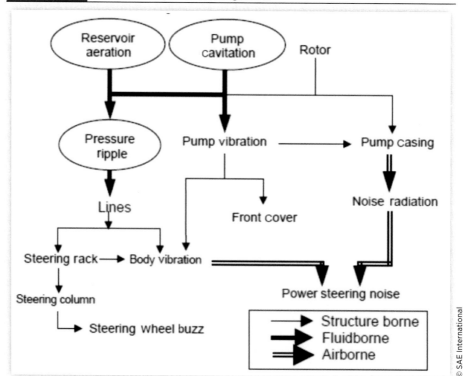

Structure-borne noises are transmitted to the driver via body structure through the pump mount, engine mounts, power steering lines, and power steering mounting brackets. Moan is the structure-borne noise. Moan frequency is driven by natural frequency and harmonics of the pump rotational vane passing frequencies of particular interest. The 10th, 20th, and 30th pump orders are the critical moan frequencies for a 10-vane pump.

Airborne noise primarily consists of the whine of the power steering pump. Whine is in the frequency range of 1–4 kHz. As with moan, whine frequency is also driven by the natural frequency and harmonics of the pump rotational group. Whine is generated when engine RPM, pump rotational natural frequency, and/or pump/bracket natural vibration frequency are aligned. The whine concern can be due to the high frequency (30, 40, 50, or 60th) orders passing by the structural resonance of the pump.

The main causes of steering pump whine consist of low steering fluid level; air trapped inside the system, foreign material trapped inside the system, and the temperature is too low.

Pump whine, as well as other noise, vibration, and harshness (NVH) issues related to the power steering system, can become customer concerns at the vehicle level. In order to avoid that, the proposed treatment of the pump structure and its installation should be performed. Most vane pumps have a wide range of excitation that can reach 1000 Hz (30th order @6000 RPM). This requires avoiding coincidence with other structure harmonics. Power steering is one of the major contributors to undesirable noise or tactile vibration.

When the vehicle powers on, the steering pump starts to work and its noise is apparent. Near the vehicle, the noise of the steering pump can be easily heard because it contains a high-frequency component. The spectrum of the steering pump is very complex. The tonal noise consists of many pure tones [80].

The hydraulic power-assist steering (HPAS) system is an example of such an issue. In HPAS, the positive displacement pump is commonly used because the system offers good value, high-speed road feel, and has a proven safety record. The pump is a powerful source of pulsation. Moan and whine is a pump's flow-induced noise, which is annoying to the consumer. Noise generated by power steering is a harmonic of the pump's fundamental frequency.

Moan is a loud humming noise, audible when the steering wheel is turned and the system is loaded. It is a low-frequency noise usually observed at engine idle RPM in the frequency range of 100–700 Hz. Moan frequency is driven by the natural frequency and harmonics of the pump rotational group.

The 10th, 20th, 30th pump orders are critical moan frequency for the 10 vane power-assist steering (PAS) pump. Whine is a general structure and airborne high-frequency noise observed in parking and city maneuvering conditions. Both noises characterize pump blade pass frequency and their harmonics.

These noises can be separated by vehicle driving condition, noise transfer path, engine RPM, frequency, and order of the pump, as shown in Figure 5.19. Moan is a pressure ripple-generated noise from the pump's rotating parts, like the rotor and vanes, due to the compression of the trapped fluid flow. The whining noise is induced from the cavity region of the pump due to pump geometrical obstruction; thereby, turbulence flows in the system. Whine is the most obvious when engine RPM, pump rotational natural frequency, and pump-mounting bracket natural frequencies align. The principle of the vane pumps is based on a pumping element, which is an aluminum alloy housing consisting of a shaft, rotor, ten vanes, a cam ring, and pressure and thrust plate. The vanes are located in the 10 radially arranged slots of the rotor.

TABLE 5.3 Pump noise generation mechanism

Mechanism of noise	Flow/pressure filed characteristic
Vane and cam interaction	1) Impingement of wake flow on cam
	2) Unsteady flow recirculation and separation at cam surface
Cam/vane and flow	1) Discontinuity in the pressure field between suction and discharge side of cam/vane
	2) Local flow separation on vane surface
Vortex interaction with the flow	Vortex near five main zoned as mentioned above

Pump cells are formed between two vanes and the cam and rotor, shaped like a crescent moon. There are five main zones in a fixed displacement pump: a) low pressure or suction zone, b) high-pressure zone, c) supercharge zone, d) control valve zone, and e) finger exit zone. Two suction zones and two pressure zones formed inside the cam ring are opposite each other. Each pump cell delivers twice its own volume for every revolution of the input shaft. In addition, this double-action arrangement of the suction and pressure zones, the hydraulic radial forces acting upon the rotor cancel each other.

Rotation of the input shaft with a rotor results in the vane force radially by centrifugal force on the track of the fixed cam ring. The pump cell draws oil into the two crescent-shaped pump chambers when the volume increases and expels it into the pressure chamber when the volume decreases. The oil flow generated in the crescent-shaped pressure chambers is directed to the control valve zone for flow relief and pressure control.

Indecent geometry of the rotating parts, excessive compression of the trapped fluid volume, and differences in pressure across leakage path lead to pressure ripple. Flow separation of said five zones due to incorrect pump housing, pressure, and thrust plate geometry results in turbulence. The pump noise generation mechanism is summarized in **Table 5.3**.

5.6.3 Air-Conditioning (A/C) System Noise and Vibration

Automotive refrigerant system-induced noise and vibration transients usually accompany A/C compressor engagement/disengagement. These transients include audible/perceivable metallic impact/slip (clink, chirp, etc.), engagement thump, delayed accumulator thump, orifice tube/thermal expansion valve (TXV)-induced hiss, and occasionally very loud slugging.

The basic noise sources of the A/C compressor are caused by trapping a definite volume of fluid and carrying it around the case to the outlet with a higher pressure. The pressure pulses from compressors are quite severe, and equivalent sound pressure levels could be very high. The noise generated from compressors is periodic with discrete tones and harmonics present in the noise spectrum.

Due to the significant improvements in electrical powertrain NVH, road NVH, wind-noise, etc., the climate control system CCS is perceived as the loudest subsystem since it resides under the instrument panel. All the CCS-induced noise can be either steady-state or transient. In addition, they may vary depending on the weather conditions and are controlled by the customer.

The refrigerant circuit of the CCS is mostly used when there is a need for A/C. The A/C compressor usually cycles on/off, 3–4 times every minute, causing the transients.

Major NVH sources of the A/C system consist of the following: A/C compressor clutch engagement noise, A/C engagement thump, delayed accumulator thump, refrigerant hiss, and A/C compressor vibration.

5.6.3.1 **A/C Compressor Clutch Engagement Noise:** The A/C compressor clutch engagement impact/slip noise is due to the impact of the clutch plate and pulley, and subsequent rotational slippage at the interface. That usually sounds like metallic engagement noise that can be heard outside and occasionally in the vehicle interior.

Under certain operating conditions and some unique compressor clutch assemblies, this engagement can also sound like a brief chirp. The severity of this phenomenon depends on the speed, compressor type (scroll or piston), the surface roughness of the clutch/pulley assembly, and its design and inherent structural characteristics.

5.6.3.2 **A/C Engagement Thump:** Engagement thump usually occurs and is perceived following the compressor clutch engagement. The major source and root cause of this occurrence is the rapid reduction of suction pressure resulting in sudden suction pulse propagation through the A/C suction line to the accumulator mounted on the dash-wall. It also results in a torsional pulse from the A/C compressor that readily propagates structurally via the belt/components to the vehicle interior via floor and steering wheel. Use of a variable capacity/displacement type A/C compressor and mounting the accumulator away from the dash-wall isolates/attenuates this thump. In addition, reducing the stiffness of the suction line by use of softer/flexible/longer hose can mitigate this issue. The use of a suction muffler and torsional damper type clutch can also help reduce the severity of this issue.

5.6.3.3 **Delayed Accumulator Thump:** This type of thump usually is not as severe as the A/C engagement thump but can be perceived by very sensitive customers. If present, this is usually perceived several (10–15) seconds after the A/C compressor engagement. It feels like a softer knock on the vehicle floor and can be perceived on the steering wheel under severe operating conditions. Only the clutch cycling orifice tube (CCOT) systems equipped with accumulators exhibit this delayed thump. The root cause of this phenomenon is the rapid boiling of the liquid refrigerant in the desiccant bag resulting in a transient burst/jolt in the accumulator. Increasing the porosity of the desiccant bag material and/or locating the accumulator away from the dash-wall can help resolve this issue.

5.6.3.4 **Refrigerant Hiss:** The refrigerant hissing noise may get induced during compressor cycling due to the throttling of the refrigerant across the TXV or OT. It readily travels through the flowing (higher velocity) refrigerant vapor and structurally

through the metallic tubing that connects the TXV to the evaporator. Remote mounting the TXV/OT via a decoupling hose length can help but is not preferred since it slightly degrades the A/C performance.

The metallic tubing and the evaporator (especially, if devoid of liquid refrigerant) are very lightly damped. Hence, this noise excites and amplifies through the evaporator that acts like a speaker. And, if the evaporator is not completely isolated from the HVAC case (under the IP), it is perceived in the vehicle interior. As the high-side (discharge) pressure increases, refrigerant vapor gets condensed to liquid due to subcooling, dissipating the refrigerant-induced hiss.

The TXV/OT hiss cannot be eliminated at the source because thermal expansion is essential for refrigeration and A/C systems. However, it can be mitigated and managed by ensuring sufficient subcooling and/or liquid refrigerant at the TXV inlet. The application of damping materials on the TXV, tubing, and the evaporator also helps. Finally, a complete structural decoupling and iso-mounting of the evaporator in the HVAC case are essential to control this issue.

Without engine noise, the cabin of an electric vehicle is quiet, but on the other hand, it becomes easy to perceive refrigerant-induced noise in the automotive A/C system [81–86]. When determining the A/C system at the design stage, it is crucial to verify whether refrigerant-induced noise occurs in the system or not before the real A/C systems are made. If refrigerant-induced noise almost never occurs during the design stage, it is difficult to evaluate by vehicle testing at the development stage.

Without the powertrain noise, the cabin of an electric vehicle becomes very quiet. On the other hand, secondary noise such as A/C noise becomes more audible due to improvements in vibration reduction techniques for other NVH sources. Therefore, it becomes more important to reduce A/C noise such as refrigerant-induced noise.

It is also important to meet the requirements for both the performance and quality of the automotive A/C system. When designing A/C components, such as the compressor and condenser, it is crucial to confirm the performance and quality of the A/C system before testing at the system level. For refrigerant-induced noise, it is important to verify whether it occurs in the system. Although refrigerant-induced noise is not generated from each A/C component, it may occur from the completed A/C system consisting of these components.

During one vehicle's development, both self-excitation noise and hiss noise were found as refrigerant-induced noise related to the TXV. From vehicle and bench test results, it was assumed that the self-excitation noise of one A/C system was generated by refrigerant pressure pulsations entering the TXV when the A/C compressor ran from high to low speed rapidly. Another hissing noise, in contrast to self-excitation noise, a sound like a long "s," was generated by the jet of refrigerant from the TXV just after turning the A/C on. These two noises were inconsistent and both sources needed noise reduction.

The TXV is one of the parts in an automotive A/C system. A schematic diagram of the TXV is shown in **Figure 5.20**. TXV-related noises include self-excitation noise and hissing noise.

An internal TXV typically consists of a temperature-sensitive cylinder with a diaphragm, shaft, ball valve, spring, and sliding friction material. The TXV generally keeps a steady superheat value by sensing the evaporator outlet temperature. When the degree of superheat is large, a downward force is generated at the diaphragm that acts in a

FIGURE 5.20   Schematic diagram of a TXV.

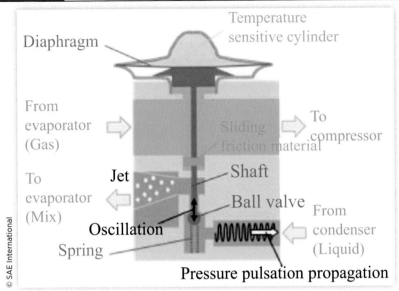

© SAE International

direction in which the ball valve opens more. Thus, the TXV adjusts the flow rate of the refrigerant in the A/C system so that the refrigerant in the evaporated state has a steady degree of superheat. The degree of superheat corresponds to the differential pressure between the temperature sensitive cylinder and at the evaporator outlet.

The refrigerant subsystem (RSS)-induced noise phenomena such as hissing, gurgling, and tones become readily audible and can result in customer complaints and concerns [87]. One of the key components that induce these noise phenomena is the TXV. The TXV throttles compressed liquid refrigerant through the evaporator that results in A/C or thermal system comfort for occupants and dehumidification for safety when needed. Under certain operating conditions, the flow of gas and/or liquid/gas refrigerant at high pressure and velocity excites audible acoustical and structural modes inherent in the tubing/evaporator/HVAC case. These modes may often get masked and sometimes enhanced by the engine harmonics and blower noise. Due to the seasonal demands on A/C compressors, cycling on/off on fixed displacement and during de-stroking of variable displacement compressors, these noises being transient can be easily perceived in the vehicle interior. This paper presents a case study with a systematic approach to excite and quantify these noises induced by the RSS at the bench-level in a laboratory environment with simulated vehicle operating conditions at a variety of thermal loads. This capability facilitates the development of cost-effective and robust countermeasures that can be validated at the bench and implemented later at the vehicle level if needed. Based on the sound quality analysis and listening studies of bench and vehicle level measurements, some sound quality metrics were investigated for their suitability. Reliable and objective targets for hiss and gurgle phenomena can be developed at an early stage of vehicle development.

Typical descriptors for the RSS generated noises are *initial hiss, continuous/steady hiss, gurgle, unsteady hiss-gurgle, and flow-induced tones/whistles.* Loudness (sone) and sharpness (acum) are quite commonly used sound quality metrics for quantifying RSS-induced transients. However, loudness also captures the lower frequency noise from powertrain including compressor and blower noise. Sharpness is a better metric for hiss and gurgle since it readily captures the high-frequency phenomena.

Initial hiss occurs and can be readily perceived for the first time after the ignition is turned on and followed by the A/C turning on, especially if the car has been parked overnight or for an extended period. Occasionally, this sudden hiss may be also preceded or accompanied by an audible whistle or tone.

Tones/whistles are audible narrowband peaks induced when acoustical cavity/length resonances and/or sometimes the resonance of the spring-loaded valve in the TXV are excited.

Continuous/steady hiss is usually observed right after the initial hiss. It is of lower amplitude and may stay steady until the A/C shuts off. Most often, the continuous hiss sounds like added airflow because of its spectral content and can readily get masked by the low blower induced air-rush noise. However, sometimes if the blower operates at a very low-duty cycle/speed, continuous hiss becomes audible during usually fixed displacement A/C compressors cycling.

Gurgle is usually due to the occurrence of unsteady and/or fluctuating hiss. This can be induced by gas-liquid refrigerant bubbles/flow impact on tubes and/or end-tanks of the evaporator which can sound like trickling and boiling (gurgling). Spectrally, the frequency bands corresponding to the inherent acoustical and structural modes get excited and suppressed due to the gas or liquid refrigerant flow through an impingement in tubes and evaporator.

Decaying gurgle can often be perceived briefly on warm days when the blower and engine are shut off following a drive with A/C on at extreme thermal loads and operating conditions. Occasionally, it can also be perceived following A/C turning off.

Fluctuating gurgle can be induced and perceived at low-thermal-load-operating conditions with extended A/C on the operation, especially on applications with the variable displacement compressors, and on systems with fixed displacement compressors during A/C cycling. The fluctuating hiss and gurgle can occur and be experienced under low ambiance with extended A/C on.

Evaporator tank width resonance frequency $F1 = c/2t$, where c is the acoustical velocity and t is the end-tank width. Sometimes, the offset charge port tubes on the high-pressure side located on the liquid line and suction line can also induce very short duration audible tone/whistle following an A/C on at frequencies corresponding to the quarter wavelengths ($F2=c/4L$) and/or the first circular mode around $F3 = 1.84c/\pi d$, where L and d stand for the pipe length and diameter, respectively.

In general, all the inherent acoustical and structural resonances in the complete refrigerant system comprise the hiss/gurgle noise sources and have the potential of getting excited under the right thermal and aeroacoustic operating conditions. If the excited resonance is of very high amplitude, it may sound like a transient brief or continuous whistle. The same resonance with lower amplitude may sound like a transient or continuous hiss. Also, when there is a coincidence of both acoustical and structural resonances, they may sound the loudest when these resonances are excited with intermittent impingement of gas-liquid drop.

Factors affecting refrigerant flow noise are briefed as follows [86–90]. The audible refrigerant hiss and flow noise is a combination of several excitations on the evaporator assembly. A typical evaporator core assembly includes: 1) TXV, 2) inlet and outlet tubes between the TXV and evaporator core body, 3) evaporator core refrigerant passages (plates or tubes) and fins, 4) top tank, and 5) bottom tank. The internal geometry of the evaporator plates or tubes, top and bottom tanks, local stiffness of the tanks, tubes, and fins contribute to the generation of refrigerant flow noise, amplification of certain frequencies and subsequent radiation from the evaporator into the cabin interior.

The major factors believed to be contributing to a higher level of refrigerant flow noise are related to the physical characteristics such as vapor density and mass flow rate.

Using scaling laws for flow-induced sound and vibration and approximating the refrigerant flow through the TXV and evaporator core passages to turbulent jet flow, we can estimate the effect of refrigerant characteristics on the sound and vibration power. The increase in flow velocity has a significant increase in sound power.

$$W \propto \frac{\rho_0 U^8 D}{c_0^5} \tag{5.13}$$

where, W = Sound Power (W), ρ_0 = density (kg/m³), U = flow velocity (m/s), D = length (m), and c_0 = speed of sound (m/s).

5.6.3.5 A/C Compressor Vibration:

The reduction of motor unit mass compared to conventional engine mass causes an upward shift of the resonance frequency of the powertrain system. In a typical internal combustion engine (ICE) application, the isolation range may begin around 25 Hz, which provides ample cover for the operating frequencies of compressors and water pumps. The reduction of mass means that it is possible for the compressor to be in the amplification range, as opposed to the isolation range causing a peak in compressor vibration by coupling it with the motor resonance.

To reduce A/C compressor vibration, two strategies for mounting and isolating the AC compressor include mounting the unit directly to the electric motor unit or mounting the unit to the body through rubber mounts [91]. Both strategies have advantages, disadvantages, and unique challenges.

Mounting the AC compressor to the motor can provide additional isolation and mass compared to mounting the unit to the body but with the comparatively small size of the electric motor, challenges arise with packaging the compressor on the motor unit. Similar to the powertrain vibration section with the reduction of motor unit mass compared to convention engine mass, we will see an upward shift of the resonance frequency of the power plant. However, the resonance shift has an adverse effect on the isolation range provided by the motor, as seen in **Figure 5.21**. In a typical ICE application, the isolation range may begin around 25 Hz which provides ample cover for the operating frequencies of compressors and water pumps, the reduction of mass means that it is possible for the compressor to be in the amplification range, as opposed to the isolation range, causing a peak in compressor vibration by coupling it with the motor resonance. A second strategy for mounting the AC compressor is to mount directly, with some

FIGURE 5.21 Effects of mass on transmissibility.

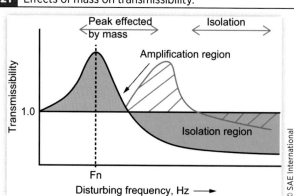

isolation layer, to the body itself. This can become a risk due to the increased vibration in the body without the added mass provided by the motor or engine in the power plant mounting strategy seen in conventional vehicles and some electric vehicles (EVs). By using this mounting strategy, it is likely to input an increased amount of vibration into the body. Therefore, it is important to optimize the mounting structure and location in order to achieve a level similar to a conventional or hybrid vehicle.

If the mounting strategy is not optimized and designed to the proper performance level, it is possible for the customer to detect elevated acoustic levels or vibration at response points, such as the steering wheel or seat, while the vehicle is at rest. During this condition, an electric vehicle has very little vibration and acoustic inputs to the customer, so components that are operating with the vehicle at rest become more perceptible.

When a vehicle attains a certain speed, a very loud noise is occasionally projected into its interior, which momentarily affects the driver's hearing. This noise is known as vehicle booming noise. Booming noise is usually similar to a pure tone below 200 Hz, and it results in the interior noise level reaching more than 3 dBA. There are many types and causes of booming noises, such as the A/C unit. A/C booming noise occurs when the pressure pulsation built up in a compressor generates and transmits vibration to a vehicle's interior through the attached pipes and chassis, which constitute a typical noise transfer path between the compressor and the interior. A/C booming noise is a typical structure-borne type of noise that occurs harmonically based on the number of cylinders. Furthermore, such noise is generated frequently at high temperatures because within the AC system, refrigerant pressure rises with temperature.

The main contribution of the A/C compressor to the overall noise generally comes from the compressor itself and its components because of connections to the vehicle body—directly or indirectly—such as pipes, brackets, and so on. These produce all kinds of aerodynamic, hydraulic, and mechanical noises [92]. **Figure 5.22** shows the schematic of acoustic propagation's coupling effect from the A/C compressor to the interior vehicle cabin.

Some electric A/C compressors in EVs must cool down not only the passenger cabin but also the battery, the inverter modules, and the electric motors. These units exhibit a high level of vibration and noise emission. The vibrations induced are minimized by

FIGURE 5.22 Schematic of acoustic propagation's coupling effect from the A/C compressor to the interior vehicle cabin.

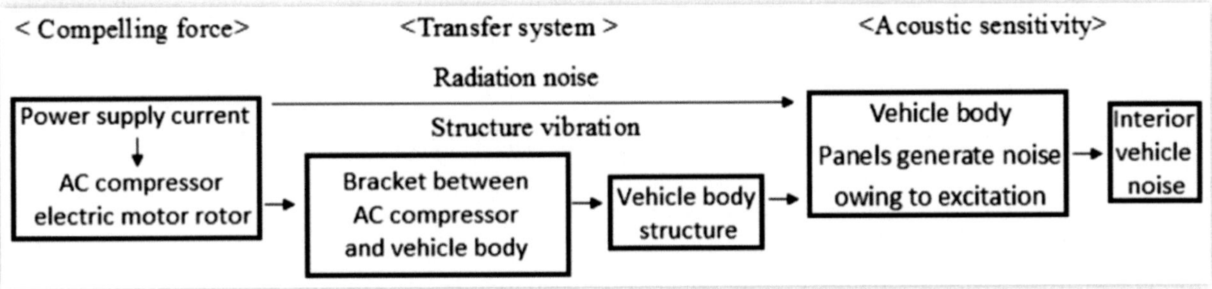

decoupling all the cooling circulation pipelines, fans, and the compressor. The A/C compressor was fitted with a new type of bracket that allows the assembly to be mounted on four soft-coupling rubber bearings. Due to the fact that the RPM range of the compressor invokes frequencies that stimulate other surfaces, the operating strategy of the compressor has been coordinated to "skip over" these critical resonance frequencies. Furthermore, the compressor is mounted at the point of the greatest mass of the electric motors and gearbox unit so that the decoupling now coincides with a mass-spring system, whereby the last mass instance represents the mass of the overall vehicle. These measures make it possible to reduce the noise transmitted in the driver area from over 20 dB in key frequency ranges.

In view of the lightweight strategy of the car, lighter absorption materials are used in place of heavy insulation to reduce the pass-through noise of the electric motors and gearboxes, as well as road noise input [93]. Only in large vibrational areas is alubutyl foil fitted to increase the inner damping of these surfaces. The noise induced at the rear of the vehicle is more critical than at the front as the front section still has a firewall typical of a combustion engine that helps keep disturbing auxiliary noise inputs at bay. In addition to the acoustic insulation provided, several auxiliary components require further optimization measures. For the break-vacuum pump, a new bracket has been designed with optimized decoupling elements. Another, much larger decoupling element has been engineered for the two powertrain units (a double electric motor gearbox coupling). By using four mountings in the rear and three in the front module, it was possible to realize a sufficiently dimensioned coupling for the powertrain units. The electric A/C compressor could not be covered with acoustic insulation due to its thermal design. A very soft decoupling and operating strategy adaption, however, that takes critical frequencies into account, could be implemented to improve vibrational acoustics. The primary cooling fan in the front features a new decoupling mechanism and a new operating strategy in relation to the driving speed. Similar to the electric steering pump, critical frequencies are skipped. In addition to decoupling, the steering pump is leveled down by a sound insulation capsule.

5.6.4 Hydraulic Pump Noise

Hydraulic pumps can play an integral role in electric vehicle noise and vibration. While a typical internal combustion engine vehicle has a single water pump, electric vehicles could have several water pumps for different components, such as the motor, battery, and the HVAC system [94–96].

Another difference for electric vehicle water pumps is that the pump itself is electric, as opposed to mechanical (belt-driven). With no engine to attach the motor to, the pump can be rigidly connected to the body by way of a mounting bracket. Picking the correct mounting strategy and location is the key to minimizing vibration in the vehicle body.

Water or coolant pumps are electronically commuted centrifugal pumps. It is suited for various applications such as hybrid and electric vehicles, where they can be used for cooling the power electronics modules and batteries. The pumps make a contribution to achieving CO_2 reduction targets, which is a centrifugal pump with an electronically commuted motor, where the stator and electronics are located completely separate from the rotor in the dry motor housing. The magnetic rotor is located in the hydraulic circuit and acts as the pump wheel. The integrated pump electronics communicate with a superordinate control unit and adapt the coolant supply volume to the respective cooling situation by regulating the speed. The main components of the pump consist of the stator with winding, electronics (PCB) for motor actuation and diagnosis, motor housing with electric connector, electromagnetic rotor with integrated vane and its bearing (glandless design), and a statically sealed pump housing with inlet and outlet connections.

Several components contribute to the noise emission of a pump, such as pressure pulsations originating from the hydraulics or electromagnetic forces in the electrical drive. The hydraulics of pumps consists of one or several rotating impellers and resting diffusers, respectively, in a guide casing. The impellers convey energy to the fluid according to the Euler equation. In the diffuser, the guide casing the flow is decelerated and the fluid is guided to the next impeller or the pipe system beyond the pump. The flow inside a pump is three-dimensional, nonstationary, and mostly very turbulent.

Several processes in the pump result in pressure pulsations and consequently excite fluid-borne sound that propagates in connected pipes, and also excite vibration of the solid boundaries that leads to the radiation of airborne sound.

Nonstationary flow: The nonstationary flow is caused by secondary flows in the impeller due to rotation, the finite number of blades and finite blade thickness, and also by effects of turbulence. The finite number of blades leads to secondary flow, caused by the asymmetric outgoing flow of the impeller. The finite blade thickness causes a notch in the wake flow. Both effects result in a time-dependent incident flow on the resting parts (guide vanes of the diffuser, volute tongue of the guide casing) and consequently excite vibration of these parts.

Pressure pulsations: Pressure pulsations are detected at discrete frequencies that are multiples of the rotating frequency and the number of blades; these frequencies are also called BPF:

$$f_{bpf} = n_b f_{rot} \qquad\qquad (5.14)$$

The amplitude of these pressure pulsations depends on a number of design parameters of the impeller and diffuser and operating parameters. One of the most important parameters is the distance between the impeller and the volute tongue. Smaller distances typically result in much higher amplitude of the BPF. However, this distance also affects the efficiency of the pump. Therefore, in industrial practice often a compromise has to be found between hydraulic and noise specification. In addition to peaks at the BPF, there are sometimes peaks to be found at lower multiples of the rotating frequency. This is often the case when the number of blades is not a prime number, for example, an impeller with eight or nine blades. In such a case, the impeller can be regarded as a superposition of several virtual impellers, each with a number of blades that corresponds to a prime factor of the actual number of blades, for instance, an impeller with nine blades can be regarded as a superposition of three virtual impellers with three blades each. Measurements of such impellers often reveal BPFs at the 3rd and 6th multiple of the rotating frequency in addition to the expected BPF at the 9th multiple. As the pressure pulsations are generated by the wake flow, their amplitude is also dependent on the operating point of the pump. In general, it is infeasible to operate a pump at a partial load far away from the best efficiency point (BEP). Especially if the pump is operated at low flow rates, this also means a higher manometric head, that is, higher differential pressure. As the overall differential pressure of the pump is the mean of the fluctuating pressure, this increase of the mean differential pressure typically results in an increase of the amplitude of pressure pulsations, caused by an unsteady separation of the flow. Therefore, it is important during the planning of a new installation to select a pump whose BEP is close to the specified operating point. In case several alternatives are available, the pump with the lower hydraulics characteristics should be selected. This is especially important for installations with a varying load profile (e.g., heating systems) where the maximum load is only rarely needed. In addition, also the radial hydraulic forces acting on the impeller are minimum at the BEP. As the radial hydraulic force is transmitted via the shaft and the bearing, it will excite vibration of the pump housing and further propagate as structure-borne sound along the connecting pipes. The amplitude of the pressure pulsations can be reduced by a design of impeller outlet and volute tongue that is optimized with respect to acoustics, as the pressure pulsations are influenced by the interaction of the wake flow and the volute tongue. If the distance between impeller outlet and volute tongue cannot be increased due to the hydraulic requirements, noise generation can be reduced by improving the geometry (for instance, the shape of the trailing edge of the blades). Sharp edges and bends at the volute tongue should be avoided. In addition, it is often advantageous to design the volute tongue and the impeller blades in such a way that trailing edges of the blades and the leading edge of the volute tongue are tilted against each other. This can either be achieved with impeller blades that are twisted in the axial direction or by a volute tongue that grows in the axial direction so that the leading edge rises toward the diffuser outlet. By this tilt of the edges, the smallest gap between the impeller and the volute tongue will occur at different angular positions with respect to the rotational axis. As a result, the pressure oscillation will be "smeared" over time, reducing the amplitude of its tonal noise. For installations with varying loads, it is recommended to use a pump that can adapt to these conditions. For example, in heating installations with thermostatic valves, the flow resistance changes with the positions of the valves. In the case of mostly closed valves, a pump without control will provide higher differential pressure that may result in unwanted whistling noises in valves and radiators. An electronically controlled pump recognizes the lower power need of the circuit and will adapt to the new state of the system.

Turbulence: The impact of turbulence on noise excitation is still a subject of basic research. Principally, the flow in pumps (including the sound field) can be calculated by solving an equation system formed by equations for the conservation of momentum (Navier-Stokes equations) and the conservation of mass [2]. The resulting system of partial differential equations has not been solved analytically until today. The numerical solution is only possible for a limited number of cases (e.g., relatively simple geometries, small Reynolds numbers) due to the extreme variety of different sizes of eddies that have to be taken into account. Therefore, for engineering purposes, often the time-averaged equations are solved numerically. By this method, the impact of turbulence is taken into account statistically. The representation of small eddies that may contain significant energy is extremely simplified. The effect of this approach on the performance of acoustic calculations is a field of research nowadays. However, it may be hypothesized that even small energy-rich eddies have a significant impact, at least on the generation of fluid-borne sound. Another major hydraulic source of noise is cavitation in the pump. Yet, as for most applications, cavitation is a forbidden state of operation; its effects are not further discussed here.

In an electric vehicle, the vacuum pump is driven by an electric motor and helps power the brake system [97]. It starts to work while pushing and releasing the brake pedal. It is not a pleasant experience to hear this noise while braking. The test shows that the interior noise rises by 2.6 dB(A) while pushing the brake pedal, in which level the driver can hear the noise and will not feel very annoyed.

The noise rises by 7.5 dB(A) outside the vehicle so that one can clearly perceive the vacuum pump when it gets to work. This may not be a paramount issue for the drivers but may affect the pedestrian's impression of the brand. With better cushions and tailor-made brackets, the vibration performance of the vacuum pump can be improved. To resolve this problem, a better pump type can be used such as a membrane pump. Therefore, the vacuum pump noise can be reduced by a great level and may not be audible in the passenger compartment. Likewise, acoustics encapsulation for the pump can also be considered.

Electrified powertrains have become one of the powertrain options along with gasoline and diesel powertrains in-vehicle lineups. Battery electric vehicles (BEVs), hybrid electric vehicles (HEVs), and plug-in hybrid electric vehicles (PHEVs) use HV battery packs with different levels of cell capacity and energy. HEVs use power cells, while PHEVs and BEVs use energy cells. PHEV battery packs are designed to have enough energy, initially, to drive in electric-only mode and provide power later on to enhance the gasoline powertrain performance, like an HEV in charge-sustain mode [98].

HV battery packs undergo extensive cyclic charging and discharging events depending on driver demand and battery SOC. During charge and discharge events, battery cells/packs generate varying amounts of heat energy that must be cooled effectively. HV batteries can be utilized efficiently only when operated within a specific temperature range. At higher temperatures the battery life span is reduced, battery performance declines, and thermal system efficiency drops.

Air (cabin climate air) and liquid (50:50 water and ethylene glycol mixture) are the most common cooling mediums used for cooling HV battery packs. When selecting and designing a cooling system for the HV battery, vehicle attributes such as cabin interior noise, driver comfort, and passenger comfort must be met. Each cooling system has advantages and shortcomings in meeting vehicle attributes and driver comforts, aside from cost and system interface complexities.

5.7 Road/Tire-Induced Vehicle Interior Noise in BEV

For gearbox and power electronic unit systems, the radiated noise spectrum is very high in frequency, so conventional acoustical materials are highly effective at minimizing this noise [99–104].

On the other hand, the noise generated by the road/tires and wind is still the two largest challenges in EV NVH refinements that must be aggressively dealt with. Normal levels of wind noise and tire/road noise will suddenly become unacceptable due to the absence of engine noise, especially at low to moderate speeds (less than 50 MPH). This means that achieving a new balance between electric motor noise, wind noise, and tire/road noise will be a fundamental task for EVs.

Tire/road noise is arguably a little less complex and involves somewhat lower energy levels but remains as one of the big challenges in EV NVH. Nonetheless, tire/road noise NVH refinement in EVs can follow the conventional approaches and methods used to treat ICE vehicles detailed in Chapter 3.

Most of an EV's life is spent operating between 25 and 50 MPH. At this speed, wind noise is still minimal and motor and gear noise are mostly masked by road and tire noise. That means that the most dominant noise experienced by drivers and passengers of EVs will mostly be road and tire noise.

So, the single most dominant noise experienced by the occupants of an EV over most of its life will be tire/road noise. Also, many EVs will use low rolling resistance tires that have historically shown to increase the force transmissibility from the tire patch to the spindle for a given road surface.

Increased forces from the tire combined with the NVH handicap of lighter weight structures in the rest of the vehicle will significantly increase the difficulty in mitigating tire/road noise. EV NVH efforts with an intense focus on tire technology, mechanical isolation in the suspension, body structure stiffness, and innovative noise control materials both inside and outside the car. This must be done without adding weight and cost to the vehicle. Clearly, this is where much of the engineering effort will be for most EVs.

Road noise generally starts to be noticeable at vehicle speeds above 30 MPH but its contribution to overall interior noise is maximum between 40 and 60 MPH, and then decreases at higher speeds, where aerodynamic noise becomes predominant. For this reason, tests for road noise are generally conducted at constant conditions, typically 50 MPH and in coast down on different road surfaces. Road noise is generated by the interaction between the tire and the road surface and excites the vehicle through both structural and airborne paths.

Tire/road noise may have significant acoustic contribution at low frequencies, and especially around 200 Hz, where tire acoustic cavity modes are present. Since the tire/road noise is generally transmitted only through structural paths (tire-to-wheel-to-tie-rod-to-suspension-to body) for frequencies up to 200 Hz, the tonal components due to tire acoustic cavity modes are typically structure-borne. Alongside the tire acoustic cavity modes, low orders of the tire rotation can affect the sound quality. In cases of strong phase alignment between tires, modulation may also occur and contribute to the overall perception. The perception of road noise is, therefore, mainly affected by tonality in the low-frequency range; broadband air-rush-type of noise in the mid-frequency range (500–1300 Hz), which can be measured by using broadband, amplitude-related parameters such as the articulation index; and A-weighted SPL or loudness. In cases where the level in the tire band is noticeable and yet it does not significantly impact a broadband parameter such as ASPL or loudness, then it is necessary to increase the resolution of the analysis and compute some spectral envelope type of metric to relate the content in the tire band to the overall content of the signal.

5.8 Wind-Induced Vehicle Interior Noise in BEV

For wind noise, the results show that electric and internal combustion engine vehicles have equivalent wind noise loudness levels at all speeds. However, at lower speeds (50–60 MPH), the EV is judged to have more wind noise even though the sound pressure level was the same as the ICE vehicle. The difference is that in the EV, there is no engine noise to mask the wind noise [105–120].

Wind noise is the predominant component of interior vehicle noise at speeds above 60 MPH. It is typically tested at steady vehicle speeds between 60 and 100 MPH, either on the road or in a wind tunnel. Wind noise refers to the following noise and conditions: aerodynamic noise is made by the vehicle as it moves at high speed through a steady air. This is related to the aerodynamic drag coefficient of the vehicle, which is a function of the vehicle shape and its cross-sectional area. Aerodynamic noise is due to turbulence through "holes," which is correlated to how tightly sealed the vehicle is (around doors, hood, windshield, etc.). Aerodynamic noise due to exterior varying wind conditions, such as cross-wind on a highway. This is different from the previous two since this type of wind noise is fluctuating. Very low-frequency (10–20 Hz) beating noise occurring when either a rear window or the sunroof is partially open. This is due to the Helmholtz resonance of the vehicle cabin, which is excited by the airflow along the boundary of the window or sunroof opening. The last two types of noise are also often referred to as wind buffeting or wind gusting noises. The frequency spectrum of steady wind noise is typically broadband and heavily biased toward the low frequencies. Gusting noise due to cross-wind, as an example, is impulsive and has content at higher frequencies (above 300 Hz or so).

Wind noise reaches the interior through a variety of mechanisms including the aeroacoustic penetration and aerodynamic excitation of vibration and its radiation as sound from the vehicle greenhouse (roof, windshield, side glasses, and Bakelite) and underbody. Vehicle design elements for low levels of wind noise include the vehicle body shape, appendix design, and seals design and manufacture. The importance of this latter element is well-established, and the modeling of seal aspiration and seal noise transmission will be left to another occasion. The flow disturbance of a moving vehicle generates turbulence, which impinges on the vehicle greenhouse. Typical turbulent flows induced by vehicle body shape include the A-pillar vortices, side glass reattached flows, windshield/roof boundary layers, and backlight separated flow. Wind noise is generated as the wall pressure fluctuations in these turbulent flow regions impinge on greenhouse body panels, causing panel vibration, and radiating sound to the cabin. Vehicle appendages such as mirrors, antenna, roof rack, etc. also generate regions of turbulence, as well as affecting the mean flow about the vehicle. For example, the mirror vortex can generate turbulence which impinges on the side glass and may also modify the intensity and position of the A-pillar vortex, with a dramatic effect on wind noise performance. In addition, mirror and antenna vortex shedding can acoustically excite vehicle body panels, radiating noise into the cabin.

Aerodynamicists and exterior designs focus on smooth flow transitions, especially in areas where flow separation can create abrupt pressure changes and turbulence. In addition, it is now well understood that much of the aeroacoustic energy generated at low frequencies inside the vehicle (<250 Hz) comes from the highly turbulent airflow underneath the vehicle and its dynamic coupling to the structure of the underbody panels. This underbody turbulence is also a significant factor in aerodynamic drag; so many EV designers will place more emphasis on a smooth, if not completely flat, underbody. All of these efforts will certainly be helpful but not entirely sufficient in minimizing wind noise. The low drag does not automatically mean low wind noise. In fact, if detailed attention is not given to the flow separation around the A-pillar, exterior mirror, and front side glass (efforts that do not generally affect drag), wind noise is almost certainly guaranteed to be a problem. Much of the high-frequency aeroacoustic energy present inside the vehicle is generated at the interface of the A-pillar and exterior mirrors.

The turbulent wake created by these airflow interruptions can impinge on the door side glass and quite efficiently reradiate noise inside the vehicle. Detailed computational fluid dynamics (CFD) and wind tunnel work must be carried out to manage the airflow in this region, where the smallest of shape changes can generate enormous changes in acoustic response. This requires a highly collaborative approach between the aerodynamicist, the exterior designer and the packaging engineer, among others. In reality, this process is no different than in conventional ICE cars, but the stakes will be higher for EVs since the need to reduce wind noise will be greater due to the absence of engine noise.

5.9 Sound Characteristics of BEV and Other NVH Issues

Masking is a well-known psychoacoustic phenomenon wherein one sound can decrease the perceived level of a second, spectrally similar, sound. Binaural masking occurs when the masker (in this case, engine noise) emanates from a different location than the signal (wind noise). In a conventional vehicle, the masking effect of the ICE noise can reduce the perceived amount of wind noise. However, in the EV, there is no engine noise and, thus, the impression is of increased wind noise, even though the actual loudness levels are unchanged. This effect is most prominent at speeds below 60 MPH, where the engine/wind noise ratio is high and, thus, the masking is large. This increased wind noise presence at low speed could be quite annoying to customers accustomed to gasoline vehicles. At higher vehicle speeds, the wind noise increases faster than the engine noise, masking is reduced, and the impression of EV and conventional vehicle wind noise is much the same [81, 106, 121–123].

Moreover, while EVs have no engine noise, they do have powertrain noise, primarily from the planetary gears used in the transmission. During acceleration and deceleration, there is significant gear whine. If reduced to tolerable levels, this gear whine can provide audible feedback as the throttle is manipulated. At idle, EVs make virtually no noise. Thus, any accessory noise (A/C, power steering pump, fan, etc.) will be quite prominent and maybe annoying due to its strong tonal quality. While electric vehicles will be relatively quiet, they still present significant sound quality challenges.

The electric motor itself is a potential noise source, although this noise is usually high frequency and can be fairly easily isolated by the sound package. For A/C motors, the inverter is also a source of noise, which can be very annoying.

The sound qualities are different for acceleration, deceleration, and cruise. At cruise, an EV emits little or no powertrain sound (other than the motor and inverter noise discussed above). However, during acceleration and deceleration, the planetary gearset, which serves as the transmission, can make a whining sound. The gear whine consists primarily of a single component whose frequency increases in proportion to the motor rotational rate. For example, the gear whine traverses a frequency range from 200 to 1800 Hz during the course of acceleration from 0 to 50 MPH. Its amplitude is initially large, fades during the middle, and increases toward the top end of the acceleration. The gear whine is a potential customer satisfaction issue. However, if the gear noise could be reduced to an acceptable level, it might be perceived positively. This is because it does provide audible feedback when the driver presses the throttle. Drivers are used to having some sort of audible feedback during acceleration and might be very uncomfortable with no such sound. A similar gear whine is heard during deceleration. The gear whine decreases in frequency (from 2300 to 1000 Hz) in proportion to the motor rotational speed. Initially, this whine is masked by wind noise (not shown), but, as vehicle speed decreases below 40 MPH, it becomes prominent. Unlike acceleration, the amplitude of the gear whine decreases more or less monotonically with decreasing vehicle speed. This may be one reason why EV deceleration is perceived as less annoying than EV accelerations.

An EV with a well-isolated motor produces virtually no powertrain noise at idle. The interior noise level is basically equal to that of the ambient. However, the powertrain accessories (A/C compressor, power steering pump, fans, and another pump) will produce noise. These accessory noises may be perceived as being excessive due to the lack of any engine idle noise to mask them. In a conventional vehicle, there is a great deal of spectral overlap between the idle and accessory sounds because their rotational rates are very similar. This leads to a large amount of masking. The EV does not benefit from such masking.

Conventionally, focus on vehicle sound and vibration control has been to reduce or eliminate the problematic noises and vibrations. Over the last decade, this focus has been extended to include the tuning of interior sound quality. While eliminating or reducing unwanted noises is still an area of focus, the quality of sound perceived in the vehicle interior has become increasingly important. Interior sound quality can be used to differentiate vehicles or brands as well as give an overall impression of the quality of the vehicle.

The interior noise levels in EVs have been reduced considerably. Yet, as the demand for higher levels of comfort and satisfaction is growing, car manufacturers and system suppliers are continually developing new methods to reduce sound generation and emission. The interior environment of EVs is acoustically dominated by aerodynamic wind, tire/road, electric motor powertrain, specific components, and accessories NVH. The following is an example of the interior noise characterization of EV.

Figure 5.23 shows the separation process of the road noise component and the wind noise component. The road noise component was separated into the structure-borne component and the airborne component to define a proper mitigation strategy for road noise.

Figure 5.24 shows the noise source contribution analysis result for the interior noise of EV at a constant speed of 100 KPH. Road noise has a large influence on interior noise under the 2 kHz frequency range. The structure-borne component influence is higher than the other component noise. The wind noise component effect becomes the principal component at over the 2 kHz frequency range. The motor noise effect appears at only 2–3 kHz range. That means that the motor rotation harmonic component noise is a high order.

FIGURE 5.23   Process of component noise extraction [74].

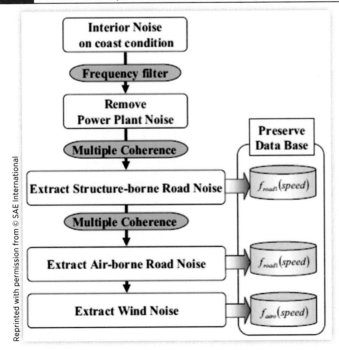

Reprinted with permission from © SAE International

FIGURE 5.24 Contribution analysis result of EV.

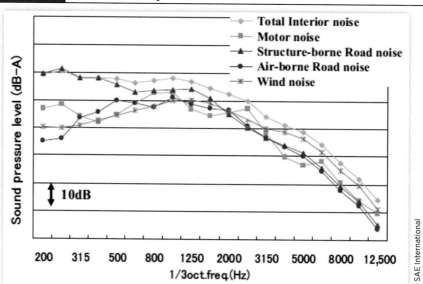

© SAE International

As another example, the EV interior noise breakdown is investigated as follows [73]. During the electrification of the drivetrain, the production transmission was retained for simplicity and used in a fixed gear. For the target-oriented sound design of the vehicle interior noise, a basic understanding of its noise composition is necessary. **Figure 5.25** shows the Campbell diagram of the vehicle interior noise under full load conditions. The interior noise consists of a set of vehicle speed proportional orders ranging from 20 Hz up to more than 10 kHz.

The vehicle makes use of the production transmission, used in a single gear configuration. Therefore, vehicle speed and e-motor speed have a fixed ratio relationship.

FIGURE 5.25 Liiondrive interior noise composition of under a full-load speed sweep.

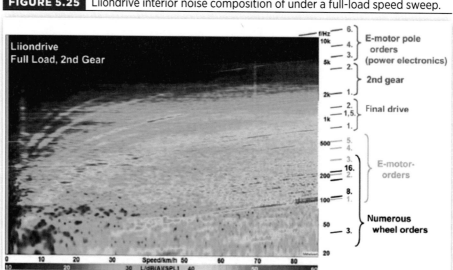

© SAE International

TABLE 5.4 Orders and frequencies of different excitation sources

Excitation process	Order number(rel. source)	Frequency at 85 km/h
Wheel orders	3rd, 8th, 16th, etc.	20–300 Hz
E-motor orders	1st–5th	100–500 Hz
Final drive orders	1st 1.5th 2nd	700–1500 Hz
2nd gear orders	1st 2nd	2000–4000 Hz
E-motor pole orders	3rd 4th 6th etc.	5000–11,000 Hz

With knowledge of the geartrain and e-motor design, interior noise excitations at various frequencies (for a fixed vehicle speed of 85 KPH) are shown below in the context of various drivetrain orders. The low-frequency range below 500 Hz is dominated by wheel and e-motor orders, the medium frequency range from 500 Hz to 4 kHz mainly consists of transmission noise from 2nd gear and final drive while the high-frequency range above 5 kHz contains pole orders from the e-motor. The medium frequency range is important for the subjective impression, as further discussed below, and excited mainly by the transmission and final drive. Due to the missing masking level of the combustion engine, the transmission noise becomes more prominent. The next generation of the vehicle will feature a smaller single-stage transmission tailored to this vehicle. The scaling of the Campbell diagram does not contribute significantly to the overall noise level. However, as described below, the e-motor pole orders are important for subjective impressions of the vehicle.

The information in **Table 5.4** is based on the knowledge of the frequencies and orders of the different excitation sources/processes. A more detailed analysis of the interior noise can be done by using time-domain transfer path techniques, such as vehicle interior noise simulation (VINS). The VINS process can separate airborne and structure-borne noise shares of the interior noise and provide details into the subcomponents of each noise share.

The main specific acoustic signature of the electric powertrain comprises orders (frequency components related to the rotational speed) due to primarily electromagnetic forces, tones, and/or orders due to DC/AC converter pulse width modulation (PWM), and finally gear mesh order(s) from the reducer [35]. All those phenomena yield tonal and high-frequency sound character, which can be perceived as annoying if prominent. The prominence of tonal components is preferably described using psychoacoustic metrics in contrast to sound pressure levels.

Quantification of the whining noise can be done by using the tone-to-noise ratio (TNR) and prominence ratio (PR) two established metrics, standardized in ECMA-74, which relate the tonal energy to adjacent broadband noise. The standard provides criteria for when a tone should be classified as prominent or not, which will be detailed in Chapter 8.

For the full-load acceleration case, in which the maximum levels were due to a very high order, the tone's frequency changes very rapidly. During the time block, this particular order has a high sound pressure level for all frequencies in the middle critical band which explains the rather low Trap–neuter–return (TNR) level. For PR, however, the effect is the opposite. The power of the middle band becomes large in comparison to the more remote (in terms of frequency) lower and upper critical bands, hence the PR level is very high displays sound pressure level, TNR and PR for *car 4* which exhibit two prominent orders for TNR and four for PR.

Based on the TPA method, the interior noise of an EV is investigated experimentally on the dynamometer in a semi-anechoic chamber [124]. The wind noise contribution to the interior noise level is excluded in such a test. The total interior noise level of the EV, in this case, is presumably the result of the total contribution from the motor assembly airborne and structure-borne noise, road structure-borne noise, and tire airborne noise.

Interior noise contribution analysis is conducted by using the TPA model to analyze the contributions from all considered noise sources to the overall interior noise level in which each noise source is obtainable from the measured data at its associated indicator locations. **Figure 5.26** shows the contributions of noise sources to the interior noise level for varied octave bands. The noise level at 80 KPH due to each noise source calculated this way is plotted in Figure 5.24, which shows the contribution ranking and energy distribution from these noise sources, respectively. From **Figure 5.27**, it is seen that the overall interior noise level is 66.8 dB(A) at 80 KPH from the synthesis analysis. The contribution from the structure-borne road noise is 64.1 dB(A), which accounts for 52.2% of the total energy, with significant contributions in the low-frequency range from 80 Hz to 315 Hz as displayed in Figure 5.27. The contribution from structure-borne power plant noise is 62.7 dB(A) that accounts for 42.5% of the total energy, with noticeable contributions around 200 Hz and 400 Hz. The contribution from airborne power plant noise is 55.0 dB(A), which is a prominent contributor for the frequency range from 1000 Hz to 1600 Hz due to the motor rotational harmonic orders. However, the total contribution is much less pronounced and accounts for merely 4.8% of the total energy. The contribution from the airborne tire noise is 44.7 dB(A) that accounts for only 0.5% of the total energy, with the most contribution in the frequency range from 2000 Hz to 4000 Hz. In conclusion, the structure-borne noise from the road and power plant contributes the most to the overall interior noise level.

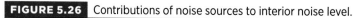

FIGURE 5.26   Contributions of noise sources to interior noise level.

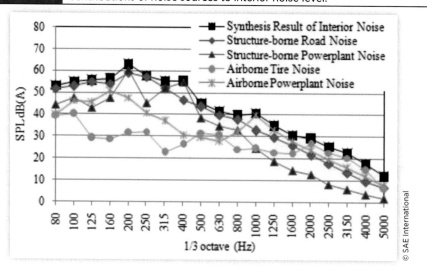

© SAE International

| FIGURE 5.27 | Rank of main contributors to the interior noise level. |

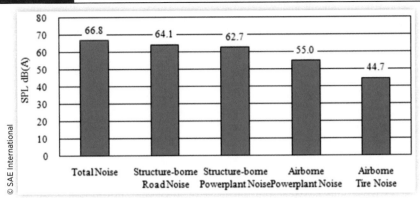

The overall interior noise levels obtained from the synthesis analysis and the test measurement are in good agreement with a consistent spectrum trend over the whole frequency gamut.

In addition to the abovementioned, two more aspects of EV NVH are also critical: buzz, squeak, and rattle (BSR) and sound quality characterization [125]. Like conventional ICE vehicles in the market today, there must be a zero-tolerance for BSRs in EVs. Experience has shown, however, that BSR problems often remain elusive and show up in the most bizarre and frustrating ways. As with the other NVH attributes, the absence of an ICE in EVs will potentially unmask many of these issues, so a great deal of focus and attention must be paid to this area to prevent the overall perception of an otherwise quiet vehicle to be damaged. This is true for any vehicle, but even more so for EVs. The BSR NVH refinement techniques are similar to those used for ICE vehicles.

Given the potential for much lower interior noise levels in EVs, the need for engineering the quality of sounds the vehicle makes will be even greater for EVs. This is due as much to the lower noise levels as it is to the expectations of the people who buy EVs. Many people view EVs as high-tech machines, so the sound field that surrounds vehicle occupants must live up to that image. Greater use of sound quality tools allows the EV NVH high-tech image can be reinforced. Chapter 8 will give comprehensive sound quality analysis with emphasis on the technology dedicated to EV/HEVs.

5.10 Exterior Warning Sound Considerations for BEV

Typically, customers' perceptions of vehicle quality closely parallel the NVH characteristics of the vehicle; therefore, huge efforts must be made for NVH refinement, as detailed in the preceding sections.

In addition to the challenges regarding customer perception of interior and exterior noise, low exterior noise levels from vehicles operating under EV-only propulsion at low vehicle speeds pose a risk to pedestrian safety. As such, a proper exterior sound is necessary for EVs.

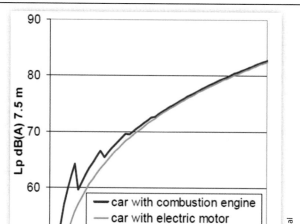

FIGURE 5.28 The noise from an electric car and an ICE car as a function of speed measured 7.5 m from the cars.

Figure 5.28 shows the noise from an electric car and an ICE car as a function of speed measured 7.5 m from the cars. The difference in sound pressure level is up to around 7 dB at low speeds. Above 40 KPH, there is hardly any difference and above 60 KPH there is no difference at all.

Due to the quietness, EV warning sounds have to be designed and implemented in vehicles to alert pedestrians to the presence of electric drive vehicles, such as EV/HEVs traveling at low speeds [125, 126]. Warning sound devices were deemed necessary by some government regulators because vehicles operating in all-electric mode produce less noise than traditional combustion engine vehicles and can make it more difficult for pedestrians, the blind, cyclists, and others, to be aware of their presence. Warning sounds may be driver-triggered (as in a horn but less urgent) or automatic at low speeds; in type, they vary from clearly artificial (beeps, chimes) to those that mimic engine sounds and those of tires moving over gravel.

Several countries have already decided that adding artificial sounds to EVs is compulsory. In the United States, it is a legislative requirement. A federal law (Public Law 111-373) was signed which directs the Secretary of Transportation to "study and establish a motor vehicle safety standard that provides for a means of alerting blind and other pedestrians of motor vehicle operation." In response to this, the US National Highway Traffic Safety Administration (NHTSA) has published guidelines for hybrid and electric vehicles (including light- and heavy-duty vehicles) to produce sounds to meet established Federal Motor Vehicle Safety Standard (FMVSS) demanding minimum requirements for exterior noise.

The US NHTSA issued its final ruling in February 2018 and requires the device to emit warning sounds at speeds less than 18.6 MPH (30 KPH) with compliance by September 2020, but 50% of "quiet" vehicles must have the warning sounds by September 2019 [127].

The European Parliament approved legislation that requires the mandatory use of "Acoustic Vehicle Alerting Systems" for all new EVs and HEVs within 5 years after the publication of the final approval of the April 2014 proposal to comply with the regulation [128].

Vehicle manufacturers will need to balance the legislative requirements with customer NVH expectations [129–156]. Several automakers have developed electric warning sound devices and made the advanced technology cars available in the market, such as Toyota Prius.

In a report released by NHTSA HVs were approximately 1.38 times more likely to be involved in a pedestrian crash than a vehicle with an ICE after completing a low-speed maneuver. As such, the NHTSA standard for exterior noise applies to all vehicles which are capable of propulsion without ICE operation at low speeds (below 30 KPH/18 MPH), where tire noise, wind noise, and other factors are deemed insufficient to alert passengers of approaching vehicles.

In the development of the NHTSA guideline for exterior noise, studies were conducted to understand the effectiveness of alert sounds, relative detectability (distance) and recognizability (as a motor vehicle). It was deemed that energy in one-third octave bands between 1600 Hz and 5000 Hz contribute most to detection, while content between 315 Hz to 1600 Hz provides an additional contribution to detection as well as pitch information. Studies indicate that participants have difficulty detecting signals containing only primary ICE firing order noise content; this was due in part to high masking by urban ambient noise below 315 Hz. Based on available research, the NHTSA guidelines provide recommendations for one-third octave levels for steady-state operation, including cruising at 10 KPH, 20 KPH, 30 KPH, backing, and stationary but activated (to alert pedestrians of potential sudden movement) condition. Further, to indicate acceleration or deceleration and to augment recognizability, the agency recommends pitch shifting of at least 1% per KPH of vehicle speed. In September 2011, the SAE published test procedure SAE-J2889-1 to provide an objective, technology-neutral means of assessing minimum sound emitted by a vehicle in specified ambient noise conditions. This standard is accepted by the NHTSA guidelines for the assessment of exterior noise levels relative to the defined standards.

Globally, regulations are already in effect in China and Japan, which mandate the use of an acoustic vehicle alert system (AVAS) for pedestrian safety at vehicle speeds below 20 KPH. The European Union allows the voluntary installation of the AVAS system, as long as it can be disabled by a control in the vehicle interior and does not produce sound louder than a similar vehicle equipped with an ICE.

The general requirement for the design of a warning sound is a low annoyance but highly detectable. Therefore, the sound should be detected easily by pedestrians despite the range of background noises. The sound should add as little noise as possible to the environment in the interest of avoiding noise pollution and annoyance, and the sound should convey information regarding the vehicle dynamics, such as speed, to the pedestrians. The warning sounds should also comply with the legislation of various countries [157–163].

The eVADER project (Electric Vehicle Alert for Detection and Emergency Response) is funded by the European Commission and aims to provide an optimal design of such warning sounds. It is suggested that these sounds should satisfy three requirements. First, they should be easily detected by the pedestrian, in spite of the background noise. During the next decade, it is expected that this background noise will be mainly due to conventional vehicles. Second, they should give some useful information to visually impaired people about the speed of the car. Third, they should not be too loud, as a major advantage of EVs and HVs is that they can dramatically reduce traffic noise level in cities.

The conducted experiment evaluated the detectability of several warning sounds. A common situation was considered: a pedestrian standing on the sidewalk, facing perpendicular to the street while waiting to cross it. An EV and a diesel one were recorded while passing close to a dummy head in that situation, from 30 m to the right of the dummy head to 30 m to its left (or vice-versa). The speed of the cars was 20 KPH. Then, various warning sounds were mixed with the recordings of the EC.

The warning sound stimuli were designed to satisfy the previously mentioned requirements and to investigate three parameters of timbre: tonal content, frequency detuning, and amplitude modulation. For each factor, three levels were considered. Tonal content represents the number of harmonics within the sound. There were three levels of tonal content: 3, 6, or 9 harmonic complexes. In all cases, the tones used were sinusoidal, and the lowest frequency of the sound was 300 Hz, due to the expected technical reasons of the loudspeakers. The second factor of frequency detuning had three levels as well. Frequency detuning could be null (purely harmonic). At a second level, a sinusoidal frequency modulation was applied to the two highest harmonics, with two different modulation frequencies (4 and 5 Hz). The third level represents a saw-tooth modulation applied to these two harmonics. These modulations periodically detuned these harmonics + or −75 Hz around these highest harmonics within any given sound. The third factor studied was amplitude modulation. As in the other factors, the first level of the amplitude modulation factor contained no amplitude modulation (the amplitude of the sound did not vary during time). Sounds containing the level 2 amplitude modulation were sinusoidally modulated (8 Hz). At the third level, complex modulation amplitude was applied to all components, which made the sound irregular. A Taguchi fractional design was applied for the design.

The warning sound engineering including design, prototype, testing verification and validation can be implemented by using psychoacoustic principles to be presented in Chapter 8. [135, 145, 164–172] presented the systematic approaches for developing the warning signal design. One example is **Figure 5.29**, with possible alternatives treated in a manner typical for decision trees.

While engineering synthetic signals for specific warning signals the possible alternatives are considered step-by-step. First, spectral features and temporal effects have to be selected. Concerning the temporal effects, the temporal envelope, that is, the gating of the signal is of utmost importance. Duration and arrangement of signals and pauses with their respective rise and decay times play an important part. In addition to these coarse temporal effects, finer grating of the time structure, such as by amplitude modulation (AM) or frequency modulation (FM), is of importance. Modulation frequency, modulation depth, and frequency deviation have to be chosen to fit the purpose of the warning signal, such as whether immediate action is necessary, or if the signal is rather of indicative nature.

FIGURE 5.29 Development of a warning signal.

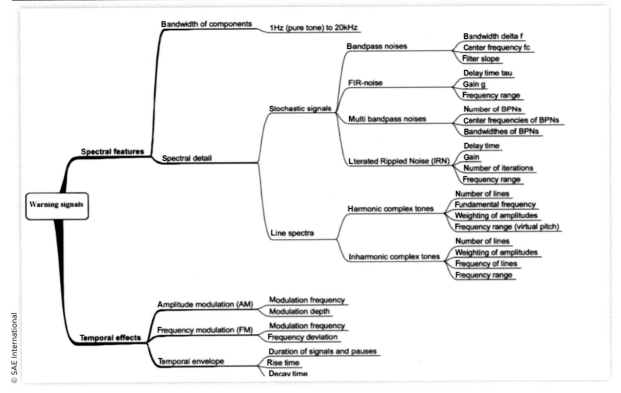

Concerning the spectral features, the bandwidth of the whole signal and how it is composed are decisive for its suitability for a specific purpose. While line spectra can be modeled to produce pleasant sounds for more indicative signals, stochastic signals can be used to tailor rather intrusive sounds. For example, with iterated rippled noise (IRN), a whole continuum from tonal, pleasant sounds to very noisy, intrusive sounds can be realized. However, with the right combination of gating and AM or FM, also with line spectra, intrusive sounds can be realized. Therefore, a general "recipe" to realize the best warning signal for a specific purpose cannot be given; rather, the choice depends both on the experience of the person engaged in the sound engineering, and the preferences of the customer.

The warning sound must have high detectability and low annoyance to pedestrians while complying with legislation for warning sound in that country. In this example, EV warning sounds were designed and tested. The AM signal is more suitable as an EV warning sound than the FM signal. As the frequency rises with the RPM of the ICE vehicle, the EV warning sounds are designed so that the frequency changes depending on the speed. The sound quality indexes such as annoyance index and detectability index of the electric vehicle warning sound were obtained with a consideration of the masking effect.

EV warning sounds are designed to comply with NHTSA regulations, which specify the frequency band between 160 Hz and 5000 Hz should be used. Depending on the driving speed, sound levels above 49 dB(A) at idle, 55 dB(A) at 10 KPH, 62 dB(A) at

20 KPH, and 66 dB(A) at 30 KPH should be generated [6]. Considering the masking effect due to the background noise, it was adjusted by 10 dB(A) higher than the level proposed by the NHTSA.

NHTSA established the necessity of additional noise for vehicles at speeds below 30 KPH [20]. Vehicle manufacturers may design their own warning sound with certain restrictions. The sound should be similar to the noise emitted by an ICE vehicle. The warning sound should be mandatory and the driver cannot turn it off manually.

A continuous sound should be emitted. It should be similar to the sound radiated by a vehicle of the same category equipped with an ICE. The sound level generated by the AVAS is limited to the sound level of an M1 category vehicle equipped with an internal combustion engine and operating under the same conditions.

According to the United Nations informal group on quiet road transport vehicles (QRTV) guidelines, warning sounds should present some requirements. Concerning the safety, sounds should be focused on the audibility, location, and directivity. Taking into account the environmental character of sounds, these should be described in terms of directivity, attenuation, and acceptability.

Audibility denotes a frequency band from 0.5 kHz to 3.5 kHz is recommended for optimal audibility. Location refers to a frequency band from 0.5 kHz to 4 kHz is recommended. The range 0.5 to 1.5 kHz provides interaural phase differences, so it indicates to the listener the angle from the center line. Frequencies up to 3 kHz provides interaural SPL differences and give an idea of the source position, left or right. Above 3 kHz, frequencies provide information related to the front or rear position as they affect the head-related transfer function.

Directivity defines, at any given frequency, how is relative sound pressure level around the source. Low directivities can negatively influence environmental behavior. On the contrary, a high directivity can be bad from the safety point of view, since pedestrians positioned in one side of the vehicle will not receive a good warning signal. At the same time, a system radiating in all directions can influence the detectability of other vehicles masking them. An SPL guideline reduction of 3 dB(A) at +/– 45° and up to 10 dB(A) at +/– 90° is suggested to provide audibility at a safe distance. In addition, sound should drop rapidly on the sides and rear of the vehicle.

Attenuation: for optimal attenuation, a frequency band from below 1 kHz to above 5 kHz is recommended.

Acceptability: in order to not increase the noise pollution and not alter the soundscapes, the warning sounds should be carefully designed.

Metrics to evaluate warning sound systems are briefed as follows. A new standard to measure noise from low noise vehicles is being prepared: ISO/CD 16254:2012 "Measurement of minimum noise emitted by road vehicles." This act proposes a method to measure the minimum noise emission of road vehicles. It also includes quantifying the characteristics of any external sound generation system installed. It proposes measuring the noise at a distance of 2 m instead of the 7.5 m that is stated in the current standards regarding pass-by measurements.

The NHTSA proposes a set of parameters and minimum requirements that EVs and HEVs must comply in order to allow pedestrians to detect the presence, direction, location, and operation of the vehicle. **Table 5.5** presents some sound parameters and requirements.

TABLE 5.5 NHTSA parameters and minimum requirements

Sound parameters	Alternative 1 (no action)	Alternative 2(preferred alternative)	Alternative 3
Min. sound required	No	Yes	Yes
Applicable speed	N/A	Idle to 30 km/h, reverse	>0–290 km/h reverse
Broadband low-frequency sounds	N/A	160–5000 Hz	N/A
One-third octave bands	N/A	Minimum sound pressure levels (SPLs) for eight specific band sets between 160 and 5000 Hz for idle, reverse, and even 10 km/h up to 30 km/h. It must include at least one tone below 400 Hz and one tone that is 6 decibels (dB) above the EV/HV's existing sound level in that band	At least two with SPL of 44 A-weight dB One band each in the ranges of 150–3000 Hz.
Pitch frequency shift with acceleration and deceleration	N/A	1% per km/h	15% monotonic shift between 5 and 20 km/h
Total minimum sound levels resulting from the individual minimum sound requirements	N/A	Idle -49 dB(A) Reverse-52 dB(A) 10 km/h-55 dB(A) 20 km/h-62 dB(A) 30 km/h-66 dB(A)	48 dB(A)

According to the technical report of NHTSA, the warning sounds could be generated by using recordings of actual ICE vehicles. A second alternative is to generate the sound by means of a digital signal processor chip programmed to emulate the sounds of an ICE. This alternative would permit a wider range of sounds taking as a reference ICE noise. Other options include the use of digital signal processors to simultaneously create both ICE noise as sounds that embody special characteristics to enhance detection.

References

1. Chan, C.C., Bouscayrol, A., and Chen, K., "Electric, Hybrid, and Fuel-Cell Vehicles: Architectures and Modeling," *IEEE Transactions on Vehicular Technology* 59, no. 2: 589–598, 2010. 10.1109/TVT.2009.2033605

2. Chan, C.C., "The State of the Art of Electric Hybrid, and Fuel Cell Vehicles," *Proceedings of the IEEE* 95, no. 4: 704–718, 2007. 10.1109/JPROC.2007.892489

3. Chan, C.C. and Wong, Y.S., "Electric Vehicles Charge Forward," *IEEE Power and Energy Magazine* 2, no. 6: 24–33, Nov. 2004.

4. Fuad, U.N., Sanjeevikumar, P., Lucian, M.P. et al., "A Comprehensive Study of Key Electric Vehicle (EV) Components, Technologies, Challenges, Impacts, and Future Direction of Development," *Energies* no. 8: 1217, 2017. 10.3390/en10081217.

5. Greg, G., "Leading the Charge - The Future of Electric Vehicle Noise Control," *Sound and Vibration* 45, no. 4: 5–8, Apr. 2011.

6. U.S. Department of Transportation National Highway Traffic Safety Administration, "Minimum Sound Requirements for Hybrid and Electric Vehicles," Docket No. NHTSA-2011-0148, 2011.

7. Lennström, D., Ågren, A., and Nykänen, A., "Sound Quality Evaluation of Electric Cars - Preferences and Influence of the Test Environment," *Proc. Aachen Acoustic Colloquium 2011*, Aachen, Germany, Nov. 2011.

8. Misdariis, N., Cera, A., Levallois, E., and Locqueteau, C., "Do Electric Cars Have to Make Noise? An Emblematic Opportunity for Designing Sounds and Soundscapes," *Acoustics Nantes*, France, Apr. 2012.

9. Konet, H., Sato, M., Schiller, T., Christensen, A. et al., "Development of Approaching Vehicle Sound for Pedestrians (VSP) for Quiet Electric Vehicles," *SAE Int. J. Engines* 4(1):1217–1224, 2011, https://doi.org/10.4271/2011-01-0928.

10. Genuit, K. and Fiebig, A., "Sound Design of Electric Vehicles - Challenges and Risks," *Inter-noise & Noise-con Congress & Conference 2014*, Melbourne, Australia, Nov. 16–19, 2014.

11. National Highway Traffic Safety Administration, "Minimum Sound Requirements for Hybrid and Electric Vehicles; Draft Environmental Assessment for Rulemaking To Establish Minimum Sound Requirements for Hybrid and Electric Vehicles," Federal Motor Vehicle Safety Standard No. 141, Apr. 27, 2018.

12. Pletschen, B., "Comfort quo vadis? In conflict between crisis and CO2 challenge," *Aachen Acoustics Colloquium 2009, Proceedings*, Aachen, Germany, May 2009.

13. Hofmann M., Lauer, M., and Engler, O., "Sound Design at Mercedes AMG," *Aachen Acoustics Colloquium 2012, Proceedings*, Aachen, Germany, Apr. 2012.

14. Wyman, O., "Elektromobilität 2025," *Powerplay beim Elektrofahrzeug*, München, Germany, May 2009.

15. Sellerbeck, P. and Nettelbeck, C., "Enhancing Noise and Vibration Comfort of Hybrid/ Electric Vehicles Using Transfer Path Models," *Aachen Acoustics Colloquium 2010, Proceedings*, Aachen, Germany, 2010, 6.

16. Kerkmann, J., "Untersuchungen zur Akzeptanz synthetischer Fahrgeräusche im Innenraum von Elektrofahrzeugen," Thesis, Technical University Berlin, Germany, 2014.

17. Küppers, T., "Results of Structure Development process for Electric Vehicle Target Sounds," *Aachen Acoustics Colloquium 2012, Proceedings*, Aachen, Germany, 2012

18. Petiot, J., Kristensen, B.G., and Maier, A.M., "How Should an Electric Vehicle Sound? User and Expert Perception," *ASME. International Design Engineering Technical Conferences and Computers and Information in Engineering Conference, Volume 5: 25th International Conference on Design Theory and Methodology*, Portland, Oregon, USA, Aug. 4–7, 2013.

19. Van der Auweraer, H. and Janssens, K., "A Source-Transfer-Receiver Approach to NVH Engineering of Hybrid/Electric Vehicles," SAE Technical Paper 2012-36-0646, 2012, https://doi.org/10.4271/2012-36-0646.

20. Albers, A., Bopp, M., and Behrendt, M., "Efficient Cause and Effect Analysis for NVH Phenomena of Electric Vehicles on an Acoustic Roller Test Bench," SAE Technical Paper 2018-01-1554, 2018, doi:https://doi.org/10.4271/2018-01-1554.

21. Wellmann, T., Govindswamy, K., and Tomazic, D., "Impact of the Future Fuel Economy Targets on Powertrain, Driveline and Vehicle NVH Development," *SAE Int. J. Veh. Dyn., Stab., and NVH* 1(2): 428–438, 2017, https://doi.org/10.4271/2017-01-1777.

22. Yu, B., Fu, Z., and Juang, T., "Analytical Study on Electric Motor Whine Radiated from Hybrid Vehicle Transmission," SAE Technical Paper 2017-01-1055, 2017, doi:https://doi.org/10.4271/2017-01-1055.

23. Tang, C., Limsuwan, N., Chandrasekhar, N., Ma, Z.et al., "Current Harmonics Impact on Torque Ripple in PM Machine Drive System," SAE Technical Paper 2017-01-1231, 2017, doi:https://doi.org/10.4271/2017-01-1231.

24. Valeri, F., Lagodzinski, J., Reilly, S., and Miller, J., "Traditional and Electronic Solutions to Mitigate Electrified Vehicle Driveline Noises," SAE Technical Paper 2017-01-1755, 2017, doi:https://doi.org/10.4271/2017-01-1755.

25. Tanabe, Y., Watanabe, M., Hara, T., Hoshino, K.et al., "Transient Vibration Simulation of Motor Gearbox Assembly Driven by a PWM Inverter," SAE Technical Paper 2017-01-1892, 2017, doi:https://doi.org/10.4271/2017-01-1892.

26. Arabi, S., Steyer, G., Sun, Z., and Nyquist, J., "Vibro -Acoustic Response Analysis of Electric Motor," SAE Technical Paper 2017-01-1850, 2017, doi:https://doi.org/10.4271/2017-01-1850.

27. Taniguchi, M., Yashiro, T., Takizawa, K., Baba, S.et al., "Development of New Hybrid Transaxle for Compact-Class Vehicles," SAE Technical Paper 2016-01-1163, 2016, doi:https://doi.org/10.4271/2016-01-1163.

28. Nakada, T., Ishikawa, S., and Oki, S., "Development of an Electric Motor for a Newly Developed Electric Vehicle," SAE Technical Paper 2014-01-1879, 2014, doi:https://doi.org/10.4271/2014-01-1879.

29. He, S., "NVH Design, Analysis and Optimization of Chevrolet Bolt Battery Electric Vehicle," SAE Technical Paper 2018-01-0994, 2018, doi:https://doi.org/10.4271/2018-01-0994.

30. Sarrazin, M., Janssens, K., and Van der Auweraer, H., "Virtual Car Sound Synthesis Technique for Brand Sound Design of Hybrid and Electric Vehicles," SAE Technical Paper 2012-36-0614, 2012, https://doi.org/10.4271/2012-36-0614.

31. Gajanan, K., Mukund, A.K., Arun, S., and Rajkumar, B., "NVH Challenges and Solutions to Mitigate Cabin Noise in Electric Vehicles," *2015 IEEE International Transportation Electrification Conference (ITEC)*, Chennai, India, Aug. 27-29, 2015.

32. Sarrazin, M., Janssens, K., Claes, W., and Van der Auweraer, H., "Electro-vibro-Acoustic Analysis of Electric Powertrain Systems," *FISITA World Automotive Congress F2014-NVH-082*, Maastricht, the Netherlands, Jun. 2014.

33. Chandrasekhar, N., Tang, C., Limsuwan, N., Hetrick, J. et al., "Current Harmonics, Torque Ripple and Whine Noise of Electric Machine in Electrified Vehicle Applications," SAE Technical Paper 2017-01-1226, 2017, doi:https://doi.org/10.4271/2017-01-1226.

34. Lennström, D. and Nykänen, A., "Interior Sound of Today's Electric Cars: Tonal Content, Levels and Frequency Distribution," SAE Technical Paper 2015-01-2367, 2015, doi:https://doi.org/10.4271/2015-01-2367.

35. Lennström, D., "Assessment and Control of Tonal Components in Electric Vehicles," PhD thesis, Luleå University of Technology, 2010.

36. Finley, W.R., Hodowanec, M.M., and Holter, W.G., "An Analytical Approach to Solving Motor Vibration Problems," *Industry Applications Society 46th Annual Petroleum and Chemical Technical Conference*, San Diego, CA, USA, Sep. 13-15, 1999, 1467–1480.

37. Yang, Z., Shang, F., Brown, I.P., and Krishnamurthy, M., "Comparative Study of Interior Permanent Magnet, Induction, and Switched Reluctance Motor Drives for EV and HEV Applications," *IEEE Transactions on Transportation Electrification* 1, no. 3: 245–254, 2015. 10.1109/TTE.2015.2470092.

38. Cao, W., Wang, F., The Min H. Kao Dong Jiang, "Variable Switching Frequency PWM Strategy for Inverter Switching Loss and System Noise Reduction in Electric/Hybrid Vehicle Motor Drives," *2013 Twenty-Eighth Annual IEEE Applied Power Electronics Conference and Exposition (APEC)*, Long Beach, CA, Mar. 17–21, 2013.

39. Cherng, J., Huang, W., Wang, Z., and Ding, P., "Integrated platform for motor stator sound power determination," *International Journal Vehicle Performance* 2, no. 3 253–274, (2016).

40. Lei, Y., Hou, L., Fu, Y., Hu, J. et al., "Transmission Gear Whine Control by Multi-Objective Optimization and Modification Design," SAE Technical Paper 2018-01-0993, 2018, doi:https://doi.org/10.4271/2018-01-0993.

41. Roche, M., Mammetti, M., and Crifaci, C., "Wavelet Analysis for the Characterization of Abrupt and Transient Phenomena in Electric Drivelines," SAE Technical Paper 2015-01-2238, 2015, doi:https://doi.org/10.4271/2015-01-2238.

42. Tuma, J., "Gearbox Noise and Vibration Prediction and Control," *International Journal of Acoustics and Vibration* 14, no. 2: 1-11, 2009.

43. Holehouse, R., Shahaj, A., Michon, M., et al., "Integrated approach to NVH analysis in electric vehicle drivetrains," *The Journal of Engineering* 17: 3842-3847, 2019, 2017-01-1226, doi:https://doi.org/10.1049/joe.2018.8247.

44. Mehrgou, M., Zieher, F., and Priestner, C., "NVH and Acoustics Analysis Solutions for Electric Drives," SAE Technical Paper 2016-01-1802, 2016, doi:https://doi.org/10.4271/2016-01-1802.

45. Zhong, Z.M. and Wei, Q., "Modeling and Torsional Vibration Control Based on State Feedback for Electric Vehicle Powertrain," *Applied Mechanics and Materials* 341: 411-417, 2013.

46. Mehrgou, M., Garcia de Madinabeitia, I., Graf, B., Zieher, F. et al., "NVH Aspects of Electric Drives-Integration of Electric Machine, Gearbox and Inverter," SAE Technical Paper 2018-01-1556, 2018, https://doi.org/10.4271/2018-01-1556.

47. Romax Tech, "Improving Noise Performance in Electric and Hybrid Vehicles," Romax discussion paper, Mar. 2015.

48. Grimmer, M., "Sideband and Sound Field Spatial Considerations in the Measurement of Gear Noise," SAE Technical Paper 2005-01-2517, 2005, https://doi.org/10.4271/2005-01-2517.

49. Durfy, J. and La, C., "Gear Whine Modulation Root Cause Analysis and Elimination," SAE Technical Paper 2007-01-2235, 2007, https://doi.org/10.4271/2007-01-2235.

50. Sun, Z., Steyer, G., and Ley, J., "Geartrain Noise Optimization in an Electrical Drive Unit," SAE Technical Paper 2015-01-2365, 2015, doi:https://doi.org/10.4271/2015-01-2365.

51. Xu, Y., Yuan, Q., Zou, J., and Li, Y., "Analysis of Triangular Periodic Carrier Frequency Modulation on Reducing Electromagnetic Noise of Permanent Magnet Synchronous Motor," *IEEE Transactions on Magnetics* 48, no. 11: 4424-4427, 2012. 10.1109/TMAG.2012.2195643

52. Chai, J-Y., Ho, Y-H., Chang, Y-C., and Liaw, C-M., "On Acoustic-Noise-Reduction Control Using Random Switching Technique for Switch-Mode Rectifiers in PMSM Drive," *IEEE Transactions on Industrial Electronics* 55, no. 3: 1295–1309, 2008. 10.1109/TIE.2007.909759

53. Capitaneanu, S.L., de Fornel, B., Fadel, M., and Jadot, F., "On the Acoustic Noise Radiated by PWM AC Motor Drives," *Automatika* 44, no. 2: 137–145, (2003).

54. Sarrazin, M., Anthonis, J., Van der Auweraer, H., and Martis, C., "Signature Analysis of Switched Reluctance and Permanent Magnet Electric Vehicle Drives," *2014 International Conference on Electrical Machines (ICEM)*, Berlin, Germany, Sep. 2–5, 2014.

55. Tsoumas, I.P and Tischmacher, H., "Influence of the Inverter's Modulation Technique on the Audible Noise of Electric Motors," *2012 XXth International Conference on Electrical Machines*, Marseille, France, Sep. 2–5, 2012

56. Iturbe, J., Fernandez, I., and Mammetti, M., "Influence of the Inverter Calibration on the NVH of an Electrified Light-Duty Truck," SAE Technical Paper 2013-01-0133, 2013, https://doi.org/10.4271/2013-01-0133.

57. Le Besnerais, J., "Reduction of Magnetic Noise in PWM-supplied Induction Machines-Low-noise Design Rules and Multi-Objective Optimisation," PhD thesis, Ecole Centrale de Lille, France, 2008

58. Numakura, K., Emori, K., Okubo, A., Shimomura, T. et al., "Silicon Carbide Inverter for EV/HEV Application featuring a Low Thermal Resistance Module and a Noise Reduction Structure," *SAE Int. J. Passeng. Cars – Electron. Electr. Syst.* 10(1):248-253, 2017, https://doi.org/10.4271/2017-01-1669.

59. Arai, K., Higashi, K., Oguchi, T., Kazutoshi, N. et al., "Development of Motor and Inverter for RWD Hybrid Vehicles," SAE Technical Paper 2011-39-7239, 2011, https://doi.org/10.4271/2011-39-7239.

60. Staunton, A.M. and Burress, C., "Evaluation of 2004 Toyota Prius Hybrid Electric Drive System," ORNL/TM-2006/423, May 2006, 10.2172/890029

61. Lennström, D., Johnsson, R., Agren, A., and Nykänen, A., "The Influence of the Acoustic Transfer Functions on the Estimated Interior Noise from an Electric Rear Axle Drive," *SAE Int. J. Passeng. Cars - Mech. Syst.* 7(1):413-422, 2014, https://doi.org/10.4271/2014-01-9124.

62. Curiac, R.S. and Singhal, S., "Magnetic Noise in Induction Motors," *Proceedings of NCAD2008 NoiseCon2008-ASME NCAD*, Dearborn, Michigan, USA, Jul. 28–30, 2008, 10.1115/NCAD2008-73077.

63. Sato, Y., Ishikawa, S., Okubo, T., Abe, M. et al., "Development of High Response Motor and Inverter System for the Nissan LEAF Electric Vehicle," SAE Technical Paper 2011-01-0350, 2011, https://doi.org/10.4271/2011-01-0350.

64. Kawamura, H., Ito, K., Karikomi, T., and Kume, T., "Highly-Responsive Acceleration Control for the Nissan LEAF Electric Vehicle," SAE Technical Paper 2011-01-0397, 2011, https://doi.org/10.4271/2011-01-0397.

65. Oki, S., Ishikawa, S., and Ikemi, T., "Development of High-Power and High-Efficiency Motor for a Newly Developed Electric Vehicle," *SAE Int. J. Alt. Power.* 1(1): 104-111, 2012, doi:https://doi.org/10.4271/2012-01-0342.

66. Shimizu, H., Okubo, T., Hirano, I., Ishikawa, S. et al., "Development of an Integrated Electrified Powertrain for a Newly Developed Electric Vehicle," SAE Technical Paper 2013-01-1759, 2013, https://doi.org/10.4271/2013-01-1759.

67. Berry, A., Gerard, A., and Masson, P., "Active Control of Multi-Harmonic Noise of Engine Cooling Fan," *INTER-NOISE and NOISE-CON Congress and Conference Proceedings*, Cleveland, OH, Jun. 23, 2003, 428–437.

68. Thawani, P., Liu, Z., and Venkatappa, S., "Objective Metrics for Automotive Refrigerant System Induced Transients," SAE Technical Paper 2005-01-2501, 2005, https://doi.org/10.4271/2005-01-2501.

69. Norisada, K., Sakai, M., Ishiguro, S., Kawaguchi, M. et al., "HVAC Blower Aeroacoustic Predictions," SAE Technical Paper 2013-01-1001, 2013, https://doi.org/10.4271/2013-01-1001.

70. John, G.C., "ME 570 Powertrain NVH of Electrified Vehicles, Chapter 5 Noise and Vibration of Cooling Fan and Accessory Components," Mechanical Engineering Department, University ofs Michigan Dearborn.

71. Otto, N., Simpson, R., and Wiederhold, J., "Electric Vehicle Sound Quality," SAE Technical Paper 1999-01-1694, 1999, doi:https://doi.org/10.4271/1999-01-1694.

72. Shin, E., Ahlswede, M., Muenzberg, C., Suh, I. et al., "Noise and Vibration Phenomena of On-Line Electric Vehicle," SAE Technical Paper 2011-01-1726, 2011, doi:https://doi.org/10.4271/2011-01-1726.

73. Govindswamy, K. and Eisele, G., "Sound Character of Electric Vehicles," SAE Technical Paper 2011-01-1728, 2011, doi:https://doi.org/10.4271/2011-01-1728.

74. Shiozaki, H., Iwanaga, Y., Ito, H., and Takahashi, Y., "Interior Noise Evaluation of Electric Vehicle: Noise Source Contribution Analysis," SAE Technical Paper 2011-39-7229, 2011, https://doi.org/10.4271/2011-39-7229.

75. Sabry, A. and Mats, Å., "Noise Reduction for Automotive Radiator Cooling Fans," *FAN 2015*, Lyon (France), Apr. 15–17, 2015.

76. Neise, W., "Review of Fan Noise Generation Mechanisms and Control Methods," *International Symposium on Fan Noise 1*, Seulis, France, 3.9.1992.

77. Longhouse, R.E., "Noise Mechanism Separation and Design Considerations for Low Tip-Speed, Axial-Flow Fans," *Journal of Sound and Vibration* 48, no. 4: 461–474, 1976. https://doi.org/10.1016/0022-460X(76)90550-2

78. Gerges, S.N.Y., Sehrndt, G.A., and Parthey, W., "Noise Sources".

79. Madani, V. and Ziada, S., "Aeroacoustic Characteristics of Automotive HVAC Systems," SAE Technical Paper 2008-01-0406, 2008, https://doi.org/10.4271/2008-01-0406.

80. Sinha, J., Kharade, A., and Matsagar, S., "Reduction of Flow Induced Noise Generated by Power Steering Pump Using Order Analysis," SAE Technical Paper 2015-26-0134, 2015, doi:https://doi.org/10.4271/2015-26-0134.

81. Thawani, P. and Liu, Z., "Flow-Induced Tones in Automotive Refrigerant Systems," SAE Technical Paper 2007-01-2294, 2007, doi:https://doi.org/10.4271/2007-01-2294.

82. Natarajan, S.S., Amaral, R., and Rahman, S., "1D Modeling of AC Refrigerant Loop and Vehicle Cabin to Simulate Soak and Cool Down," SAE Technical Paper 2013-01-1502, 2013, https://doi.org/10.4271/2013-01-1502.

83. Liu, Z., Wozniak, D., Koberstein, M., Jones, C. et al., "Flow-Induced Gurgling Noise in Automotive Refrigerant Systems," *SAE Int. J. Passeng. Cars - Mech. Syst.* 8(3):977-981, 2015, doi:https://doi.org/10.4271/2015-01-2276.

84. Thawani, P., Sinadinos, S., and Black, J., "Automotive AC System Induced Refrigerant Hiss and Gurgle," *SAE Int. J. Passeng. Cars - Mech. Syst.* 6(2):1115-1119, 2013, doi:https://doi.org/10.4271/2013-01-1890.

85. Thawani, P., Venkatappa, S., and Liu, Z., "Automotive Refrigerant System Induced Evaporator Hoot," SAE Technical Paper 2005-01-2509, 2005, doi:https://doi.org/10.4271/2005-01-2509.

86. Itoh, A., Wang, Z., Nosaka, T., and Wada, K., "1D sModeling of Thermal Expansion Valve for the Assessment of Refrigerant-Induced Noise," SAE Technical Paper 2016-01-1295, 2016, doi:https://doi.org/10.4271/2016-01-1295.

87. Thawani, P., Sinadinos, S., and Zvonek, J., "Automotive Refrigerant System Induced Phenomena - Bench to Vehicle Correlation," SAE Technical Paper 2017-01-0448, 2017, doi:https://doi.org/10.4271/2017-01-0448.

88. Koberstein, M., Liu, Z., Jones, C., and Venkatappa, S., "Flow-Induced Whistle in the Joint of Thermal Expansion Valve and Suction Tube in Automotive Refrigerant System," *SAE Int. J. Passeng. Cars - Mech. Syst.* 8(3):973–976, 2015, doi:https://doi.org/10.4271/2015-01-2275.

89. Venkatappa, S., Koberstein, M., and Liu, Z., "NVH Challenges with Introduction of New Refrigerant HFO-1234yf," SAE Technical Paper 2017-01-0172, 2017, doi:https://doi.org/10.4271/2017-01-0172.

90. Liu, Z., Jones, C., and Koberstein, M., "Refrigerant Sub-system Level Assessment on Flow-Induced TVX /Evaporator Noises," *2012 Ford Global Noise & Vibration Conference*, Leuven, Belgium, Jun. 12–14, 2012

91. Freeman, T., Thom, B., and Smith, S., "Noise and Vibration Development for Adapting a Conventional Vehicle Platform for an Electric Powertrain," SAE Technical Paper 2013-01-2003, 2013, https://doi.org/10.4271/2013-01-2003.

92. Cheng, Z., and Lu, Y., "Diagnosis and Analysis of Abnormal Noise in the Pure Electric Vehicle's Air Condition Compressor at Idle," *Journal of Low Frequency Noise, Vibration and Active Control*, 37: 1–14, 2018, 10.1177/1461348418765950

93. Mercedes-Benz SLS, AMG Coupé Electric Drive: NVH Development and Sound Design of an Electric Sports Car," SAE Technical Paper 2016-01-1783, 2016, doi:https://doi.org/10.4271/2016-01-1783.

94. Dürrer, B. and Wurm, F.H., "Noise Sources in Centrifugal Pumps," *Proceedings of the 2nd WSEAS Int. Conference on Applied and Theoretical Mechanics*, Venice, Italy, Nov. 20-22, 2006.

95. Bosch Mobility Solutions, "PAD2 Electric Coolant Pump," 2018.

96. Williams, D.J., "Avoiding Cavitation in Engine Cooling Pumps," *MTZ Worldwide* 70, no. 2: 40–44, 2009.

97. Guo, R., Cao, C., and Mi, Y., "NVH Performance of Accessories in Range-Extended Electric Vehicle," SAE Technical Paper 2015-01-0040, 2015, doi:https://doi.org/10.4271/2015-01-0040.

98. Janarthanam, S., Burrows, N., and Boddakayala, B., "Factors Influencing Liquid over Air Cooling of High Voltage Battery Packs in an Electrified Vehicle," SAE Technical Paper 2017-01-1171, 2017, doi:https://doi.org/10.4271/2017-01-1171, May 22, 2003.

99. European Standard - EN 60268-16:2003, "Sound System Equipment -Part 16: Objective Rating of Speech Intelligibility by Speech Transmission Index," 2003-05-22.

100. Lee, S.J., Park J.B., and Kang, H.S., "Study of Evaluation Method for Tire Noise by Applying Sound Quality Metrics," *Inter-Noise 2004*, Prague, Czech Republic, Aug. 22–25, 2004.

101. Onusic, H., Baptista, E., and Hage, M., "Using SIL/PSIL to estimate Speech Intelligibility in Vehicles," SAE Technical Paper 2005-01-3973, 2005, https://doi.org/10.4271/2005-01-3973.

102. Viktorovitch, M., "Implementation of a New Metric for Assessing and Optimizing the Speech Intelligibility Inside Cars," SAE Technical Paper 2005-01-2478, 2005, https://doi.org/10.4271/2005-01-2478.

103. Lee, S.D., "Characterisation of Multiple Interior Noise Metrics and Translation of the Voice of the Customer," *International Journal of Vehicle Noise and Vibration* 2, no. 4: 341–356, 2006.

104. Frank, E., Pickering, D., and Raglin, C., "In-Vehicle Tire Sound Quality Prediction from Tire Noise Data," SAE Technical Paper 2007-01-2253, 2007, https://doi.org/10.4271/2007-01-2253.

105. Barlow, J.B., William, H.R., and Pope, A., *Low-Speed Wind Tunnel Testing*, 3rd edn. (New York: Wiley, 1999) ISBN: 978-0-471-55774-6

106. Eisele, G., Genender, P., and Wolff, K., "Electric Vehicle Sound Design - Just Wishful Thinking? http://inhabitat.com/2010/08/24/prius-getsoptional-sound-enhancer-to-protect-pedestrians/, Aug. 24, 2010.

107. Cerrato, G., "Automotive Sound Quality - Powertrain, Road and Wind Noise," *Sound & Vibration* 43: 16–24, Apr. 2009.

108. Otto, N. and Feng, B., "Wind Noise Sound Quality," SAE Technical Paper 951369, 1995, https://doi.org/10.4271/951369.

109. Hoshino, H. and Katoh, H., "Evaluation of Wind Noise in Passenger Car Compartment in Consideration of Auditory Masking and Sound Localization," SAE Technical Paper 1999-01-1125, 1999, https://doi.org/10.4271/1999-01-1125.

110. Amman, S., Greenberg, J., Gulker, B., and Abhyankar, S., "Subjective Quantification of Wind Buffeting Noise," SAE Technical Paper 1999-01-1821, 1999, https://doi.org/10.4271/1999-01-1821.

111. Rossi, F. and Nicolini, A., "Theoretical and Experimental Investigation of the Aerodynamic Noise Generated by Air Flows Through Car Windows," *EuroNoise 2003*, Naples, Italy, May 2003.

112. Blommer, M., Amman, S., Abhyankar, S., and Dedecker, B., "Sound Quality Metric Development for Wind Buffeting and Gusting Noise," SAE Technical Paper 2003-01-1509, 2003, https://doi.org/10.4271/2003-01-1509.

113. Bodden, M., Booz, G., and Heinrichs, R., "Interior Vehicle Sound Composition: Wind Noise Perception," *Proceedings of the joint congress, Congres Français d'Acoustique/ Tagung der Deutschen Arbeitsgemeinschaft für Akustik (CFA/DAGA), Strasbourg, France*, Société Française d'Acoustique, Paris, France, 2004.

114. Jen, Y.H. and Coney, W.B., "Wind Noise Challenge in Automobile Industry," *The Journal of the Acoustical Society of America* 103: 2850, 1998. https://doi.org/10.1121/1.421980

115. Her, J., Lian, M., Lee, J., and Moore, J., "Experimental Assessment of Wind Noise Contributors to Interior Noise," SAE Technical Paper 971922, 1997, https://doi.org/10.4271/971922.

116. Her, J., Wallis, S., Chen, R., and Lee, W., "Vehicle Flow Measurement and CFD Analysis for Wind Noise Assessment," SAE Technical Paper 970403, 1997, https://doi.org/10.4271/970403.

117. Strumolo, G. and Babu, V., *New Directions in Computational Aerodynamics* (Bristol UK: Physics World, Aug. 1997), doi: 10.1088/2058-7058/10/8/28.

118. Strumolo, G., "The Wind Noise Modeller," SAE Technical Paper 971921, 1997, https://doi.org/10.4271/971921.

119. Coney, W., Her, J., and Moore, J., "Characterization of the Wind Noise Loading of Production Automobile Greenhouse Surfaces," *ASME International Engineering Congress& Exposition*, Dallas, Texas, 1997.

120. Coney, W., Her, J., and Moore, J., "An Experiment Based Modeling Procedure for a Primary Contributor to Automobile Wind Noise," manuscript submitted to the Journal of Sound and Vibration, September 1997.

121. BASSETT; Simon TATE; Matthew MAUNDER, "Study of High Frequency Noise from Electric Machines in Hybrid and Electric Vehicles Timothy Whitehead", Inter-noise, 2014.

122. Timothy Whitehead, B., Tate, S., and Maunder, M., "Study of High Frequency Noise from Electric Machines in Hybrid and Electric Vehicles," *INTER-NOISE and NOISE-CON Congress and Conference Proceedings*, Melbourne, Australia, Nov. 16-19, 2014.

123. Zeng, X., Liette, J., Noll, S., and Singh, R., "Analysis of Motor Vibration Isolation System with Focus on Mount Resonances for Application to Electric Vehicles," *SAE Int. J. Alt. Power.* 4(2):370-377, 2015, https://doi.org/10.4271/2015-01-2364.

124. Cao, Y., Wang, D., Zhao, T., Liu, X. et al., "Electric Vehicle Interior Noise Contribution Analysis" SAE Technical Paper 2016-01-1296, 2016, doi:https://doi.org/10.4271/2016-01-1296.

125. Graaff, E.D. and Blokland, G.V., "Stimulation of Low Noise Road Vehicles in the Netherlands," *Inter noise*, Osaka, Japan, Sep. 4–7, 2011.

126. Marbjerg, G., "Noise from Electric Vehicles, *Vejdirektoratet*, 2014

127. Shepardson, David (2018-02-26), "U.S. Finalizes Long-Delayed 'quiet cars' Rule, Extending Deadline," https://www.reuters.com/article/us-autos-regulations-sounds/u-s-finalizes-long-delayed-quiet-cars-rule-extending-deadline-idUSKCN1GA2GV, Reuters. Retrieved 2018-03-04.

128. Ray Massey (2014-04-02), "Silent But Deadly: EU Rules All Electric Cars Must Make Artificial Engine Noise," Daily Mail. Retrieved 2014-04-03.

129. Tousignant, T., Govindswamy, K., Eisele, G., Steffens, C. et al., "Optimization of Electric Vehicle Exterior Noise for Pedestrian Safety and Sound Quality," SAE Technical Paper 2017-01-1889, 2017, https://doi.org/10.4271/2017-01-1889.

130. Pedestrian Safety Act of 2010, Public Law 111-373 Stat. *4086*, Jan. 4, 2011.

131. Hanna, R., "Incidence Rates of Pedestrian and Bicyclist Crashes by Hybrid Electric Passenger Vehicles," Report No. DOT HS 811 204, U.S. Dept. of Transportation, 2009.

132. U.S. Department of Transportation National Highway Traffic Safety Administration, "Minimum Sound Requirements for Hybrid and Electric Vehicles," FMVSS 141, Jan. 5, 2013.

133. Wu, J., Austin, R., and Chen, C. L., "Incidence Rates of Pedestrian and Bicyclist Crashes by Hybrid Electric Passenger Vehicles: An Update (No. DOT HS 811 526)," U.S. Dept. of Transportation, 2011

134. Goodes, P., Bai, Y., and Meyer, E., "Investigation into the Detection of a Quiet Vehicle by the Blind Community and the Application of an External Noise Emitting System," SAE Technical Paper 2009-01-2189, 2009, doi:https://doi.org/10.4271/2009-01-2189.

135. Hastings, A, Pollard, J.K., Garay-Vega, L., et al., "Quieter Cars and the Safety of Blind Pedestrians, Phase 2: Development of Potential Specifications for Vehicle Countermeasure Sounds," (No. DOT-VNTSC-NHTSA-11-04) United States, National Highway Traffic Safety Administration, 2011.

136. SAE International Surface Vehicle Standard, "Measurement of Minimum Noise Emitted by Road Vehicles," SAE Standard J2889-1, Iss., Sep. 2011.

137. Vegt, E., "Designing Sound for Quiet Cars," SAE Technical Paper 2016-01-1839, 2016, doi:https://doi.org/10.4271/2016-01-1839.

138. Chinese General Administration of Quality Supervision, Inspection and Quarantine of the Peoples Republic of China (AQSIQ) and Standardization of China (SAC), GM/T 28382-2012, 2012

139. UNECE, "UN Regulation for QRTV," Draft MINUTES of 4th Meeting of Informal Working Group, 2015

140. Japan Engineering and Safety Department, Land Transport Bureau, MLIT, Japanese

141. Massey, R., "EU Rules All Electric Cars Must Make Artificial Engine Noise," http://www.dailymail.co.uk/news/ article-2595451/Silent-deadly-EU-rules-electric-carsmake-artificial-engine-noise-fears-kill-unsuspectingpedestrians.html, Accessed Apr. 2014.

142. Transport and Tourism (MLIT), "Guideline for the Approaching Vehicle Audible System," Ministry of Land, Infrastructure, Transport and Tourism (MLIT), 2010.

143. Fahy, F. and Walker, J., *Advanced Applications in Acoustics, Noise and Vibration* (London: SponPress, 2004), ISBN: 9780415237291.

144. Hoogeveen, L.V.J., "Road Traffic Safety of Silent Electric Vehicles," Master's thesis, Utrecht University, The Netherlands, 2010.

145. Han, M. and Lee, S.K., "Objective Evaluation of the Sound Quality for Accelerating Warning Sound of Electric Vehicle Based on Whine Index," *24th International Congress on Sound and Vibration*, London, Jul. 23–27, 2017.

146. Gustav, N.J., Jeong-Guon, I.H., Wookeun, S., and Ewen, N.M., "Predicting Detectability and Annoyance of EV Warning Sounds Using Partial Loudness," *Internoise 2016*, Humburg, 2016.

147. Altinsoy, M.E., "The Detectability of Conventional, Hybrid and Electric Vehicle Sounds by Sighted, Visually Impaired and Blind Pedestrians," *Inter-Noise 2013*, Sep. 2013.

148. Altinsoy, M.E., Landgraf, J., and Lachmann, M., "Investigations on the Detectability of Synthesized Electric Vehicle Sounds - Vehicle Operation: Approaching at 10 km/h," *Forum Acusticum*, Krakow, Poland, Sep. 2014.

149. Parizet, E., Robart, R., Ellermeier, W., Janssens, K. et al., "Additional Efficient Warning Sounds for Electric and Hybrid Vehicles," *Forum Acusticum*, Krakow, Poland, Sept. 2014.

150. GENESIS, "Loudness Toolbox," A MATLAB toolbox provided by sound company GENESIS.

151. Singh, S., Payne, S.R., and Jennings, P.A., "Toward a Methodology for Assessing Electric Vehicle Exterior Sounds," *IEEE Transactions on Intelligent Transportation Systems* 15, no. 4: 1790–1800, 2014.

152. Lee, S.K., Lee, S.M., Shin, T., and Han, M., "Objective Evaluation of the Sound Quality of the Warning Sound of Electric Vehicles with a Consideration of the Masking Effect: Annoyance and Detectability," *International Journal of Automotive Technology* 18, no. 4: 699–705, 2017.

153. Robart, R., Parizet, E., Chamard, J. C., Janssens, K., and Biancardi, F., "A Perceptual Approach to Finding Minimum Warning Sound Requirements for Quiet Cars," *AIA-DAGA 2013 Conf. Acoustics*, Merano, Italy, 2013.

154. Nyeste, P., and Wogalter, M.S., "On Adding Sound to Quiet Vehicles," *Proceedings of the Human Factors and Ergonomics Society Annual Meeting*, Sage CA: Los Angeles, CA: Sage Publications, Sep. 2008, 52, 1747–1750.

155. Yasui, N. and Miura, M. "Effect of Amplitude Envelope on Detectability of Warning Sound for Quiet Vehicle," *Inter-Noise 2016*, 2016, 1698–1705.

156. Jacobsen, G., Ih, J. G., Song, W. and Macdonald, E., "Predicting Detectability and Annoyance of EV Warning Sounds Using Partial Loudness," *Proc. 45th Int. Congress and Exposition on Noise Control Engineering*, Hamburg, Germany, 2016.

157. Fastl, H., "Psychoacoustic Basis of Sound Quality Evaluation and Sound Quality Evaluation and Sound Engineering," *The Thirteenth International Congress on Sound and Vibration Vienna*, Austria, Jul. 2–6, 2006.

158. Fastl, H., *Psycho-Acoustics and Sound Quality* (Berlin: Heidelberg, Springer, 2005), 139–162, 10.1007/3-540-27437-5_6.

159. Fastl, H., "Recent Developments in Sound Quality Evaluation," *Proc. ForumAcusticum 2005*, CD-ROM (2005b), available from AG Technische Akustik, MMK, TU München, Germany, 1647–1653, 2005.

160. Fastl, H., "Advanced Procedures for Psychoacoustic Noise Evaluation," *Proc. Euronoise 2006*, Tampere, Finland, 2006.

161. Fastl, H., Menzel, D., and Maier, W., "Entwicklung und Verifikation eines Lautheits-Thermometers," *Fortschritte der Akustik, DAGA 2006*, DEGA Berlin, 2006.

162. Hellbrück, J. and Ellermeier, W., "Hören. Physiologie, Psychologie und Pathologie (2. Auflage)," Hogrefe, Götitingen, 2004.

163. Etienne, P., Ryan, R., Jean-Christophe, C. et al., "Detectability and Annoyance of Warning Sounds for Electric Vehicles," *ICA 2013 Montreal*, Montreal, Canada, Jun. 2–7, 2013.

164. Konet, H., Sato, M., Schiller, T., Christensen, A., Tabata, T., and Kanuma, T. "Development of Approaching Vehicle Sound for Pedestrians (VSP) for Quiet Electric Vehicles", SAE Technical Paper 2011-01-0928, 2011, https://doi.org/10.4271/2011-01-0928.

165. Hanna, R., "Incidence of Pedestrian and Bicyclist Crashes by Hybrid Electric Passenger Vehicles," NHTSA Technical Report. No. HS-811 204, 2009.

166. Congress, U. S., "S. 841: Pedestrian Safety Enhancement Act of 2010," 2010.

167. Dalrymple, G., "Minimum Sound Requirements for Hybrid and Electric Vehicles: Draft Environmental Assessment," NHTSA, Washington, DC, Document number: NHTSA-2011-0100, 2013.

168. UNECE Regulation 28, "Uniform Provisions Concerning the Approval of Audible Warning Devices and of Motor Vehicles with Regard to Their Audible Signals".

169. Regulation U. 28, "Uniform Provisions Concerning the Approval of Audible Warning Devices and of Motor Vehicles with Regard to Their Audible Signals," UNECE E/ECE/324-E/ECE/Trans/505, Rev. 1/add27. 1972 and Amendments 1984 and, 2001.

170. International Organization for Standardization, "Measurement of Minimum Noise Emitted by Road Vehicles," ISO/CD 16254:2012, 2012.

171. Report No. DOT HS 811 496. Washington, DC: National Highway Traffic Safety Administration.

172. Campillo-Davo, N. and Rassili, A., *NVH Analysis Techniques for Design and Optimization of Hybrid and Electric Vehicles* (Germany: Shaker Verlag Publications, 2016), ISBN: 978-3-8440-4356-3.

NVH of HEV

6.1 **NVH Development of HEV**

To fulfill the requirements of customer satisfaction on noise, vibration, and harshness (NVH), the NVH refinement of hybrid vehicles is important. The vibroacoustic behavior of internal combustion engine (ICE) and electric motors should be well-controlled for hybrid electric vehicles (HEV). HEVs inherit the NVH issues of conventional ICE vehicles plus the new NVH source of electrical components. To improve the total efficiency of the vehicle, its engine tends to have a higher compression ratio and friction loss accompanied by lowered maximum engine speed. The engine is operated in a high torque range, even though it runs at low engine speed. These results in vibrations during start and stop, power shifting, body vibration, and torque fluctuation. The vehicles with a hybrid powertrain introduce some new NVH features [1, 2], such as no masking by low-order ICE noise, tire noise dominant component, increased impact of wind noise, secondary sources (e.g., HVAC, steering, etc.), new noise sources, electric powertrain components, electric motor, inverter, current-control strategy, new secondary sources, battery cooling, and complex gears in HEV.

The absence of masking noise from an ICE can make the noise in an HEV from other sources more annoying, such as the noise from the tire/road and accessories. Even though the noise levels in HEVs are generally lower, new challenges, such as startup/stop NVH, moreover, the electric powertrain consists of several components apart from the motor that emits noise. The cooling system emits airborne flow-noise and the inverter emits a high-frequency whining noise. To provide the customer with refined noise and vibration performance in a hybrid vehicle, the typical efforts to solve NVH issues in HEVs include (1) conventional aspects, such as rigid body modes/forced response issues, powertrain signature, acoustic package, accessory noise and vibration, wind and road/tire noise, customer expectations and (2) new NVH aspects, such as electric motor/inverter NVH issues, NVH effects from highly transient operating conditions of HEVs complex systems, new customer expectations, and necessary warning sound design [1, 3–24].

There are many new phenomena such as (1) The start/stop operation of the ICE and the switching noise of the power-cooling unit, (2) The simultaneous operation and interactions of engines in one hybrid vehicle yield new harmonic structures. For example, the order components of the electric motor(s) and the ICE can cross each other while several engines are operating simultaneously at an unrelated speed. This may cause unsuspected modulations and those very closely spaced harmonics are difficult to handle in a frequency domain analysis, (3) In a hybrid vehicle equipped with a continuously variable transmission (CVT), a typical "motor-boat effect" with a very characteristic noise can be detected. It describes the effect of the fast, almost abrupt changes of the engine rotational speed during full load acceleration, and the slow increase of the vehicle speed. **Table 6.1** shows certain typical NVH phenomena in HEV applications [25].

TABLE 6.1 Selection of typical NVH phenomena in HEVs applications.

Noise	Description range	Excitation	Frequency modulation	Sources	Spectrum (Hz)
Humming/ Buzzing	low-frequency, narrow	Periodic, tonal	Low or mid-frequency stochastic modulation	low speed rotating parts	20–180
Rumble	low-frequency, narrow	Periodic, impulsive and stochastic parts	Low-frequency modulation with stochastic parts	Relatively low speed impulses, e.g., chassis components. Start-Stop	20–120
Howling	Mid-frequency, narrow	Tonal	Mid-trequency modulation possible	Medium speed rotating parts	180–500
Whistling	High-frequency, narrow, harmonics possible	Tonal	High-frequency modulation possible	High speed rotating parts, structural resonance of gaseous medium	3,000–10,000
Squealing	High-frequency, narrow	Tonal	High-frequency modulation	Stick-Sup effect on friction pairs, e.g., brakes, slide bearings	1000–7000
Tacking	Mid- or high-frequency, broad	Impulsive	High-frequency modulation	Single impulses, e.g., caused, e.g., by endstops or valves	300–5000
Clacking	low-frequency, broad	Impulsive	Low-frequency modulation	Lose parts or play between parts, end steps	300–8000
Rattle	High-frequency, broad	Impulsive and stochastic	High-frequency modulation	Rotational play in toothings, chaindrtves	1000–8000
Hissing	High-frequency broad	Stochstic, tonal parts possible	Low-frequency modulation	Pneumatic components, grinding with high relative speed	6,000–12,000
Droning	All frequencies possible, arrow	Tonal	Heavy, low- or mid-frequency modulation	Belt drives, unsteady rotation, pitch error in gearings	300–10,000
Scrubbing	Mid or high-frequencie, broad	Stochstic	High-frequency modulation possible	friction between moving surfaces	1000–8000

NVH characteristics of HEVs are highly varied, and sufficient tests under varied operational conditions are needed to characterize the NVH properties. For example, the Ford Company considered following operations to quantify HEV NVH [8].

> *Idle is operated for testing purposes. Idle is used for improved fuel economy in an HEV, which idles the engine only as long as necessary and then the engine shuts off. Neutral operation is used when an HEV shift lever is in the neutral position and the vehicle control unit understands there to be no demand. HEVs are "drive by wire." The throttle position is controlled by the strategy and is not directly related to the pedal to gain maximum fuel efficiency. Wide-open throttle (WOT) cannot be achieved without manual control of the throttle. Similar to WOT, the part throttle position cannot be controlled to a percentage of the opening either. The battery state of charge (SOC) determines how much of the engine output is assisting the wheel torque and how much is recharging the battery. The regenerative braking test is important since it can potentially increase gear and chain noise in the transmission. The startup/shut off test is critical. One of the biggest NVH concerns for the HEV is the transient condition where the engine is starting or shutting off while the vehicle is in motion. This causes a torsional input to the system due to the change in the combustion process, which travels along with the crankshaft and is delivered to the rest of the system. Tests could be done as a slow run-up and coast down in the rpm range where the condition occurs, which is achieved by sweeping both the pedal position and dyno speed simultaneously, through low ranges. Other testing includes road load simulation, powertrain mapping to elaborate the component interactions, using specific instrumentation for specific NVH, and so on.*

In contrast to ICE vehicles and electric vehicles (EVs), hybrid vehicles have different operation modes that are essentially logically coupled to the driving conditions. **Figure 6.1** shows a map of various HEV system NVH under different operation modes for a hybrid vehicle versus some driver maneuvers. Overlaid are some critical noise and vibration phenomena [7].

Due to the different types of HEVs and associated unique operating conditions, NVH benchmarking and development is more complicated than ICE vehicles. Many efforts have been made to support competitive NVH target definition, as well as comparative analysis for investigation of relative strengths and weaknesses between vehicles with varied HEV configurations. These assessments were conducted for vehicle speed run-ups with and without ICE operation (as applicable), and during regenerative braking. Simulation and testing have been performed to get information regarding powertrain source and vehicle transfer function differences among many optional hybrid-electric vehicles. In the process, upon definition of targets for vehicle interior noise, airborne, and structure-borne path behavior was utilized to define targets for radiated noise and active side mount vibration, as would be measured in a hemi-anechoic powertrain test cell. **Figure 6.2** shows a two-dimensional landscape of typical customer vehicle maneuvers mapped against HEV operational modes [1]. Overlaid on this map are NVH issues, such as those associated with global powertrain vibration, driveline vibration, HEV component-specific noise, motor/generator whine, accessory noise, gear rattle, and noise pattern changeover.

FIGURE 6.1 HEV system NVH issues under varied operation modes with driver actions [7].

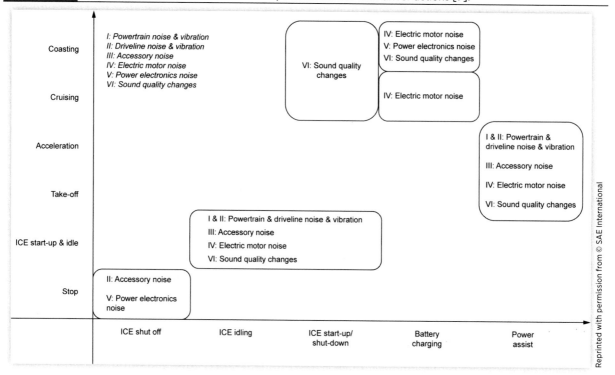

FIGURE 6.2 Unique HEV NVH challenges [1].

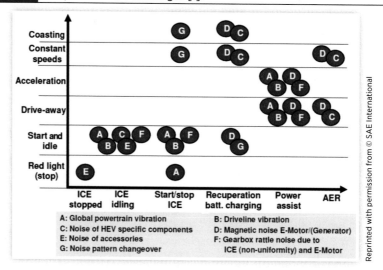

For conventional ICE vehicles, comparative benchmarking typically involves a combination of stationary data points (e.g., idle, city cruising speeds, highway cruising speeds) and vehicle speed sweeps (with the transmission in the drive) at multiple loads for assessment of customer perceived interior noise and vibration. This is then typically augmented with engine speed sweeps (with the transmission in fixed gear) to facilitate characterization of the powertrain, as perceived at the vehicle interior.

The multiple operating conditions experienced during the normal operation of various HEVs complicates the comparative benchmarking process. Given that powertrains are operated very differently, performance must be assessed based on customer perception; hence, these comparisons are made primarily based on similar vehicle operating conditions. Therefore, the hybrid system strategy and calibration play a significant role in the NVH performance. Secondary assessments of ICE performance (relative to engine speed), can then be conducted to understand the core powertrain NVH behavior and the potential for reduction of radiated noise through calibration and engine hardware.

In the development, the synthesis process has been used for evaluating interior noise, based on measured source and vibration, in which the process can be effectively used for reverse cascading from interior noise targets to powertrain level targets.

The expectation for interior noise content from an ICE depends highly on the vehicle class by following conventional vehicle development procedures. However, the tonal noise from electric machines (motor/generator) is universally considered annoying; hence, any perceived motor whine (in-vehicle interior) is considered unacceptable in a refined vehicle. Therefore, although interior noise targets can be defined based on EV whine orders from competitive data sets, it is perhaps more appropriate to develop targets based on the ability to perceive whine noise in a given vehicle application. Masking band analysis can be used to assess the ability to perceive tonal noise in the context of other noise sources in the same frequency range. The noise targets can be set for the primary and secondary orders from the electric machine. Once the interior noise order targets are established, the following steps can be taken to derive powertrain level targets. Overall interior noise contribution must be split into defined airborne and structure-borne interior noise share targets of the main excitation sources. Each of the main noise shares identified above must be broken down to their sub-components. Once targets are defined for each noise contribution, frequency-based vehicle sensitivity information can be used to cascade these interior noise share targets to powertrain level noise and vibration targets.

For the electric motor (EM) applications, the excitation sources are generally limited to powertrain airborne and structure-borne noise inputs. Airborne sources, defined as powertrain radiated noise, are broken down into multiple locations around the powertrain to account for variations in radiated noise directivity and vehicle sensitivity (acoustic attenuation between each source location and vehicle interior). The dominant structure-borne noise sources are defined as active side powertrain mount vibrations, broken down into vibration content in each direction of each mount.

A significant difference between electrical and combustion engines is the torque-characteristic. For the ICE, the drivers demand acceleration is transformed from the gas-pedal and actual gear position by stationary engine characteristics and in-stationary parameters to the torque on the drive-train resulting in a smooth, rotational speed-dependent parabolic curve with low torque at low rotational speeds. Electrical machines can deliver the full moment at very low rotational speeds, as shown in Chapter 2, which results in high agility. This high agility is a big challenge for the engineer, as it is necessary to design the control parameters in a way that no comfort-reducing load alteration vibrations occur.

6.2 **ICE Start/Stop Vibration and Noise**

In HEVs, ICE start/stop operations cause severe noise and vibration issues [12, 13, 23, 26–31]. This is similar to conventional vehicles where shake is a kind of low-frequency vibration, including steering wheel, seat track vibrations while HEV ICE starts/stop operations cause structure-borne NVH issues.

Figure 6.3 illustrates the engine start dynamic responses of the powertrain block and driveline, as well as vehicle motion. It is seen from the figure that the seat track acceleration captures the engine start dynamic responses from the powertrain to the vehicle or the responses from the NVH source to the driver/passenger. However, it is difficult to gain more insights into the dynamic event and the interaction among the powertrain block, driveline, and vehicle by examining these responses. The Fast Fourier Transform (FFT) results of these responses show that there are four vibration modes in the seat track acceleration response at the following frequencies: 6, 10, 15, and 20 Hz. Based on computer-aided engineering (CAE) analysis results, these modes are driveline

FIGURE 6.3 HEV vibrations during engine startup. Top: seat track acceleration; Middle: half shaft torque; Bottom: powertrain block top fore/aft acceleration [13].

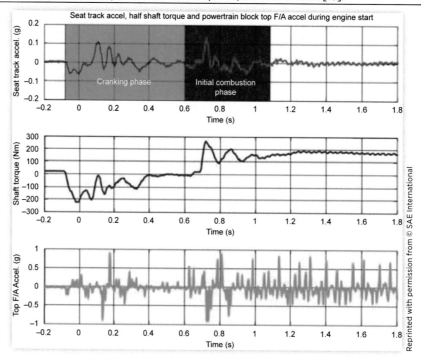

torsional mode, powertrain block roll mode, engine damper torsional mode, and vehicle mode, respectively. During engine cranking, the generator provides a cranking torque to ramp up engine speed for engine start/ignition, which results in in-cylinder compression due to the reciprocating motion of the engine piston. This compression can excite the powertrain block roll mode.

The engine/damper torsional mode can also be excited during the engine speed ramp-up if the rise of the engine speed is not controlled properly (e.g., slowly passing through the engine torsional resonance range). Besides, the generator's cranking torque is being transmitted to the driveline due to the permanent mechanical connection in this powertrain. This transmitted torque can excite the driveline torsional mode and cause undesired vehicle motions if it is not compensated properly by the traction motor.

For ICE vehicles, key start-stop is initiated by the driver and occurs only a few times a day, and the start-stop NVH is not a main concern for typical drivers. However, it is not the case on HEVs where frequent auxiliary power unit (engine, motor and controls– auxiliary power unit [APUs]) start-stop may deteriorate the NVH performance. The APU might be required to start and stop many times depending on the driving cycle, the configuration of the powertrain, and the control strategy, as shown in **Table 6.2**.

Several milliseconds elapse between the initial ignition and the completion of combustion, so the ignition spark must be triggered early enough to ensure the main combustion (the combustion-pressure peak) occurs shortly after the piston reaches top dead center (TDC), and the ignition angle should move farther in the advance direction along with increasing engine speed. It is also known that the correct ignition and valve timing significantly affect the performance of an engine, fuel economy, and emissions. They also impact the pressure of the combustion chamber, which therefore results in vibration of the powertrain.

The APU start-stop system, which is an integrated starter generator (ISG), consists of a tuned 4-cylinder engine, a clutch that transmits engine torque and attenuates the bidirectional vibration instead of engaging and disengaging powertrain, and a motor that allows cranking at the restart and also functions as a generator when APU stops.

TABLE 6.2 Engine/APU starts/stops induced NVH [31].

Working condition	Item	Subjective feeling	Average score
APU start transients	Unexpected noise	Unsatisfactory/Rated disturbing by all consumers	5.50
	Floor vibration		5.67
	Steering wheel vibration	Poor/Rated as failure by all consumers	4.83
	Synthetical comfort	Unsatisfactory/Rated disturbing by all consumers	5.17
APU stop transients	Unexpected noise	Very poor/Complained as failure by all consumers	3.67
	Floor vibration	Poor/Rated as failure by all consumers	4.50
	Steering wheel vibration	Very poor/Complained as failure by all consumers	3.50
	Synthetical comfort		3.67

For conventional vehicles, key start-stop is initiated by the driver and occurs only a few times a day. The start-stop NVH is not a main concern for typical drivers, but it is not the case on a ReEV that frequent APU start-stop may deteriorate the NVH performance. Since frequent APU start-stop will degrade the NVH performance, more emphasis should be placed on it [10, 12, 13, 19, 23, 31, 33–39].

Figure 6.2 shows a two-dimensional landscape of typical customer vehicle maneuvers mapped against HEV operational modes, from which the engine start/stops NVH effect is specified. From the customer acceptance of the end product perspective, the plot should be customized to a given vehicle program, so that a focused plan can be developed to address each NVH item systematically the engine needs to start/stop behavior to be transparent to the driver from both a tactile (vibrations) and acoustic (noise) standpoint.

Figure 6.3 shows an example of seat track vibrations measured during ICE start-up conditions on an HEV. The start-up transient event can be split into the engine cranking phase and the initial start of combustion.

The accelerations are recorded as a vibration index. Usually, an acceleration RMS value is used to quantify system vibrations. The acceleration RMS value calculates the average acceleration during a certain period.

$$\text{RMS} = \sqrt{\int_{t_0}^{t_f} \tilde{a}^2 \, dt \, / \left(t_f - t_0\right)} \tag{6.1}$$

where

a is vehicle acceleration in m/s²
t_0 is the starting time in s
t_f is the final time in s

Table 6.3 shows the human comfort levels for varied vibration accelerations.

A well-established metric to describe vibration during transient events of ICE start-up is the vibration dose value (VDV).

$$\text{VDV} - \sqrt[4]{\int_{ts}^{te} a^4 \left(t\right)\left(dt\right)} \tag{6.2}$$

where a is the acceleration of seat track or steering wheel vibration. Although the vibration behavior might seem generally similar to the starting phase of a conventional ICE-powered vehicle, the ICE start-up vibration for HEV can be more important.

In a conventional ICE-powered vehicle, the ICE start-up occurs only at the beginning of vehicle operation and the resulting vibration feedback (in response to cranking the engine to start the vehicle) is expected from the driver of the vehicle. For HEV, the start-up of the ICE is linked to factors such as the SOC of the battery and driver torque demand, which can result in unexpected vehicle vibration. Depending on the layout of the driveline, the start-up vibrations for HEV can be further complicated by the fact that the start-up of the ICE does not occur in neutral, causing excitation of driveline torsional modes (in addition to powertrain rigid body modes).

TABLE 6.3 Comfort reaction to vibration environments (ISO 2631-1 1997).

Acceleration RMS value (m/s²)	Comfort reaction
≤0.315	Not uncomfortable
0.315–0.63	A little uncomfortable
0.5–1	Fairly uncomfortable
0.8–1.6	Uncomfortable
1.25–2.5	Very uncomfortable
≥2	Extremely uncomfortable

© SAE International

The vibration dose value (VDV) is typically based on the low-frequency content of the vehicle's fore-aft acceleration signal during the transient event of interest [3]. The VDV results in a single number metric, which can be used for vehicle benchmarking, target setting, and to track the ICE start/stop performance (the seat track vibration [longitudinal direction]) during ICE start-up of a vehicle during the HEV development phase.

ISO 2631-1 recommends several measures that are appropriate to use when investigating human vibration excitation. These are root mean square value (RMS value), maximum transient vibration value (MTVV), and VDV.

The MTVV is defined as the maximum running RMS value with an integration time τ,

$$a_w(t_0) = \left\{ \frac{1}{\tau} \int_{t_0-\tau}^{t_0} \left[a_w(t) \right]^2 dt \right\}^{\frac{1}{2}}$$
$$\mathrm{MTVV} = \max\left[a_w(t_0) \right] \tag{6.3}$$

The integration time is set to 0.25 s.

There were efforts made to correlate subjectively graded vibration levels of discomfort to measurable quantities and it was concluded that the MTVV yielded the smallest deviations from a curve fitted relationship between these two domains.

While the VDV is useful for benchmarking, target setting and assessment of the ICE start/stop the behavior of a given HEV, it does not provide additional insights that might be needed to focus development efforts. Another useful metric that can help with ICE start/stop refinement is the energy spectral density (ESD) of vehicle seat track fore-aft (or steering wheel) vibration. The ESD is defined as

$$\mathrm{ESD} = \frac{G_s(\omega)}{\Delta f} \Delta T \tag{6.4}$$

where
    $G_s(\omega)$ is the auto power spectrum of seat track or steering wheel vibration
    Δf and ΔT are the frequency resolution and period for the ESD calculation, respectively

The ESD metric allows for the development of scattering bands based on benchmarking the state-of-the-art HEV for their ICE start/stop the behavior. Besides, the ESD-based analysis can provide insights into the ICE start/stop measurements on a development vehicle and assist with focusing refinement efforts.

The ESD analysis is normally conducted on a time window that includes the ICE cranking phase and the initial combustion phase. These phases can contribute to the excitation of different frequencies, such as those corresponding to ICE rigid body modes, torsional modes of the driveline, and vehicle body modes (vehicle sensitivity). Therefore, for further investigation and optimization of the ICE start-up vibration, it is beneficial to have a good understanding of all relevant modes in the frequency range of interest (<50 Hz). A significant contributor to the vehicle vibrations during an ICE start-up (or shut down) event is the excitation of the engine rigid body modes.

Specifically, the roll mode of the ICE (motion about the crankshaft axis) is excited under ICE start-up and shut down events. To minimize the transmission of vibration to the passenger compartment during these transient events, it is essential to optimize the mounting layout of the engine. This is accomplished by using a multi-body systems-based analysis of the ICE installed in a simplified vehicle model. Specifically, the model includes a rigid body representation of the ICE (and transmission or motor/generator, as appropriate) and vehicle body with appropriate values for the location of the center of gravity, masses, and mass moments of inertia. A simplified wheel and suspension assembly are often used with appropriate mass, geometry, stiffness, and damping information. The powerplant mounts are represented using specific mount models that capture the static nonlinearities and frequency-dependent dynamic (stiffness and damping) behavior.

The multi-body systems-based model is utilized to decouple the ICE rigid body modes. Specifically, it is important to decouple the ICE roll mode from the other rigid body modes as much as possible.

Having achieved the desired degree of separation from the ICE rigid body modes, the engine calibration and engine start procedure should be optimized for reducing the vibration excitation. This can affect both the cranking phase and the initial combustion phase of the engine start procedure.

Typical engine control-related parameters which influence the ICE start/stop vibration are injection time, amount of initial injection, compression rate at cranking (which can be influenced by valve opening time and throttle position), crank speed rise rate, and engine speed at which the combustion process starts. The crank angle positions are defined for engine stop. The position of the crankshaft when starting influences the vibration levels.

For conventional powertrains, the engine start/stop is usually performed with the transmission in neutral. Hence, torsional vibrations during the start/stop events are decoupled from the vehicle driveline. However, HEV drivetrains do not necessarily have this decoupling, as a result of which the torsional vibrations from the ICE cranking and combustion phase can excite low-frequency driveline torsional modes, such as driveline shuffle. As part of the start/stop calibration, it is important to ensure that these torsional modes are not excited by engine torsional during these transient events. As indicated previously, seat track acceleration ESD analysis can be used to diagnose the presence of resonances excited and focus development effort to refine the ICE start/stop the behavior.

Finally, the use of the EM to actively damp the engine torsional vibrations should be explored. In such a scenario, the EM applies a torque fluctuation with a 180° phase offset relative to the torque fluctuation caused by the ICE cranking, so that the torsional excitation of the driveline can be minimized. For the implementation of active damping, the crankshaft position must be known, and a good understanding of the cranking torque is mandatory.

It is equally important to minimize the increase of powertrain noise in a hybrid electric vehicle when the ICE starts up. **Figure 6.4** illustrates the increase in overall noise level caused by the start-up of the ICE. The ICE start-up event is noticeable in the overall noise levels, as indicated by the encircled area in Figure 6.4. The interior noise level increased by ~8 dB(A) when the ICE started up. In addition to the measurement data, trend lines are shown in this figure for the vehicle speed dependency of the overall interior noise level while operating in all-electric range (AER) mode and with the fired

FIGURE 6.4   Interior noise during ICE start-up [23].

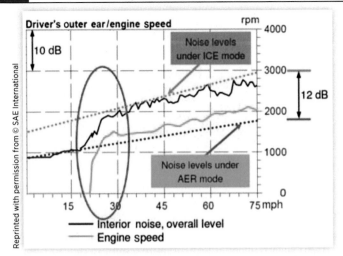

combustion engine (ICE mode). These lines show that the interior noise level difference between the AER mode and ICE mode can be up to 12 dB(A).

Therefore, it is extremely important to carefully manage the transitions between the ICE mode and the AER mode on an HEV.

To reduce customer annoyance with ICE start/stop noise issues, it is important to reduce the overall powertrain noise share in the interior noise signature of the vehicle. This becomes especially important, as ICE in HEV applications are typically downsized and can have significantly increased torsional fluctuations, noise and vibration levels in comparison to ICE in conventional vehicles.

Optimization of the ICE-related noise and the vibration is necessary to minimize significant level differences between AER and ICE modes of operation, as well as to smooth out the interior noise pattern changes during transitions. For successful NVH optimization, this process needs to start early in the development process of the combustion engine. Specifically, the following NVH related issues should be addressed during development.

- NVH effects from highly transient operating conditions, including change of torque flow and activation/deactivation of components. The most important operating conditions are start/stop, ICE engagement, and transition between operating conditions [3].
- NVH issues of hybrids are related to engine start-stop conditions, which deteriorate passenger comfort during vehicle standstill. In particular, the engine start is very critical in terms of NVH.

In the subjective evaluation test, a series of HEV interior sounds, including decelerating, stop and accelerating conditions were recorded and presented to participants and the uneasiness of the sound only at the stop condition were evaluated. The evaluation term "uneasiness" was employed in addition to "loudness" as the fundamental psychoacoustic metric. In the loudness evaluation, the participants rated the loudness of the sound using an integer scale from 0 to 100. In the uneasiness evaluation, four categories, "very uneasy," "uneasy," "slightly uneasy," and "not uneasy" were given [40].

Based on the results, uneasiness was found to be increased by the unexpected engine start sound. Subsequently, they considered a method to relieve the uneasiness feeling by changing the engine starting duration for battery charge at the vehicle stop condition. Three patterns of engine starting durations at the automatic start conditions were prepared. In the original, middle, and long patterns, the engine started from stop to 2500 rpm for 2, 3, and 6 s, respectively. The subjective evaluation result indicated that the uneasiness was relieved when the duration was in the middle at 3 s. On the contrary, the ratio was largest at the duration of 6 s. This revealed that the uneasiness to the engine automatic start sound varies depending on the engine starting duration and there is a suitable duration for engine start to relieve the feeling.

The start event, in this case, can be described in two discrete phases: (a) the initial application of torque by the starting machine, and (b) the firing of the engine as it accelerates. In this example, both of these phases are visible at the seat rail. Generally, this distinction of phases may not be so readily discernible due to the modal response of the power unit and the vehicle. The seat rail response is, therefore, a complex response resulting from the excitation of the starter torque and the compressive gas forces, in combination with vehicle transfer functions, which might be determined by vehicle attributes other than NVH (e.g., ride, steering precision) [41].

Optimization of the start-up vibration event requires careful design of all these factors. But an improvement is gained if the initial torque application is smoothed to reduce shock input and the crank oscillations controlled to avoid exciting sensitivities in the vehicle transfer functions.

Electric motors will generally provide maximum torque at low speed, which is precisely the opposite of what the NVH engineer would like. Ideally, torque should be applied progressively, rather than with a harsh initial step. This characteristic is not feasible with a standard, low-cost starter motor with minimal control electronics. A significant challenge for the NVH engineer is to provide a sufficient approximation to the ideal characteristic by either modifying the existing start system at minimum cost, or utilizing the traction motor present on all hybrid vehicles without compromising other attributes, or by employing some alternative start device.

For shut-down vibration, the shut-down process is much simpler – especially where there is no involvement of the starting machine. It becomes clear that the vehicle response during shut-down is excited only by the compression-derived accelerations of the engine. The low-frequency region (where the engine is in its final 1 or 2 cycles) is the most important. This is partly because the vehicle and powertrain modal responses peak in these frequencies. This low-frequency response is also driven by the fact that, for near-constant acceleration, the displacement is maximum at low frequencies, and consequently high displacement at the engine mounts results in high forcing into the vehicle body.

Active control was used to suppress engine start vibrations [12]. Generally, in vehicles with longitudinal engines, vertical and lateral vibrations of the vehicle body are generated during engine start in conjunction with the roll and yaw resonance of the power plant. For example, the same sort of vibration occurs in the hybrid luxury sedan with THS II.

Because it has no clutch mechanism between the engine and the drive train, the torque fluctuation of the engine excites torsional resonance in the drive train, generating longitudinal vibration in the vehicle body. The compelling forces of these forms of vibration are the motor torque reaction force and engine compression pressure during motoring, as well as the combustion pressure during the explosion. The following section describes technologies for reducing these compelling forces, as well as technologies for reducing vibration in the power plant mounting system and drive train.

A variety of countermeasures that were developed for the THS II for front-wheel drive (FWD) vehicles have been implemented in the HEV. The compelling forces were reduced by changing the intake valve closing timing, by controlling the piston stop position to reduce the engine compression pressure, and by adjusting the injected fuel volume and ignition timing.

The lateral vibration of the vehicle body was reduced by shortening the distance between the principal elastic axis and the center of the gravity of the power plant.

The excitation of torsional resonance in the drive train and resonance in the power plant mounting system was inhibited by operating MG1 at high torque during engine start. In the hybrid luxury sedan, in addition to the countermeasures described above, vibration-reducing motor control and the torsional damper, which has two-stage hysteresis characteristics, were adopted to reduce the drive train torsional vibration.

Moreover, active vibration-reducing motor control is implemented in the form of feed-forward control by MG1 and feedback control by MG2. The THS II power split device (PSD) uses a planetary gear, the revolution speeds of MG1, the engine, and the drivetrain axis. **Figure 6.5** shows the motor control system for vibration reduction and **Figure 6.6** shows the vibration reduction with motor control.

FIGURE 6.5   Motor control system for vibration reduction.

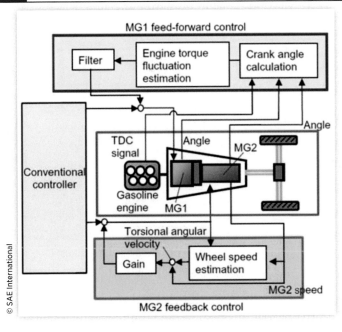

FIGURE 6.6 Vibration reduction with motor control.

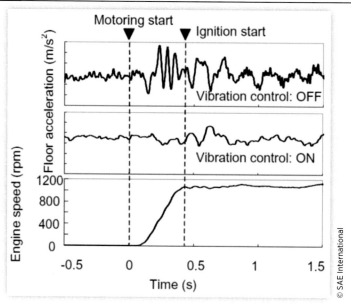

The torsional vibration level of the drive train can also be reduced by increasing the damping of the torsional damper. Increasing the hysteresis torque is an effective way to increase damping, but it raises the concern of a worse booming noise from the engine explosion first-order component because of the torque fluctuation to the drive train increases. In consideration of the different torsional angles of these two phenomena, a damper with two-stage hysteresis characteristics was adopted. During engine start, the hysteresis torque is high in the high-amplitude region, but the hysteresis torque is low in the very low-amplitude region where booming noise tends to be a problem.

6.3 **HEV Driveline NVH**

The HEV NVH has been widely investigated [10, 12, 14, 42–48]. **Figure 6.7** shows a typical driveline architecture for HEVs. In hybrid transmissions, the torque converter in a traditional automatic transmission is normally replaced by a motor. Most hybrid

FIGURE 6.7 Dynamical system of HEV driveline.

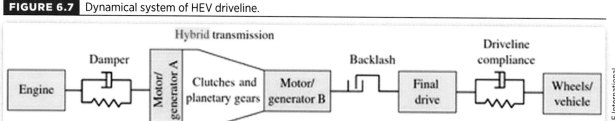

vehicles can be driven by the motor/generator B alone in the electric drive mode, by the engine alone in mechanical drive mode, or by all the motor/generators A and B and the engine together in hybrid mode. These drive modes are achieved through controlling and corresponding clutches in the transmission to meet the high-level requirements of fuel economy and drivability. In most hybrid vehicle systems, the engine and a motor/ generator are directly connected through a planetary gear set as a result of removing the torque converter. While this improves system responsiveness and efficiency, it also easily causes twisting of the driveshaft and produces torsional vibration, and this becomes even worse during engine start and stop.

In HEVs, engine booming noise has been a critical issue. In a vehicle powered by a V6 engine, in addition to the third-order component of engine revolution, the booming noise of the second-order component of engine revolution sometimes becomes a problem. The second-order excitation force of the revolution is mainly a second-order couple of the reciprocating inertia of the piston.

The vibration is amplified by the bending resonance of the power plant and is transmitted to the body from the mounts at the rear of the transmission, generating a booming noise. A two-speed reduction gear was adopted for the RWD hybrid sedan, which made it possible to reduce the motor diameter and weight. But the overall length of the THS II transmission is 50 mm greater than that of a 6-speed automatic transmission (A/T) in an ordinary gasoline-powered vehicle, and the mass is approximately 35 kg greater.

There was concern that the increase in mass and overall length would cause the power plant resonance to drop into the normal engine speed range. An increase in the transmitted force was also predicted because the increase in the load allocated to the mounts would require that the spring constant for the mounts be raised to 1.5 times that of the mounts for a 6-speed A/T.

The deformation mode of the power plant resonance is such as to cause a large deformation in the transmission. A study was undertaken on how to improve the resonant frequency to separate it from the normal engine speed range. Finite element analysis (FEM) was used to optimize the shape of the transmission case by smoothing the outline, reinforcing the ribs, and so on.

As a result, the resonant frequency was raised to 180 Hz, equivalent to that of a 6-speed A/T. Also, the mounting position was shifted 80 mm farther forward than in the original plan to set it at a nodal point of the vibration mode. Changing the mounting position raised new issues of installation space and the separation of principal elastic axis and the center of gravity, which strongly influences the engine start vibration performance. To ensure the adequate installation space, the mounting is attached to the cross member, a major change from the structure used for a 6-speed A/T. The distance of the principal elastic axis and the center of gravity was reduced by optimizing the lateral-to-vertical ratio of the mount spring constants. These countermeasures successfully addressed the issues of booming noise and engine start vibration [10, 13, 49–52].

The HEV NVH has many issues, one of which is the idle vibrations as investigated in THS system as follows [53]. Although idling vibration is usually caused by 1st order of engine combustion force, other engine forces also occur at frequencies lower than the

1st order of combustion (low frequency idling vibration). The driveline of the Toyota Hybrid System II (THS II) has different torsional vibration characteristics compared to a conventional gasoline engine vehicle with an A/T. Nonlinear characteristics caused by the state of the backlash of pinions and splines influence changes in the torsional resonance frequency. The torsional resonance frequency of the driveline can be controlled utilizing the hybrid system controls of the THS II.

A wide variety of hybrid vehicles have been proposed and commercialized by automakers around the world. Generally, these vehicles shut down the engine while the vehicle is stopped to improve fuel economy and lower emissions, resulting in an extremely quiet interior environment. However, hybrid vehicles will engage the engine while stopped to charge the main battery or to warm up the engine or catalytic converter. In these cases, idling vibration is generated by 1st order of engine combustion force. However, other engine forces occur at frequencies lower than the 1st order of combustion. Low-frequency idling vibration in vehicles installed with the THS II has a unique generation mechanism based on the characteristics of the driveline and the engine speed during idling.

Next, the mechanism of this low-frequency idling vibration is described. Idling vibration is generally described as a phenomenon caused by 1st order of engine combustion force. Low-frequency idling vibration is the vibration that occurs at frequencies lower than the 1st order of combustion. There are two types of low-frequency idling vibration: steady-state vibration felt like a swaying motion and intermittent unsteady-state vibration felt as a rattling motion. The former is mainly caused by differences in combustion pressure between each cylinder in the engine or imbalances in engine revolution. In contrast, the latter is caused by unstable combustion pressure in each combustion cycle within the same cylinder. **Table 6.4** shows steady-state low frequency idling vibration.

Table 6.5 shows the specification of HEV compared to a similar ICE vehicle. Although the idling engine speed of the hybrid vehicle and conventional vehicle are different, the results were compared after changing the idling engine speed of the conventional vehicle to 1000 rpm, the same as the HV. In both vehicles, the gear lever was placed in the P range and the air conditioning and all other accessory equipment were switched off.

The operating states of the hybrid vehicle include the application of high engine torque for charging the main battery (operation with load) and very low engine torque

TABLE 6.4 Idling vibrations of HEV with a V6 engine.

V6 Engine	Idling vibration	Low frequency idling vibration
Engine torque fluctuation	1st Order of engine combustion (3rd order of revolution)	0.5th to 1st Order of engine revelution
Frequency (1000 rpm)	50 Hz	8.3–16.7 Hz
Sensory felling	Shudder (steady-state)	Sway (steady-state)
		Rattle (unsteady-state)

© SAE International

TABLE 6.5 Test vehicle specifications.

		Hybrid Vehicle	Gasoline Engine Vehicle
Configuration	Engine	2GR-FXE	2GR-FSE
	Engine Mounts	Hydraulic Mounts	←
	Transmission	THS II	6-speed AT
	Suspension	Multi-Link	←
	Wheels	19-Inch	←
	Body	N-Platform	←
Operating Conditions	Engine Speed	1,000 rpm	←
	Shift Position	P	←
	Engine Torque	Low	←

for warming up the engine or the catalytic converter (operation without load). However, to set an engine torque level similar to that of the conventional vehicle, both vehicles were operated without load. Under these conditions, the fuel injection quantity was adjusted to establish the same 0.5th order of engine revolution forces. The other specifications of the test vehicles were the same. As shown in **Table 6.5**, the only difference between the two vehicles is the transmission.

Figure 6.8 shows the frequency analysis results for the floor vibration level at an engine speed of 1000 rpm. The normal peak of idling vibration peak is at 50 Hz, which is the 1st order of engine combustion (i.e., the 3rd order of engine revolution). However, there are multiple peaks at lower frequencies that are felt like a swaying motion. These peaks are examples of steady-state low-frequency idling vibration. Vibration at the 0.5th and the 1.5th order of engine revolution are mainly caused by the difference in combustion pressure between each cylinder. The 1st order of engine revolution is mainly caused by engine revolution imbalance. The level of vibration at 8.3 Hz, which corresponds to the 0.5th order of engine revolutions, is particularly large. The following sections focus on this frequency phenomenon.

Of the forces generating steady-state low-frequency idling vibration, the primary cause of the 0.5th order of engine revolution is fluctuations in combustion pressure between each cylinder. The fluctuations are generally the result of differences in the air intake volume into each cylinder and differences in the quantity of fuel injected into each cylinder. **Figure 6.9** shows the measurement results of engine combustion pressure for each cylinder during idling.

The combustion pressure increases cyclically in cylinder 5 and decreases cyclically in cylinders 3 and 4. This generates torque fluctuations in a cycle consisting of two engine revolutions, which acts as the 0.5th order of engine revolution forces. Steady-state low-frequency idling vibration is the result of force generated by differences in combustion pressure between each cylinder. The force is then transferred from the engine to the engine mounts and causes the floor to vibrate. **Figure 6.10** compares the measurement

FIGURE 6.8 Floor vibration level under idling conditions.

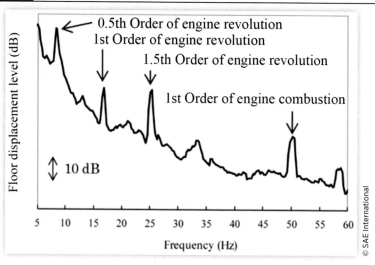

FIGURE 6.9 Cylinder pressure of the engine.

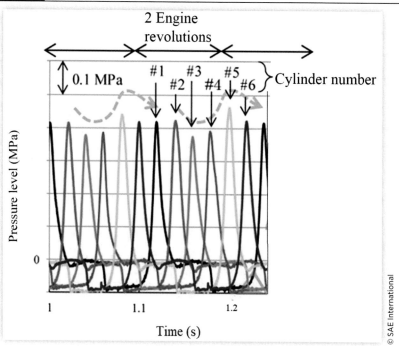

results of crankshaft angular fluctuation in a V6 HV equipped with the THS II and a conventional ICE vehicle.

Therefore, it was concluded that the THS II driveline amplifies the 0.5th order of engine revolution component. At the 8.3 Hz peak for steady-state low-frequency idling vibration, the hybrid vehicle was approximately 10 dB higher than the conventional vehicle.

In addition to the shake discussed in the last section [53–55], driveline boom is another structure-borne NVH issue in HEVs [41].

FIGURE 6.10 Comparison of the measurement results of crankshaft angular fluctuation in a V6 HV equipped with the THS II and a conventional ICE vehicle.

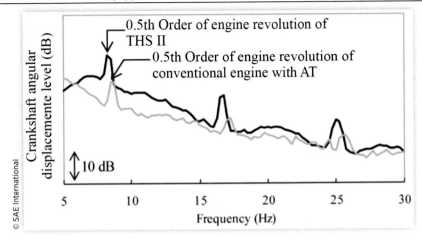

The installation of an electric machine can generate new torsional modes, or modify existing modes of the driveline, as shown in **Figure 6.11**. Care needs to be taken that this inertia does not act upon the compliance of other driveline elements (shafts, direct manual focus [DMF] springs) to form a dynamically unstable system and generate interior noise booms or vibration.

It is normally possible to deal with driveline modes by the use of a clutch slip (or torque converter slip) to increase driveline damping, or by application of absorbers. However, neither of these palliative solutions is ideal since they involve inefficiencies – clutch slip wastes power and absorbers add weight. In any case, it is impossible to fit absorbers at the antinode of the torsion mode when this is the rotor itself. A better solution is to ensure that additional inertias in the driveline are accommodated without introducing torsional modes, or that such modes are absorbed by other dynamic elements within the driveline. These actions must be taken by the transmission supplier at the concept stage – they cannot be accommodated later in the vehicle integration stages.

FIGURE 6.11 New driveline dynamics result from the addition of a new inertia.

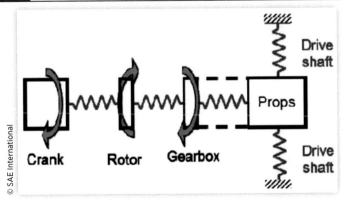

During the start and stop of the combustion engine and load alterations, vibrations can occur that the driver may perceive as subjectively annoying. Particularly during the starting process, vibrations are caused by the excitation of eigenmodes of the powertrain due to the low engine speed and by an impulse-like excitation due to the sudden increase of torque. Hybrid vehicles are especially problematic in this regard since the starting process of the engine occurs often. On the other hand, the new components of the powertrain also offer possibilities to counteract these comfort-impairing effects.

The following example illustrates this in more detail for a power-split hybrid vehicle. After a description of the powertrain, its dynamic behavior will be analyzed using a multi-body simulation model. It will be shown how EM control can help to reduce vibrations.

A full hybrid vehicle has a power-split powertrain and an all-wheel-drive without the mechanical coupling of the axles. A V6 cylinder gasoline engine and two electric motors propel the front axle. A third EM (mechanically decoupled) propels the rear axle.

Figure 6.12 shows the assembly of the transmission and ICE of the vehicle. The crankshaft is connected to the planetary gear carrier of the first planetary gear (PG1). The motor/generator (MG1) drives the central gear and the annulus gear is fixed to the annulus gear of second planetary gear (PG2). Motor/generator 2 (MG2) drives the central gear of the second planetary gear (PG2) and the planetary gear carrier of PG2 is fixed to the housing. This results in a fixed transmission ratio from MG2 to the annulus gear and thus also to the wheel.

MG2 works mainly as an engine, as it propels the wheels directly and controls the powertrain torque. Only during recuperation, it is used as a generator.

FIGURE 6.12 Transmission and combustion engine of the test vehicle [57].

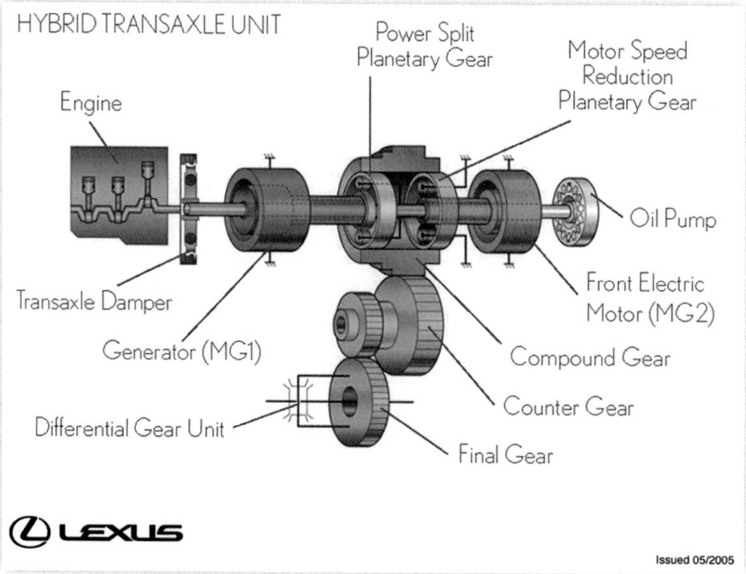

On the other hand, MG1 works mainly as a generator providing the electric power for MG2 and it charges the high-voltage battery. Moreover, MG1 has to propel the combustion engine at the start. MG1 counters the torque of the combustion engine and controls the transmission ratio of the combustion engine and the annulus gear. Through a variable adjustment of the transmission ratio, the combustion engine can be increasingly operated within the fuel-efficient range. PG2 has to reduce the speed of the electric motor, while PG1 serves as a power distributor.

Figure 6.13 shows the model which was used for the torsional vibration simulation. The ICE and the two electric motors (MG1 and MG2) act as excitation sources on the transmission. The very rigid hybrid planetary gear is depicted as a purely kinematic module without stiffness. Its inertias are reduced to the differential gear.

Torsion spring-damper elements model the driveshaft and the tires. The model has several rotational degrees of freedom and one longitudinal degree of freedom in the driving direction. The vehicle is represented within the vibrational system as rotational inertia.

The inertia of MG1 is small, so small torque fluctuations of the combustion engine would lead to considerable speed changes of MG1 at almost constant annulus gear speed due to the transmission ratio. To achieve a stable speed ratio, the electric counter-torque of MG1 needs to be controlled. Thus, to simulate the powertrain, the set-up and the integration of a controller is necessary. Moreover, further control concepts are used to reduce powertrain vibrations. This will be discussed as follows.

The modal analysis of the powertrain model shows that four characteristic values in the investigated low-frequency range below 250 Hz: (1) jerking mode (10 Hz), (2) torsional-vibration damper mode (17 Hz), (3) tire mode – asymmetrical (29 Hz), and (4) tire mode – symmetrical (35.6 Hz).

Using impact tests and operational measurements, the jerking mode at 10 Hz and the torsional-vibration damper mode at 17 Hz are verified, which matches well with the simulation.

FIGURE 6.13   Sketch of the multibody simulation model of the hybrid drivetrain.

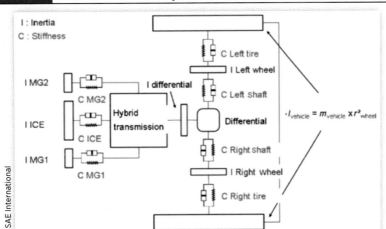

The eigenmodes of the hybrid concept are similar to the eigenmodes of a conventional powertrain with a dual-mass flywheel. In case of a conventional vehicle with a single-mass flywheel, the second eigenmode lies between 40 to 80 Hz. High vibrational amplitudes occur mainly in the proximity of the gearbox and can lead to transmission rattling. In case of a powertrain with a dual-mass flywheel, a part of the flywheel mass is shifted behind the torsional-vibration damper. The jerking mode is thus nearly unchanged, but the second eigenmode is shifted below the idle range between 10 and 15 Hz. This is also the case for the hybrid concept. The rotor of MG2 is comparable to the secondary mass of a dual-mass flywheel and the flywheel corresponds to the primary mass. In the torsional-vibration damper mode of this hybrid powertrain, the MG2 mass vibrates against the combustion engine, and in the DMF mode of a conventional engine, the DMF secondary mass vibrates against the DMF primary mass.

In the jerking mode, the entire powertrain vibrates against the vehicle and is particularly excited during start/stop. In the torsional-vibration damper mode, the combustion engine vibrates against the powertrain, which corresponds to major tensions in the torsional vibration damper. With a 6-cylinder engine, the 3rd engine order meets this 17 Hz resonance at the critical speed of approximately 340 rpm. This critical speed is only reached during the start and stop process of the combustion engine, which occurs much more often in hybrid vehicles than in conventional ICE vehicles.

Both tire modes are highly damped and thus hardly noticeable. The investigation of the parameter dependency yields that the inertia of MG2 and the crankshaft stiffness are the relevant parameters for the jerking mode. The parameter with the dominant influence on the torsional-vibration damper mode is the torsional-vibration damper stiffness.

With hybrid vehicle concepts, start and stop occur very often. To reduce the vibrations caused by this, control methods of the electric components are used in addition to measures at the engine mounts, as well as combustion engine control.

The starting of the combustion engine can be divided into two phases. First is the starting phase, where MG1 speeds up the combustion engine to about 1000 rpm; combustion has not yet started. The excitation occurs through the third engine order at low frequencies. Second is the combustion start at about 1000 rpm. The sudden torque increase acts as an excitation on the powertrain.

During the start, the powertrain vibrations are caused by the excitation of the torsional-vibration damper mode when the ignition order of the combustion engine meets its eigenfrequency. For this purpose, starting torque control is applied to minimize the rotational angle in the torsional vibration damper by controlling the additional torque of MG1, as the largest vibrational amplitudes are found at the torsional vibration damper.

The jerking mode is excited by a rapid torque increase or decrease of the combustion engine, especially during start/stop, sudden acceleration, or deceleration. To take care of this problem, jerking control is applied. Its task is to minimize the difference between the current dynamic wheel speed and the (according to the transmission ratio converted) speed of MG2 by control of MG2 torque. This suppresses the jerking mode, as it features considerable angle and/or speed deviations between MG2 and the converted wheel speed.

To achieve a stable transmission ratio, the electric counter-torque of MG1 needs to be controlled. This is why a speed regulator is necessary for this type of hybrid concept, even for standard operation. The torque of MG1 is not an input value of the total system, but it is determined by the speed regulator. The starting process module yields the torque of the combustion engine while allowing for the start of combustion. Starting torque control accesses MG1; jerking control accesses MG2. The controllers receive the speeds (n) as input values and the corresponding torques (T) are yielded as output values.

The starting process of the combustion engine during driving is simulated with and without control. Without control (gray curves), MG2 yields the constant driving torque for vehicle acceleration. MG1 becomes active twice. At the start, it revs up the combustion engine to approximately 1000 rpm. At the beginning of combustion, the torque of the combustion engine needs to be countered.

With control, the dynamic intervention of the starting torque control into MG1 can be noticed (black curve). The maximum amplitude of the drive shaft irregularity and vehicle acceleration are reduced to a third.

All in all, without control, the excitation during the starting process is more critical than at the beginning of combustion. With active control, both vibration phenomena are significantly reduced and on a similar level.

The starting process was simulated with a delayed beginning of combustion compared to the real behavior to facilitate the separate investigation of the phenomena in both areas.

One important criterion for vibration reduction mechanisms is their robustness against disturbances that occur in reality, for instance, through inaccuracies or production variance. As expected, the vibration reduction through MG2 jerking control at the beginning of combustion proves to be considerably more robust against the interferences that were built into the model than MG1 starting torque control during a start.

It can be seen that through the specific use of electric motors in the form of pre-regulated or controlled dynamic torques, the powertrain vibrations can be reduced to a third. Regarding the regulation and control algorithms, it is necessary to ensure sufficient robustness against disturbance.

Possible causes of booming noise include the following: poor transmission mount isolation; excessive induced floor vibration; vibration transferred through exhaust hangers, inducing floor vibration; excessive tailpipe noise, transmitted into the cabin; and excessive muffler shell noise.

The powertrain motion mainly consists of rotation due to the torque imbalance at start-up and shut down. This may also be valid for engine idle and low engine speeds.

Driveline boom is affected by the installation of an electric machine, which can generate new torsional modes, or modify existing modes of the driveline. Care needs to be taken that this inertia does not act upon the compliance of other driveline elements (e.g., shafts, DMF springs) to form a dynamically unstable system and so generate interior noise booms or vibration [41].

It is normally possible to deal with driveline modes by the use of clutch slip (or torque converter slip) to increase driveline damping, or by application of absorbers. However, neither of these palliative solutions is ideal since they involve inefficiencies – clutch slip wastes power and absorbers add weight. In any case, it is impossible to fit absorbers at the antinode of the torsion mode when this is the rotor itself. A better solution is to ensure that additional inertias in the driveline are accommodated without introducing torsional modes, or that such modes are absorbed by other dynamic elements within the driveline. These actions must be taken by the transmission supplier at the concept stage – they cannot be accommodated later in the vehicle integration stages.

Active booming controls have also been explored [58]. In HEVs, pressure variation during engine combustion generates torque fluctuation that is delivered through the driveline. Torque fluctuation delivered to the tire shakes the vehicle body and causes the body components to vibrate, resulting in a booming noise. Hyundai Kia Motor Company's (HKMC) transmission mounted electric device (TMED) type HEV generates booming noises due to increased weight from the addition of customized hybrid parts and the absence of a torque converter. Some of the improvements needed to overcome this weakness include reducing the torsion-damper stiffness, adding dynamic dampers, and moving the operation point of the engine from the optimized point. The method of reducing lock-up booming noise in the HEV at low engine speed is employed as follows. Generating a torque profile via the propulsion motor, which has the antiphase profile of the driveline vibration, can, therefore, suppress driveline fluctuation. Engine vibration extracted from the hybrid starter generator (HSG) is used for the desired input signal. The generated signal, based upon the rotation angle of the driving motor, is used as a reference input for the adaptive filter. The filter updates its coefficients to minimize the phase difference between the reference signal and the desired input signal. Frequency response characteristics of the powertrain driveline components are used to generate the antiphase vibration profile of the point causing booming. Effectiveness was verified with development vehicles under various driving conditions. The results show an 8 dB(C) improvement in booming noise, and 9 and 4 dB improvements in floor panel vibration and steering wheel vibration, respectively.

TMED is comprised of a propulsion motor, hybrid starter generator, 6-speed A/T, engine clutch, combustion engine, and high-voltage main battery. The primary feature of the system is the inclusion of two motors: one for propulsion and one for generating energy. The engine power can be used for the generation or propulsion via engine clutch control. The TMED system has no torque converter, unlike the conventional ICE; and while this allows for minimal power loss, the driveline vibrations cannot be sufficiently attenuated, leading to poor NVH characteristics. Anti-jerk technology is applied to reduce free vibration jerk in the drivetrain that occurs with the tipping of the pedal either in or out, gear shifting, and engine clutch jointing. Besides, the booming noise generated by the excitation force from variations in engine combustion pressure causes the vehicle body system to resonate at particular engine speed.

Lowering the torsion damper stiffness, adding dynamic dampers on the resonating body component, and moving the engine operation point from the optimized point can reduce this booming noise, but the resulting increased cost and sacrificed fuel economy may degrade the vehicle's market value. While some additional advanced technologies, such as active noise cancelling have been developed, it needs components such as speakers, microphones, an amp, and control devices.

An active method is introduced to overcome the disadvantages of the passive methods mentioned previously. The active booming control method uses the frequency response characteristics of the powertrain driveline components to determine the anti-phase vibration of the booming and the point causing the vibration. Inserting the anti-phase torque of the booming causative torque reduces the noise very effectively, as verified using HKMC's 2016 Sonata Hybrid under various driving conditions.

Booming noise can be divided in accordance with the rotational speed of the engine as follows: low speed (under 1300 rpm), medium speed (1800–2200 rpm), and high speed booming noise (over 3000 rpm). Low-speed booming is a result of the resonance that occurs in body components, such as the trunk lid and roof panel, and is caused by engine torque fluctuation transmitted through the driveline. This is called "low-speed lock-up booming noise" in the TMED system because it only happens when the engine clutch has locked up.

Booming noise, caused by driveline vibration from the excitation force generated at the ignition stroke of the engine, has two pathways, as shown in **Figure 6.14**. One path is attenuated via the torsional damper and then travels through the motor, A/T, driveshaft, and tires, thereby causing the vehicle to vibrate.

In the case of a 4-cylinder, 4-stroke engine, the engine ignition occurs twice per one engine rotation, which means that the excitation force is two times faster than the engine speed. When analyzing the frequency of low-speed lock up booming noise, the C2 component is the primary contributor, as seen in **Figure 6.15**.

The vibration caused by the engine excitation force is transmitted through the driveline including the propulsion motor. Therefore, if we identify the vibration phase, and include antiphase vibration torque in the motor, as shown in **Figure 6.16**. We can reduce the driveline vibration downstream of the motor. To this effect, we developed a combination of a phase estimation method based on frequency response and a feed-forward antiphase torque control method.

FIGURE 6.14   Delivery path of booming in the TMED system.

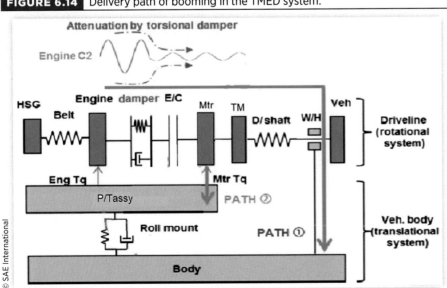

FIGURE 6.15 Frequency analysis result of booming noise in a 4-stroke, 4-cylinder engine. The RPM (left) and the magnitude of the engine speed components via FFT (right).

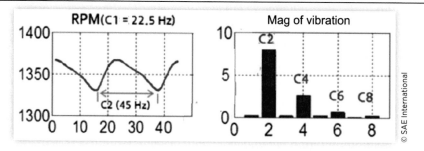

FIGURE 6.16 Main idea of antiphase torque reduction in the driveline.

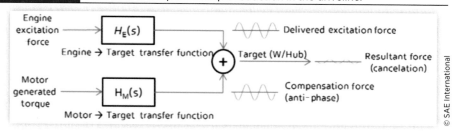

6.4 Transmission and Gear NVH

In particular in hybrid vehicles, gear whine noise from the complex transmission is one of the dominant noise sources and is perceived as highly unpleasant because of its tonal character. Gear transmissions are used in hybrid vehicles to allow a mechanical power flow between different components: ICE, electric motor/generator, and wheels. An example of such a gear system for commercial hybrid vehicles is shown in **Figure 6.17**.

The design and modeling of gear transmissions are even more challenging for hybrid vehicles compared to conventional ICE-powered vehicles because reverse and mixed flows of mechanical power occur during operation. Even with the ICE being switched off, the gear transmission is still used to transmit power from the EM to the wheels, or vice-versa for regenerative braking and will continue to generate noise because of the gear whine excitation mechanism. Since the gear shine excitation mechanism is coupled with the dynamics of the driveline, an accurate prediction of gear whine excitation can only be guaranteed by accurate multi-body modeling of the entire gear transmission, including flexibility of the gear teeth, bearing, shafts, and so on.

The Toyota Prius of the third generation was analyzed for a WOT condition on a chassis dynamometer (dyno). The microphone was positioned on the driver seat. Multiple engines, two electric engines, and an ICE with different rotation speeds constitute the hybrid powertrain and make a sound synthesis of the interior noise, engine noise, and environmental noise more complex. The independent rotational speed that creates new sound features is caused by the PSD. This complex planetary gear set can be considered

FIGURE 6.17   Driveline of a Toyota Prius [56].

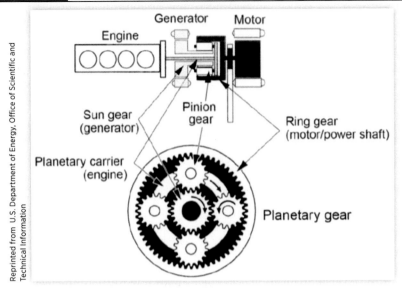

as a planetary continuous variable transmission (CVT) and connects the ICE with the two electric motors. A hybrid synergy drive (HSD) controls the drive of the electric engine and the PSD. Practically, the position of the gas pedal selects with a drive-by-wire system (HSD) the optimal gear ratio required for any desired speed [56].

Figure 6.17 demonstrates a detailed, mutual connection between the three motors. The gear wheel system comprises several elementary parts. A first gear wheel, a sun gear, is connected to the electric motor/generator MG1. A second gear wheel, the ring gear wheel, is connected to the wheels and the EM MG2. At last, a planetary carrier can be noticed with four planet gears and is connected to the ICE.

For hybrid vehicles with a CVT, an algorithm has been developed that can deal with crossing orders. Each group of order harmonics had to be tracked and resynthesized one by one to obtain a good sound approach.

The calculation of the rotational speed of the electric motors was done with the help of the gear ratios in between the various gears and the measured parameters, RPM_{ICE} and RPM_{wheel}. The teeth of the various sprockets (**Table 6.6**) inside the Toyota Prius were used to calculate the transmission ratios.

The electric engine speed RPM_{MG2} can be determined after calculating the corresponding gear ratios:

$$n_{\text{total final drive}} = n_{CG} \cdot n_{FG}$$

$$n_{MSR} = \frac{58}{22} \tag{6.6}$$

$$RPM_{MG2} = n_{\text{total final drive}} \cdot n_{MSR} \cdot RPM_{wheel}$$

A similar procedure is followed to find the rotational speed RPM_{MG1} of the other electric engine.

TABLE 6.6 Teeth of gears.

	$N_{Follower}$	N_{Driver}
Counter gear(CG)	55	54
Final gear (KG)	77	24
Teeth of power split device (PSD)		
$N_{Ring\ gear\ wheel\ (MG2)}$	78	
$N_{Planetary\ carrier\ (ICE)}$	24	
$N_{Sun\ gear(MGI)}$	30	
Teeth of motor speed reduction (MSR)		
$N_{Ring\ gear\ wheel\ (PSD\ ring)}$	58	
$N_{Planetary\ carrier\ (fixed)}$	18	
$N_{Sun\ gear\ (MG2)}$	22	

© SAE International

$$RPM_{MG1} = RPM_{ICE} + \left(RPM_{ICE} - RPM_{MG2} \cdot a_1 \cdot a_2\right)$$
$$a_1 = \frac{22}{58}, \quad a_2 = \frac{78}{30} \tag{6.7}$$

Based on the last two equations and the measured tacho signals, a noise spectrogram can be established in function of the rotational speeds RPM_{MG1}, RPM_{MG2}, and RPM_{ICE}. In this way, the order tracking algorithm can be correctly applied to the orders of the various engines.

In addition to whine, transmission system rattles also need to be reduced to attain superior performance. For example, in the development of Toyota Multi-Stage Hybrid Transmission, the technology was developed to reduce rattling noise [59].

The system has adopted a special EM mounting mechanism where the motor rotor constitutes the torque transmission path. By locating the motor, the largest inertia body, within the torque transmission path, inertia unloading is reduced during the light cruise, resulting in improved NVH characteristics. To reduce the rattling noise generated between the motor rotor and the shift device, the tolerance ring is inserted in the subject region. The tolerance ring is inserted into a groove in the shift device input shaft, located at the fitting surface between the motor rotor which enables the tolerance ring to transmit torque when the torque level is below the retention torque of the tolerance ring, whereas normal driving torque is transmitted through the spline. Hence, when the shift device input torque is near 0 Nm, the spline does not contact any torque, resulting in the elimination of spline rattling. Moreover, even when transmitting normal driving torque, the tolerance ring can act as a dampening mechanism when the direction of the torque is reversed, resulting in reduced impact force at the spline and enhanced NVH characteristics. Moreover, engine torque fluctuation is absorbed by carefully tuned O-rings between shafts, as well as the hybrid vehicle damper. The amount of backlash is optimized at all splines, while rattling of non-engaged components is reduced through the addition of retainer springs and leaf springs.

Since the ICE is not their only power source, typical HEVs use smaller ICEs with fewer cylinders compared to conventional vehicles. Such ICEs suffer from higher torque fluctuations that lower the NVH performance. Mechanically dampening torque fluctuations is one way to enhance NVH performance. However, the energy loss in such approaches can be large, and that can affect the efficiency of the powertrain. Previous studies of NVH in HEV powertrains focus mostly on vibrations which occur during the transient period upon mode shifting, clutch engaging, or disengaging. The novel powertrain architecture has been studied using electric power to suppress torque and power fluctuations without additional damping, which saves energy [59, 60, 61, 62, 63].

6.5 Motor and Power Electronics NVH

In addition to the shake and boom type of structure-borne NVH issue in HEVs, the motor whine is another critical issue in HEVs [41]. Reference 64 contains example based on the development of the Jaguar Land Rover and it discusses one way of resolving the problem. The most common electric machine currently in use for hybrid applications is a permanent magnet synchronous machine. In this machine, the rotating part includes a set of permanent magnets and the stationary part applies a rotating magnetic field using electromagnets.

The simple design has only two magnetic poles on the rotor and three pairs of coils (or teeth) on the stator. An alignment of pole and teeth occurs once every 60° or six times per revolution, so it is intrinsic to know how the motor operates so that it generates a 6th order excitation. In a real-world scenario, an alignment between poles and teeth occurs more often, since traction motors have higher numbers of poles and teeth.

In the case of a development system at Jaguar Land Rover, an alignment happens every few degrees, resulting in a high order whine that we can hear as a whistle in the 2-4 kHz region. Lower orders exist too, driven by imbalances in the magnetic fields and poles. The characteristics of electric driving sounds will be discussed later, but it is worth noting that whines in this frequency range would traditionally be associated with transmission whines, and so would be thought of as an error-state to be eradicated.

As another example, in the development of the new "THS II," compared to the previous THS, the EM had NVH problems and the countermeasures were taken against the NVH problems [10].

THS II is provided with the Motor Generator No. 1 (MG1), which mainly performs generation and engine start, and the Motor Generator No. 2 (MG2), which mainly performs drive and regeneration. The power performance of the THS II was boosted by the following improvements made to the MG1 and MG2. For the MG1, the maximum speed of the rotor revolution was raised from 6,500 rpm to 10,000 rpm to increase the power supply to the MG2 and to accommodate higher engine revolutions.

Therefore, acceleration performance was improved in the middle to high vehicle speed range. Moreover, the molding material was used for mass reduction. For the MG2, the arrangement of the permanent magnets was modified, enabling an approximate 1.5× increase in output. As a consequence of these improvements for MG1 and MG2, it was a concern that electromagnetic noise would become worse during acceleration in the middle to high vehicle speed range, as well as during regeneration.

Modifying the rotor shape of MG1 enabled simultaneous achievement of an increase in rotor speed and a decrease in torque ripple. But, at the rotor speed range from

FIGURE 6.18 (a) Rearrangement of permanent magnets in MG2 rotor; (b) Reduction of torque ripple by permanent magnet rearrangement in MG2.

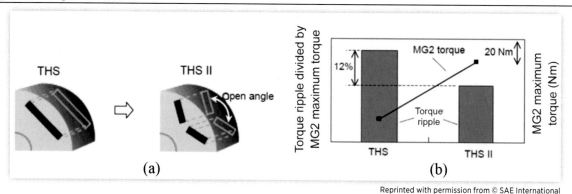

6,500 rpm to 10,000 rpm, it was a concern that MG1 electromagnetic noise becomes worse, moreover, using the molding material makes the vibration of the motor housing increase. As a result of structural FEM analysis including a magnetic field, it was confirmed that housing vibration did not get worse because of the decrease in torque ripple and the separation between the normal rotational speed of the MG1 and the resonance frequency of the rotor, stator, and transaxle housing.

For MG2, to simultaneously increase reluctance torque and to reduce the harmonic component in magnetic flux, the arrangement of the permanent magnets was optimized into a V-shape (**Figure 6.18a**).

Optimizing the open-angle of the permanent magnets in the rotor enables a reduction of approximately 12% in torque ripple for the same MG2 torque (**Figure 6.18b**).

Table 6.7 shows a summary of the principles for improving motor/generator noise as applied to the development of both the THS and the THS II (3). Motor/generator noise can be decreased by reducing the torque ripple and improving the vibration characteristics of the rotor, stator, and transaxle case. However, it is simultaneously necessary to design the system considering factors relating to motor efficiencies, such as output-power and heat radiation performance.

The example of EM NVH refinement is provided in the development of the Toyota Lexus HEV [12]. The overall length of the transmission of the new hybrid luxury sedan

TABLE 6.7 Countermeasures for motor/generator noise.

	Adopted countermeasure items	THS	THSII
Reduction of torque ripple	–Superimpose high order components to basic current	O	O
	–Optimization of rotor/stator shape	O	O
	–Optimization of permanent magnet shape and arrangement		O
Improvement of vibration characteristics	–Resonance frequency optimization of rotor/stator/transaxle case	O	O
	–Increase transaxle case stiffness	O	O
	–Separation MG1/MG2 resonance frequency from usual rotor speed		O

is greater than that of the THS II, so the resonance that deforms the entire transmission is generated at a comparatively low frequency. Moreover, since the MG2 reduction gear ratio is low, the resonance is generated by the 24th order component of MG2 speed in the low vehicle speed range where there is less background noise, which means that the motor noise is readily audible.

Moreover, the transmission is installed in the center tunnel, which makes the acoustic transfer function from the transmission to the occupants higher than in FWD hybrid vehicles, where the transmission is mounted in the engine compartment. The following section explains countermeasures that were employed.

To reduce the 24th order electromagnetic force of revolution that is generated in MG2, the permanent magnets were arranged in a V shape and the angle of the magnet arrangement was optimized.

To improve the radiated noise characteristics of the transmission, FEM analysis was used to analyze in detail the vibration modes that contribute to radiate noise. The results indicated that two vibration modes exist. One vibration mode couples the bending resonance of the transmission case with a resonance in which the MG1 and MG2 rotors serve as mass elements and the support bearings serve as spring elements. The other vibration mode couples the bending resonance of the transmission case.

To dissipate these resonant frequencies, the resonant frequency of the transmission case was changed, thereby improving the radiated noise characteristics. The radiated noise characteristics were also improved by installing a dynamic damper on the back end of the transmission, where the amplitude was high, and by adding ribs in radiating areas of the transmission case.

In the Lexus LS600h, a sound-proof cover was added on the surface of the transmission. To improve the acoustic transfer function, sound-absorbing, and insulating material was added around the center tunnel and the dash panel. These countermeasures made it possible to achieve motor noise performance suitable for a hybrid luxury sedan.

The motor noise was improved by reducing the 24th order electromagnetic force of MG2 revolution, changing the resonance of the transmission case, installing a dynamic damper to the transmission case, adding ribs in radiating areas, and improving the acoustic transfer function from the transmission to the occupants by adding sound absorbing and insulating material to the vehicle.

The overview to provide a detailed description of the noise-generating mechanisms of induction machines switched reluctance machines, and permanent magnet machines are given in Chapter 4. An example is the use of a permanent magnet synchronous machine for hybrid applications.

An acoustic camera may be used to identify which parts of the transmission give the biggest concerns. In this example, sealing actions were taken on the machine casing to reduce airborne noise, and a palliative acoustic absorber was fitted to outside of the casing. An alternative could be to improve the acoustic transfer function of the vehicle. However, careful design of the rotor, stator, driving waveforms and transmission casing are the best way of curing the problem at the source. Acoustic palliatives are tantamount to an admission of failure, and best avoided.

The electromagnetic noise stemming from the inverter is a high-frequency noise generated on the conversion of DC to AC. The paths through which this noise propagates into the interior of the car include airborne noise and structure-borne noise from the inverter case, and airborne noise from the transmission. The airborne noise was addressed by improving the vehicle's sound-absorbing performance, and the structure-borne noise was handled by improving the vibrational characteristics of the inverter case and brackets. As a result of these countermeasures, the hybrid luxury sedan achieves a level of quietness vastly superior to that of ordinary gasoline-powered vehicles in the same class

Similar to the cases in EVs, inverter noise, circuit switching, or pulse width modulation (PWM) causes high-frequency noise from 6 to 20 kHz. It is critical and annoying to the customer. The best method is to integrate it with a motor isolation unit and/or wrap the inverter with absorptive layers.

Sideband noise is an artefact of the PWM technique used to control motor speed. The PWM waveform is generated by solid-state switches in the inverter, resulting in a pulse-train whose duty cycle is defined by the instantaneous pulse width, which is designed to generate a sine-like flux density in the magnetic circuit of the motor.

Figure 6.16 shows why PWM generates sideband energy which is equally disposed about the fundamental carrier frequency (since $T_1 + T_2 = T$). Since a PWM signal consists of pulses of varying width, the distribution of sideband energy depends on the ratio of the pulse frequency to the modulation frequency. Rather than a single pair of sidebands, a real PWM signal will contain multiple harmonic sidebands.

Figure 6.19 shows how the PWM signal is formed from the combination of a triangular carrier waveform (of a fixed, high frequency) and a sinusoidal reference signal (much lower frequency) which relates to the motor speed. This sinusoid is called the electrical speed of the motor and it is a fixed ratio of the actual motor speed depending on the motor design. The Fourier analysis in **Figure 6.20** only extends to just above the fundamental frequency of the triangle waveform (the carrier frequency) and does not show the harmonics of the triangle waveform fundamentals [41].

Real-world test data measured on the surface of an inverter during a run-up clearly shows diverging sidebands (**Figure 6.21**). Sideband noise is unpleasant because of the rising and falling orders. If the falling order is most obvious, it is distinctly counter-intuitive when the motor is accelerating. It is never a desirable noise and is best avoided by setting the inverter in a position which has low noise sensitivity, and possibly via small vibration isolators.

FIGURE 6.19 Formation of sidebands due to varying pulse widths.

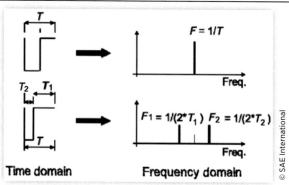

© SAE International

6.6 **Ancillary Noise**

Similar to the accessories of NVH in EVs, HEVs have many accessory NVH issues. HEV has water pump noise because it may have several water pumps for different cooling purposes, and they are electrically driven by a motor. Mounting strategy and location are critical for NVH reduction. HEV has fan noise as it may either use the vehicle cooling fan or dedicated cooling fan for HEV components. It must be masked by other noise sources and noise levels need to be controlled at low speed and stop conditions. HEV has an HVAC compressor noise because it has an A/C system that generates pulsate waves and can create both structure-borne and airborne noises. It can be either mounted directly to the motor or mounted separately. It is important to optimize the mounting structure and location to achieve a level similar to a conventional vehicle.

For example, Lexus GS450h and LS600h vehicles feature a newly developed version of the THS II for longitudinal power trains [12]. These vehicles have class-leading power performance while achieving NV and environmental performance superior to conventional gasoline vehicles.

To maintain this superiority, an active effort was made to eliminate the noise of the water pump in the inverter cooling system, as well as the electromagnetic noise of the motor, the inverter, and other units.

Figure 6.20 shows an example of the noise improvement of the water pump in the inverter cooling system. During development, the noise caused by the rotation of the pump's impeller and the motor became a problem when the engine was stopped, and the background noise was low. The compelling force was reduced by reducing imbalance in the impeller, changing the bearing structure, and changing motor structure. To reduce the input force to the body, the rubber isolation on the pump was installed, and the rigidity of the pump brackets attached to the body was optimized. These countermeasures reduced the noise level significantly.

Another example is the Toyota Prius HEV, for which the specific component operational limits and key HEV operations include idle (ICE off), low speed with the only EM operation, and initial ICE boost when ICE begins to turn on to give additional assists; normal operation with ICE dominant mode; high acceleration with ICE and assist from

FIGURE 6.20 Reduction of water pump noise when the vehicle is stopped.

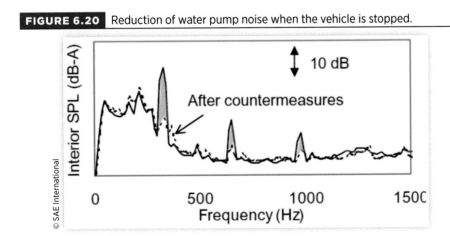

FIGURE 6.21 Noise and vibration phenomena of the Toyota Hybrid System in different driving conditions.

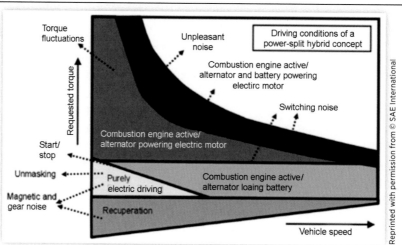

the electric motor; high torque/power requirements with ICE and assist from the electric motor; and regenerative braking with separate generator operates to recapture electricity to store in a battery.

At first, the EM operates as a function of wheel speed, hence lying on the same axis as the vehicle speed. Then, the ICE begins turning at 17 mph (EM is rotating at 1000 rpm). The ICE will cycle on and off as needed up to 42 mph, entering *normal operation* (ICE dominant mode) where the ICE will always turn, and providing power to drive the vehicle due to the limits of the EM being reached. The ICE will continue to be the dominant APU under normal conditions up to 65 mph, beyond this the EM can provide additional power required for higher speed/torque demands.

Figure 6.21 gives an overview of the NVH phenomena occurring in a power-split hybrid system in different operating modes [57]. The transitions between the different operating modes are generally critical since transient phenomena are to be expected. For instance, during the transition from a purely electric mode to a mixed-mode with the combustion engine turned on, the starting process should not produce unpleasant noise and vibration phenomena. In both operating ranges, with the combustion engine active either in combination with an alternator or with an electrical motor, intelligent regulation procedures are necessary to avoid sudden torque changes and torque fluctuations. The relevant NVH aspects of hybrid vehicles fall into three categories: dominant noises due to the absence of masking effects, unexpected acoustic behavior, and specific acoustic phenomena. Some of these phenomena are presented in more detail here.

For example, **Figure 6.22** shows the interior noise spectra of the ventilation system at different fan settings.

In driving conditions where the combustion engine is turned off, acoustic phenomena become prominent, which in conventional engines are masked by the combustion engine noise. Typical examples for this are the pump noise of the electric water pump and the vacuum pump, the ventilator, and the rolling noise, as well as ambient noise.

FIGURE 6.22   Interior noise spectra of the ventilation system at different fan settings.

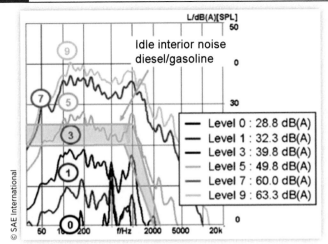

Figure 6.22 depicts the interior noise spectrum of a ventilation system in different fan settings. The gray area indicates the interior noise of typical gasoline and diesel engines at idle. It is obvious that the fan noise up to ventilation level 3 is masked by the engine noise. At idle, the fan noise becomes more prominent due to the low engine noise. This has been a steady trend for several years. With hybrid vehicles, a low blower noise becomes even more important since the combustion engine is often turned off.

Dealing with low masking noise is a major new problem in the new generation of quiet vehicles. We need to consider the magnitude of this problem and a suitable method for dealing with it, before describing how component targets may be set for ancillaries around the vehicle. We will also look at some problems arising in hybrid and electric vehicle interior noise.

Considering first a typical idle condition with an ICE running, there is a significant level of noise present in the car. This is illustrated in **Figure 6.25** with the graph showing a smoothed 1/3 octave masking spectrum. Any noises under this line (cooling pumps, vac pumps, air suspension compressors, motors, etc.) are inaudible [41].

If we turn the engine off, then effectively our "masking noise" reduces to the audibility threshold. For much of the spectrum, this would require our various ancillaries to be 30 dB quieter, or even more. This is simply infeasible in the vast majority of cases and would add considerable cost and engineering effort to the vehicle design. We should bear in mind that these ancillaries would have to be engineered to be 30 dB quieter on the donor base vehicle, not simply on the relatively low volumes of the hybrid variant.

A pragmatic solution is to consider that most customers will have an HVAC system running in the vehicle. If we take this as a guideline, then we can set a more realistic set of goals for interior noise. This will allow us to develop a hybrid or electric vehicle which has "silent" ancillaries for the vast majority of used cases. This would be the toughest condition for most ancillaries here, a stationary vehicle with no wind and road noise and no engine running. However, most customer experience relates to driving the vehicle, and for this, we need to use a more complex process for defining what the customer hears.

6.7 Interior and External Noise and NVH Characterization

HEVs acoustic phenomena are complicated. Compared to a conventional powertrain, a hybrid powertrain features additional components such as electric engines, electronic control units, and a high-voltage battery. This results in different new interactions between these components which are not found in this form in conventional engines. Additional components and the resulting interactions can lead to acoustic problems with a negative effect on comfort. In the following, NVH phenomena due to hybrid-specific components and their interactions are listed as follows: low-frequency vibrations of the powertrain during start/stop of the combustion engine at load change, modified moments of inertia and eigenfrequencies in the powertrain, "streetcar noise" (magnetic noise of the engine/generator during electric driving and regenerative braking), aerodynamic noises of the battery cooling system, and switching noise of the power control unit.

One of the essential aspects to be considered in the development of hybrid powertrains is the frequent starting and stopping of the combustion engine. With a start/stop system, the engine is shut down automatically at a standstill, such as at a traffic signal or in traffic congestion, and it gets started again when the driver wants to move. This can reduce fuel consumption in urban traffic considerably. If the hybrid powertrain is built in such a way—which is the case for instance for the power-split hybrid—that combustion engine speed is decoupled from vehicle speed, the combustion engine can be turned off also in driving conditions with little load demand.

Since the start/stop event happens frequently, its influence on the noise and vibration behavior should be thoroughly investigated and analyzed. The main tasks regarding the improvement of hybrid vehicle NVH behavior are the optimization of the vibrations in the vehicle during start and shut off of the engine and the prevention of disturbing noises due to the start/stop system. Typical problems are the toothing noise due to the use of a pinion starter and the gas forces of the piston engine, which leads to severe shaking of the engine by excitation of the roll eigenfrequency.

With a power-split hybrid, a frequent change between electric and combustion engine propulsion can occur during constant-speed driving at low speeds. It has been shown the example where the driver hardly notices the change between electric and combustion engine propulsion. It shows a period from constant driving in which the combustion engine is operated during three short time intervals. Each time the 3rd engine order (6-cylinder) appears on the level diagram.

However, the level of this order is 15 dB(A) below the total level and thus it is only perceived as slight background noise. The reason for not providing an acoustic response to the driver in this situation is that the operation of the combustion engine depends on the charge state of the battery and not of the driver's load demand. If during constant driving, suddenly and without any recognizable reason, the combustion engine started up and dominated the vehicle interior noise, it would only serve to confuse the driver.

Aside from the above-described acoustic effects due to the operation of the combustion engine, the starting process can also lead to disturbing vibrations. Control methods can be applied to reduce unwanted vibrations. The interior noise needs to be minimized since the driver does not expect a whining noise during electric start or braking.

Figure 6.2 shows a two-dimensional landscape of typical customer vehicle maneuvers mapped against range vehicle operational modes. Overlaid on this map are NVH issues, such as those associated with global powertrain vibration, driveline vibration, HEV component-specific noise, motor/generator whine, accessory noise, gear rattle, and noise pattern changeover. From the point-of-view of customer acceptance of the end product, the plot shown in Figure 6.1 should be customized to a given vehicle program, so that a focused plan can be developed to address each NVH item systematically.

The expectation for interior noise content from an ICE (i.e., powertrain presence) depends highly on the vehicle class and target demographic. While luxury cars target low interior noise content, performance vehicles demand some level of powertrain noise feedback (with an emphasis on the development of the desired "brand character"). Conversely, the tonal noise from electric machines (motor/generator) is universally considered annoying; hence, any perceived motor whine (in-vehicle interior) is considered unacceptable in a refined vehicle. Therefore, although interior noise targets can be defined based on EV whine orders from competitive data sets, it is perhaps more appropriate to develop targets based on the ability to perceive whine noise in a given vehicle application. Once the interior noise order targets are established, the following steps can be taken to derive powertrain level targets: overall interior noise contribution must be split into defined airborne and structure-borne interior noise share targets of the main excitation sources; each of the main noise shares identified above must be broken down to their sub-components; once targets are defined for each noise contribution, frequency-based vehicle sensitivity information can be used to cascade these interior noise share targets to powertrain level noise and vibration targets.

For the EM applications, the excitation sources are generally limited to powertrain airborne and structure-borne noise inputs. Airborne sources defined as powertrain radiated noise, broken down into multiple locations around the powertrain to account for variations in radiated noise directivity and vehicle sensitivity (acoustic attenuation between each source location and vehicle interior). The dominant structure-borne noise sources are defined as active side powertrain mount vibrations, broken down into vibration content in each direction of each mount.

The NVH refinement of road/tire noise, wind noise, and the powertrain-induced NVH, bending/torsional are similar to the treatments in preceding chapters for ICE and EVs. Powertrain sound quality can be described in terms of both powerful and refined factors [3, 48, 63, 65, 66].

The most straightforward method to address a noise and vibration concern is to break it into separate parts using the source-path-receiver model shown in **Figure 6.23**. What the customer complains about, what we set targets for, and what we initially measure to represent the noise or vibration concern is the receiver position. This measurement is the most appropriate for representing the voice of the customer but is generally not enough to provide good engineering insight to all of the design opportunities available to address the issue.

To understand the options available to resolve the issue, the sources and paths should be identified and then quantified. Breaking the problem into these categories leads to

FIGURE 6.23 Source-path-receiver model in HEVs.

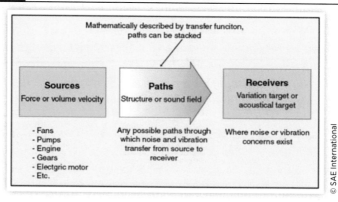

the development of a test work plan and allows for a better understanding of the issue. For example, a vehicle boom issue was measured in the vehicle interior and was determined to be related to second-order engine firing. This was done by measuring engine speed during the objectionable noise condition and determining that the frequency of the noise concern was equal to twice the rpm of the engine. The first decision in the source-path-receiver model is what to bound as the source and whether to treat it as a black box or to break it down even further. This decision depends on your engineering responsibility in the program.

For a vehicle-level engineer with responsibility for full-vehicle NV performance, then the engine can be considered a black box source that produces second-order pulsations that are input to the exhaust. It may be that the energy from the source needs to be reduced, but someone else is responsible for that, and you are just considering it a black box. On the other hand, for the engine manufacturer, the engine itself needs to be broken into a source-path-receiver model to reduce the amount of energy it produces. The source would be the engine combustion, the paths would be the engine design and structure, and the receiver would be the engine second-order pulsations output to the exhaust. An example of the source-path-receiver model for a vehicle-level engineer working on an engine-related boom issue is shown in Figure 6.23.

The NVH simulators have been widely developed to enable driving sounds to be engineered in the lab. The system usually combines a powerful signal processing and manipulating tool with an interactive display that simulates various driving environments. Road and wind noise components can be extracted and analyzed using advanced algorithms and in this way, we can determine the "threshold of audibility" for additional noise signatures. It is, therefore, possible to fully simulate the interior noise environment of a hybrid or electric vehicle with speed-dependent road and wind noise, with optional engine sounds, and with optional resonances, whines, booms, and other error states. Once we have constructed a target sound and achieved an appropriate level of sign-off we then need to cascade that sound to component targets.

Given a defined level of interior noise (that is a set of spectra dependent on road speed), we then apply these sounds to a set of transfer functions which then generates a set of targets which may be applied to the component. A series of transfer functions is required to enable us to specify sound pressure levels and excitation force levels for various zones around the vehicle.

This process may be used for the hybrid transmission and driveline components enabling rig-testing of noise levels using various development levels of software and hardware. Pressure variation during engine combustion generates torque fluctuation that is delivered through the driveline. Torque fluctuation delivered to the tire shakes the vehicle body and causes the body components to vibrate, resulting in a booming noise. HKMC's TMED type generates booming noises due to increased weight from the addition of customized hybrid parts and the absence of a torque converter. Some of the improvements needed to overcome this weakness include reducing the torsion-damper stiffness, adding dynamic dampers, and moving the operation point of the engine from the optimized point. These modifications have some potentially negative impacts, such as increased cost and sacrificed fuel economy. Herein, we introduce a method of reducing lock-up booming noise in an HEV at low engine speed. Generating a torque profile via the propulsion motor, which has the antiphase profile of the driveline vibration can, therefore, suppress driveline fluctuation.

Similar to EVs, HEVs need to consider the warning sound design, as HV/HEVs tend to be quieter than ICE vehicles [58]. Electric vehicle warning sounds are a series of sounds designed to alert pedestrians to the presence of electric drive vehicles, such as EVs/HEVs travelling at low speeds.

The National Highway Traffic Safety Administration in the USA performed some measurements in which the noise from hybrid vehicles was compared to the noise from some similar ICE cars [67]. In the reference, these are called "twin" cars. The two couples of "twin" cars were a Toyota Matrix (ICE) and a Toyota Prius (hybrid) along with an ICE Toyota Highlander and a hybrid Toyota Highlander. The Matrix and the Prius are midsize passenger cars and the Highlanders are midsize SUVs. The results presented here only include the hybrid cars driven in electric mode. The main aim of the study was to investigate whether or not artificial sound needed to be added to electric or hybrid vehicles and what these sounds could be. The measurements were taken as pass-by measurements 3.7 m from the center of the track and 1.5 m above the ground. The speeds studied were 6 mph (9.7 km/h), 10 mph (16.1 km/h), 20 mph (32.2 km/h), 30 mph (48.3 km/h), and 40 mph (64.4 km/h). The results can be seen in **Figure 6.24**, which show that the Prius is more silent than its "twin," the Matrix, at 6 mph. From Figure 6.24, it can be said that at 30 mph and above, the noise from the hybrid cars and the ICE cars converge.

Under the law Pedestrian Safety Enhancement Act (PSEA), National Highway Traffic Safety Administration (NHTSA) issued a performance standard for EVs and hybrid vehicles (HVs) to ensure that they emit a sound that meets certain minimum requirements to aid visually-impaired and other pedestrians in detecting vehicle presence, direction, location, and operation. EVs and HVs pose a greater potential risk to pedestrians while operating under electric propulsion at slow speeds when tire and wind noise are less dominant. The new performance requirement that NHTSA mandates must enable a pedestrian to reasonably detect a nearby EV or HV operating at a constant speed, accelerating, decelerating, and operating in any other scenarios that NHTSA deems appropriate. Under the PSEA, the added sound must also

FIGURE 6.24 Noise from a Toyota Prius (hybrid car is driven in electric mode) and its "twin" Toyota Matrix (ICE car) measured as pass-by 3.7 m from the center of the track [68].

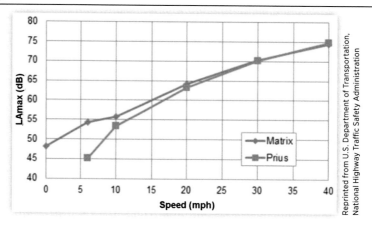

Reprinted from U.S. Department of Transportation, National Highway Traffic Safety Administration

be "recognizable" as that of a "motor vehicle" in operation. The agency's proposed rule is projected to reduce the number of incidents in which EVs and HVs strike pedestrians [1, 69, 70].

To address the quietness issue of HEVs, Toyota has explored the vehicle proximity notification sound designed for its Prius vehicles. It is explained that Toyota's notification sound activates only at speeds below about 15 mph, uses a combination of high- and low-pitched tones to make it easily audible over city sounds, and rises and falls in pitch with the vehicle speed to give pedestrians a sense of whether the approaching Prius is accelerating or decelerating. The tone is generated by externally mounted speakers, so the driver potentially would not be able to hear it with the windows secured.

References

1. Tousignant, T., Govindswamy, K., Stickler, M., and Lee, M., "Vehicle NVH Evaluations and NVH Target Cascading Considerations for Hybrid Electric Vehicles," SAE Technical Paper 2015-01-2362, 2015, https://doi.org/10.4271/2015-01-2362.

2. Goetchius, G., "Leading the Charge - The Future of Electric Vehicle Noise Control," *Sound & Vibration*, 45, no. 4: 5-11, Apr. 2011.

3. Brandl, S., Biermayer, W., Graf, B., and Resch, T., "Hybrid Vehicle's NVH Challenges and Influences on the NVH Development," SAE Technical Paper 2016-01-1837, 2016, doi:https://doi.org/10.4271/2016-01-1837.

4. Gover, J., "A Tutorial on Hybrid Electric Vehicles: EV, HEV, PHEV and FCEV," UofM, 2015.

5. Franco-Jorge, M., "Sound Quality in Hybrid Vehicles," *European Conference on Vehicle Noise and Vibration 2002 Whole Vehicle*, London, Jun. 11, 2002, 167 (10p, 7 fig, 3 ref), Doc No. 198876.

6. Masao, I., "Automotive Noise and Vibration Technology in near Future," *JSAE Review* 25: 43-49, 2004.

7. Sturesson, P.-O., Svensson, C., Weckner, J., Karlsson, R. et al., "N&V Integration and Optimization of Driveline Using Transfer Path Analysis," SAE Technical Paper 2012-01-1531, 2012, https://doi.org/10.4271/2012-01-1531.

8. Volinski, B., "NVH Testing of Hybrid Electric Powertrains," SAE Technical Paper 2001-01-1543, 2001, https://doi.org/10.4271/2001-01-1543.

9. Yoshioka, T. and Sugita, H., "Noise and Vibration Reduction Technology in Hybrid Vehicle Development," SAE Technical Paper 2001-01-1415, 2001, https://doi.org/10.4271/2001-01-1415.

10. Komada, M. and Yoshioka, T., "Noise and Vibration Reduction Technology in New Generation Hybrid Vehicle Development," SAE Technical Paper 2005-01-2294, 2005, https://doi.org/10.4271/2005-01-2294.

11. Allam, E., Ahmed, I., Hammad, N., and Abouel-Seoud, S., "Noise Characteristics for Hybrid Electric Vehicle Induction Motor," SAE Technical Paper 2007-01-2261, 2007, https://doi.org/10.4271/2007-01-2261.

12. Kawabata, N., Komada, M., and Yoshioka, T., "Noise and Vibration Reduction Technology in the Development of Hybrid Luxury Sedan with Series/Parallel Hybrid System," SAE Technical Paper 2007-01-2232, 2007, https://doi.org/10.4271/2007-01-2232.

13. Kuang, M.L., "An Investigation of Engine Start-Stop NVH in A Power Split Powertrain Hybrid Electric Vehicle," SAE Technical Paper 2006-01-1500, 2006, https://doi.org/10.4271/2006-01-1500.

14. Tomura, S., Ito, Y., Kamichi, K. and Yamanaka, A., "Development of Vibration Reduction Motor Control for Series-Parallel Hybrid System," SAE Technical Paper 2006-01-1125, 2006, https://doi.org/10.4271/2006-01-1125.

15. Beuschel, M., Rau, M., and Schröder, D., "Adaptive Damping of Torque Pulsation Using a Starter Generator Opportunities and Boundaries," *Industry Applications Conference* 3, no. 8: 1403-1408, Oct. 2000.

16. Davis, R.I. and Lorenz, R.D., "Engine Torque Ripple Cancellation with an Integrated Starter Alternator in a Hybrid Electric Vehicle: Implementation and Control," *IEEE Transactions on Industry Applications* 39, no. 6: 1765-1774, Nov./Dec. 2003.

17. Shulz, M., "Low-Frequency Torsional Vibrations of a Power Split Hybrid Electric Vehicle Drive Train," *Journal of Vibration and Control* 11: 749-780, 2005.

18. Takeuchi, T., Choi, K.G., Kann, S., and Togai, K., "Sensibility Analysis of Vibration Transfer Path and control of Input Force for Reduction of Acceleration and Deceleration Shock," Mitsubishi Technical Papers No.15, 2003.

19. Alt, N.W., Wiehagen, N., and Schlitzer, M.W., "Interior Noise Simulation for Improved Vehicle Sound," SAE Technical Paper 2001-01-1539, 2001, https://doi.org/10.4271/2001-01-1539.

20. Alt, N., Wiehagen, N., and Schlitzer, M.W., "Vehicle Interior Noise Simulation for Evaluating Prototype Powertrains in the Vehicle (Pt 1 & 2)," *ATZ* 103: 5-6, 2001.

21. Albers, A., Schille, F., and Behrendt, M., "Method for Measuring and Analyzing Transient Powertrain Vibrations of Hybrid Electric Vehicles on an Acoustic Roller Test Bench," SAE Technical Paper 2016-01-1835, 2016, doi:https://doi.org/10.4271/2016-01-1835.

22. Conico, B., Barth, M, Hua, C , Lyons, C. et al., "Development of Hybrid-Electric Propulsion System for 2016 Chevrolet Malibu," *SAF Int. J. Alt. Power.* 5, no. 2: 259-271, 2016. doi:https://doi.org/10.4271/2016-01-1169.

23. Govindswamy, K., Wellmann, T., and Eisele, G., "Aspects of NVH Integration in Hybrid Vehicles," *SAE Int. J. Passeng. Cars - Mech. Syst.* 2, no. 1: 1396-1405, 2009. doi:https://doi.org/10.4271/2009-01-2085.

24. Arvanitis, A., Orzechowski, J., Tousignant, T., and Govindswamy, K., "Automobile Powertrain Sound Quality Development Using a Design for Six Sigma (DFSS) Approach," *SAE Int. J. Passeng. Cars - Mech. Syst.* 8, no. 3: 1110-1119, 2015. doi:https://doi.org/10.4271/2015-01-2336.

25. Albers, A., Bopp, M., and Behrendt, M., "Efficient Cause and Effect Analysis for NVH Phenomena of Electric Vehicles on an Acoustic Roller Test Bench," SAE Technical Paper 2018-01-1554, 2018, doi:https://doi.org/10.4271/2018-01-1554.

26. Affi, S., "NVH Study of Stop & Start System and Optimized Solutions for Hybrid Vehicles," SAE Technical Paper 2014-01-2068, 2014, doi:https://doi.org/10.4271/2014-01-2068.

27. Li, L. and Singh, R., "Start-Up Transient Vibration Analysis of a Vehicle Powertrain System Equipped with a Nonlinear Clutch Damper," *SAE Int. J. Passeng. Cars - Mech. Syst.* 8, no. 2: 726-732, 2015. doi:https://doi.org/10.4271/2015-01-2179.

28. Sugimura, H., Takeda, M., Takei, M., Yamaoka, H. et al., "Development of HEV Engine Start-Shock Prediction Technique Combining Motor Generator System Control and Multi-Body Dynamics (MBD) Models," *SAE Int. J. Passeng. Cars - Mech. Syst.* 6, no. 2: 1363-1370, 2013. doi:https://doi.org/10.4271/2013-01-2007.

29. Wellmann, T., Govindswamy, K., and Tomazic, D., "Integration of Engine Start/Stop Systems with Emphasis on NVH and Launch Behavior," *SAE Int. J. Engines* 6, no. 2: 1368-1378, 2013. doi:https://doi.org/10.4271/2013-01-1899.

30. Chen, J.S. and Hwang, H.Y., "Engine Automatic Start-Stop Dynamic Analysis and Vibration Reduction for a Two-Mode Hybrid Vehicle," *Proc IMechE, Part D: J Automobile Engineering* 227: 1303-1312, 2013.

31. Guo, R., Mi, Y., and Cao, C., "Subjective and Objective Evaluation of APU Start-Stop NVH for a Range-Extended Electric Vehicle," SAE Technical Paper 2015-01-0047, 2015, doi:https://doi.org/10.4271/2015-01-0047.

32. Rust, A. and Graf, B., "NVH of Electric Vehicles with Range Extender," *SAE Int. J. Passeng. Cars - Mech. Syst.* 3, no. 1: 860-867, 2010. doi:https://doi.org/10.4271/2010-01-1404.

33. Bang, J., Yoon, H., and Won, K., "Experiment and Simulation to Improve Key ON/OFF Vehicle Vibration Quality," SAE Technical Paper 2007-01-2363, 2007, doi:https://doi.org/10.4271/2007-01-2363.

34. Ng, H., Anderson, J., Duoba, M., and Larsen, R., "Engine Start Characteristics of Two Hybrid Electric Vehicles (HEVs) - Honda Insight and Toyota Prius," SAE Technical Paper 2001-01-2492, 2001, doi:https://doi.org/10.4271/2001-01-2492.

35. Henein, N., Taraza, D., Chalhoub, N., Lai, M. et al., "Exploration of the Contribution of the Start/Stop Transients in HEV Operation and Emissions," SAE Technical Paper 2000-01-3086, 2000, doi:https://doi.org/10.4271/2000-01-3086.

36. Guo, R., Cao, C., and Mi, Y., "Experimental Research on Powertrain NVH of Range-extended Electric Vehicle," SAE Technical Paper 2015-01-0043, 2015, doi:https://doi.org/10.4271/2015-01-0043.

37. D'Anna, T., Govindswamy, K., Wolter, F., and Janssen, P., "Aspects of Shift Quality With Emphasis on Powertrain Integration and Vehicle Sensitivity", SAE Technical Paper 2005-01-2303, 2005, https://doi.org/10.4271/2005-01-2303.

38. Horste, K., "Objective Measurement of Automatic Transmission Shift Feel Using Vibration Dose Value", SAE Technical Paper 951373, 1995, https://doi.org/10.4271/951373.

39. Eisele, G. Wolff, K. Alt, N., and Hüser, M., "Application of Vehicle Interior Noise Simulation (VINS) for NVH Analysis of a Passenger Car", SAE Technical Paper 2005-01-2514, 2005, https://doi.org/10.4271/2005-01-2514.

40. Yoshida, J., Ueno, T., and Nitta, S., "Influence of Starting Duration on the Uneasi-Ness to Engine Automatic Start Sound of HEV," *The 23rd International Congress on Sound and Vibration*, Athens, Greece, 2016.

41. Holton, T., Bullock, L., and Gillibrand, A., "New NVH Challenges within Hybrid and Electric Vehicle Technologies," *5th CTi Conference* Auburn Hills, May 18, 2011.

42. Kanai, H., Ueda, K., and Yamaguchi, K., "Reduction of the Engine Starting Vibration for the Parallel Hybrid System," *JSAE Spring Convention Proceedings* 983 1998-5, 9833467.

43. Yoshioka, T. and Sugita, H., "Noise and Vibration Technology in Hybrid Vehicle Development," SAE Technical Paper 2001-01-1415, 2001, https://doi.org/10.4271/2001-01-1415.

44. Ito, Y., "The Vibration Control Using the Traction Motor in Hybrid Electric Vehicle," *The Japan Society of Mechanical Engineers D&D Proceedings* 2004.

45. Tomura, S. and Ito, Y., "Vibration Reduction Control at the Time of Engine Starting in Hybrid Vehicles," *JSAE Convention Proceedings*, 122-04 20045752.

46. La, C.,Poggi, M., Murphy, P., and Zitko, O., "NVH Considerations for Zero Emissions Vehicle Driveline Design," SAE Technical Paper 2011-01-1545, 2011, https://doi.org/10.4271/2011-01-1545.

47. Tomura, S., Ito, Y., Kamichi, K., and Yamanaka, A., "Development of Vibration Reduction Motor Control for Series-Parallel Hybrid System," SAE Technical Paper 2006-01-1125, 2006, https://doi.org/10.4271/2006-01-1125.

48. Zhao, T., Liu, X., Cao, Y., Li, C. et al., "Noise Control during Idle Charging for Hybrid Vehicles," SAE Technical Paper 2016-01-1322, 2016, doi:https://doi.org/10.4271/2016-01-1322.

49. Yoshioka, T., "Noise and Vibration Reduction Technology in Hybrid Vehicle Development," SAE Technical Paper 2001-01-1415, 2001, https://doi.org/10.4271/2001-01-1415.

50. Allam, E. and Ahmed, I., "Noise Characteristics for Hybrid Electric Vehicle Induction Motor," SAE Technical Paper 2007-01-2261, 2007, https://doi.org/10.4271/2007-01-2261.

51. Kawabata, N., "Noise and Vibration Reduction Technology in the Development of Hybrid Luxury Sedan with Series/Parallel Hybrid," SAE Technical Paper 2007-01-2232, 2007, https://doi.org/10.4271/2007-01-2232.

52. Kokaji, J., Komada, M., Takei, M., and Takeda, M., "Mechanism of Low Frequency Idling Vibration in Rear-Wheel Drive Hybrid Vehicle Equipped with THS II," *SAE Int. J. Passeng. Cars - Mech. Syst.* 8, no. 3: 910-915, 2015. https://doi.org/10.4271/2015-01-2255.

53. Lennström, D., Johnsson, R., Agren, A., and Nykänen, A., "The Influence of the Acoustic Transfer Functions on the Estimated Interior Noise from an Electric Rear Axle Drive," *SAE Int. J. Passeng. Cars - Mech. Syst.* 7, no. 1: 413-422, 2014. https://doi.org/10.4271/2014-01-9124.

54. Senousy, M., Larsen, P., and Ding, P., "Electromagnetics, Structural Harmonics and Acoustics Coupled Simulation on the Stator of an Electric Motor," *SAE Int. J. Passeng. Cars - Mech. Syst.* 7, no. 2: 822-828, 2014. https://doi.org/10.4271/2014-01-0933.

55. Otokawa, K., Hayasaki, K., Abe, T., and Gunji, K., "Performance Evolution of a One-motor Two-Clutch Parallel Full Hybrid System" *SAE Int. J. Engines* 7, no. 3: 1555-1562, 2014. https://doi.org/10.4271/2014-01-1797.

56. Sarrazin, M., Janssens, K., and Van der Auweraer, H., "Virtual Car Sound Synthesis Technique for Brand Sound Design of Hybrid and Electric Vehicles," SAE Technical Paper 2012-36-0614, 2012, https://doi.org/10.4271/2012-36-0614

57. Eisele, G., Wolff, K., Wittler, M., Abtahi, R. et al., "Acoustics of Hybrid Vehicles," SAE Technical Paper 2010-01-1402, 2010, https://doi.org/10.4271/2010-01-1402.

58. Kang, H., Chung, T., Lee, H., and Ihm, H., "Active Booming Noise Control for Hybrid Vehicles," *SAE Int. J. Passeng. Cars - Mech. Syst.* 9, no. 1: 167-173, 2016, https://doi.org/10.4271/2016-01-1122.

59. Tateno, H., Yasuda, Y., Adachi, M., Suzuki, H. et al., "Rattling Noise Reduction Technology for Multi Stage Hybrid Transmission," SAE Technical Paper 2017-01-1157, 2017, doi:https://doi.org/10.4271/2017-01-1157.

60. Walker, P.D. and Zhang, N., "Active Damping of Transient Vibration in Dual Clutch Transmission Equipped Powertrains: A Comparison of Conventional and Hybrid Electric Vehicles," *Mech Mach Theory* 77: 1-12, 2014, http://dx.doi.org/10.1016/.

61. Hwang, H.S., Yang, D.H., Choi, H.K., Kim, H.S. et al., "Torque Control of Engine Clutch to Improve the Driving Quality of Hybrid Electric Vehicles," *Int J Automot Technol* 12, no. 5: 763-768, 2011.

62. Kim, H., Kim, J., and Lee, H., "Mode Transition Control Using Disturbance Compensation for a Parallel Hybrid Electric Vehicle," *IProceedings of the Institution of Mechanical Engineers Part D Journal of Automobile Engineering,* 225: 150-166, 2010.

63. Yi, C., Epureanu, B.I., Hong, S.-K., Ge, T. et al., "Modeling, Control, and Performance of a Novel Architecture of Hybrid Electric Powertrain System," *Applied Energy* 178: 454-467, 2016.

64. Yu, B., Fu, Z., and Juang, T., "Analytical Study on Electric Motor Whine Radiated from Hybrid Vehicle Transmission," SAE Technical Paper 2017-01-1055, 2017, doi:https://doi.org/10.4271/2017-01-1055.

65. Joslin, A., Henderson, M., Suffield, I., and Kerber, S., "Active Noise Cancellation System to Tackle Charge Sustain Idle Noise in a PHEV Vehicle," SAE Technical Paper 2018-01-1562, 2018, doi:https://doi.org/10.4271/2018-01-1562.

66. De Hesselle, E., Grozde, M., Adamski, R., Rolewicz, T. et al., "Hybrid Powertrain Operation Optimization Considering Cross Attribute Performance Metrics," SAE Technical Paper 2017-01-1145, 2017, doi:https://doi.org/10.4271/2017-01-1145.

67. Garay-Vega, L., Hastings, A., Pollard, J.K., Zuschlag, M., and Stearns, M.D., "Quieter Cars and the Safety of Blind Pedestrians: Phase I, Report DOT HS 811 304," National Highway Traffic Safety Administration, U.S. Department of Transportation, April 2010.

68. U.S. Department of Transportation National Highway Traffic Safety Administration, "Minimum Sound Requirements for Hybrid and Electric Vehicles," 459 CFR Part 571, Docket No. NHTSA-2011-0148, 2011.

69. U.S. Department of Transportation National Highway Traffic Safety Association, "Minimum Sound Requirements for Hybrid and Electric Vehicles," NHTSA 1-13, Jan. 2013, 5.

70. Khan, R., Ali, M., and Frank, E., "Analysis of Vehicle Voice Recognition Performance in Response to Background Noise and Gender Based Frequency," SAE Technical Paper 2017-01-1888, 2017, doi:https://doi.org/10.4271/2017-01-1888.

NVH of FCEV

7.1 NVH Development of FCEV

A fuel cell electric vehicle (FCEV) is a type of electric vehicle (EV) that uses a fuel cell instead of a battery to power its on-board electric motor. Fuel cells in vehicles generate electricity to power the motor, generally using oxygen from the air and compressed hydrogen [1–4].

The key systems of a typical FCEV are presented as follows. An auxiliary battery provides electricity to start the car before the traction battery is engaged and also powers the vehicle accessories. A battery pack stores energy generated from regenerative braking and provides power to the electric traction motor. A direct current (DC)/DC converter converts higher-voltage DC power from the traction battery pack to the lower-voltage DC power needed to run vehicle accessories and recharge the auxiliary battery. An electric traction motor uses power from the fuel cell and the traction battery pack to drive the vehicle's wheels. Some vehicles use motor generators that perform both the drive and regeneration functions. A fuel cell stack is an assembly of individual membrane electrodes that use hydrogen and oxygen to produce electricity. The fuel filler is a filler or "nozzle" used to add fuel to the tank. A fuel tank (hydrogen) stores hydrogen gas on board the vehicle until it is needed by the fuel cell. A power electronics controller manages the flow of electrical energy delivered by the fuel cell and the traction battery, controlling the speed of the electric traction motor, and the torque it produces. The thermal system (cooling) maintains a proper operating temperature range of the fuel cell, electric motor, power electronics, and other components. The transmission transfers mechanical power from the electric traction motor to drive the wheels.

The first commercially produced hydrogen fuel cell automobile, the Toyota Mirai, was introduced in 2015, after which Hyundai and Honda entered the market. The Mirai is based on the Toyota fuel cell vehicle (FCV) concept car. The FCV concept also uses portions of Toyota's Hybrid Synergy Drive (HSD) technology for the hybrid car drivetrain technology used in vehicles. As a refinement of the original Toyota Hybrid System (THS), HSD includes the electric motor, power control unit, and other parts and components from its hybrid vehicles to improve reliability and minimize cost. The hybrid technology is also used to work with the fuel cell. At low speeds, such as city driving, the FCV runs just like any all-electric car by using the energy stored in its battery, which is charged through regenerative braking. At higher speeds, the hydrogen fuel cell alone powers the electric motor. When more power is needed, for example during sudden acceleration, the battery supports the fuel cell system as both work together to provide propulsion.

The Mirai, which is the production version of the FCEV Concept, is built around a 114-kW fuel-cell stack that converts compressed hydrogen gas into electricity. That electricity is stored in a nickel-metal hydride battery pack that powers electric motors, which drive the front wheels.

The car's proton-exchange membrane fuel cell stack, which is encased in a protective carbon fiber composite housing, has a power density of 3.1 kW/L. Unlike many previous units, it requires no humidifier to keep the polymer membranes damp to promote proton transport. Instead, the system circulates the water that the stack produces. The compressor and blower whirls and whines that have accompanied most earlier fuel-cell systems have been mostly eliminated to help keep the passenger cabin quiet. The Mirai contains a pair of 10,000-psi hydrogen storage tanks: one under the rear seats and another just behind them. The tanks, which have 62.4-L (16.5-gal) and 60-L (15.9-gal) capacities, respectively, feature outer protective shells composed of filament-wound fiber/resin composites that are lined with impermeable membranes. At 1,850 kg (4,100 lb), the car is still 100-150 kg (220-330 lb) lighter than a Camry hybrid. Engineers placed all the fuel-cell system components low in the frame to create a low center of gravity for good handling performance. The underbody of the car is completely covered to cut road noise and aerodynamic drag because the fuel cell's heat output is less than that of a traditional car engine. The huge air intake grilles are there not only to help cool the fuel-cell system's three radiators but also to feed the vehicle's air-management system.

The vehicles powered solely by fuel cells have some disadvantages, such as a heavy and bulky power unit caused by the low power density of the fuel cell system, long start-up time, and slow power response. Furthermore, in propulsion applications, the extremely large power output in sharp acceleration and the extremely low power output in low-speed driving lead to low efficiency. Hybridization of the fuel cell system with a peaking power source (PPS) is an effective technology to overcome the disadvantages of the vehicles powered only by the fuel cell. The fuel cell hybrid electric vehicle (HEV) is totally different from conventional internal combustion engine (ICE)-powered vehicles and ICE-based hybrid drive trains. The fuel cell-powered hybrid drivetrain is usually constructed as shown in **Figure 7.1** [2].

It mainly consists of a fuel cell system as the primary power source, PPS, electric motor drive (motor and its controller), vehicle controller, and an electronic interface

FIGURE 7.1 Structures of a fuel cell hybrid electric vehicle drive train (ESS indicates another energy storage system, either battery or ultracapacitor).

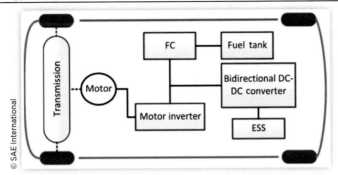

© SAE International

between the fuel cell system and the PPS. According to the power or torque command received from the accelerator or the brake pedal and other operating signals, the vehicle controller controls the motor power or torque output and the energy flows between the fuel cell system, PPS, and the drive train. For peak power demand, for instance, in a sharp acceleration, both the fuel cell system and the PPS supply propulsion power to the electric motor drive. In braking, the electric motor, working as a generator, converts part of the braking energy into electric energy and stores it in the PPS. The PPS can also restore its energy from the fuel cell system when the load power is less than the rated power of the fuel cell system. Thus, with proper design and control strategy, the PPS will never need to be charged from outside the vehicle [3–10].

FCVs have a series hybrid configuration. They are often fitted with a battery to deliver peak acceleration power and to reduce the size and power constraints on the fuel cell. A fuel cell hybrid electric always has a series configuration, with the engine-generator combination replaced by a fuel cell.

7.2 **NVH of Blowers/Compressor/ Pump in FCEV**

In practice, fuel cells need auxiliaries to support their operation [11–13]. The auxiliaries, such as fan, blower, compressor or pump, are used for air circulating, coolant circulating, ventilation, fuel supply, and electrical control, as shown in the example in **Figure 7.2a**. For a fuel cell system, two types of air feeding systems, the blower-type and the compressor-type, have been widely used. The available machines for auxiliaries are highly varied. **Figure 7.2b** shows the types of compressors [11].

A fan moves large amounts of gas with a low increase in pressure. A blower is a machine used for moving gas with a moderate increase of pressure like a more powerful fan. A compressor is a machine for raising gas to a higher level of pressure actually making the air denser by cramming air into a small space like a more powerful blower. The compressors are like the fan and blower in terms of noise mechanisms. Similar to the compressor, a pump is a machine for raising liquid (or gas) to a higher level of pressure.

FIGURE 7.2 (a) Hydrogen-air fuel cell system; (b) Types of compressors.

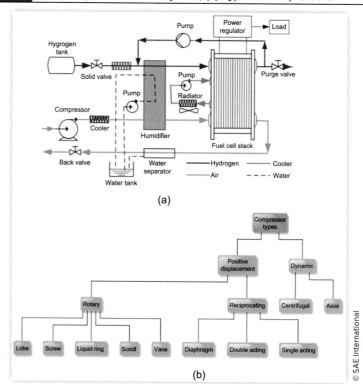

(a)

(b)

© SAE International

The fan/blower/compressor/pumps are currently used in FCEVs, such as fans and blowers for oxygen and air supply for fuel cells with reactive air; fans and blowers for transporting heat generated by the fuel cell; fans and pumps for cooling the fuel cells; fans and blowers for transporting steam; fans, blowers, and pumps for conveying reactive gas, hydrogen, or natural gas; blowers for reformer burners; blowers and compressors for hydrogen recirculation; and blowers with a sensor for mass flow measurement.

7.3 Noise and Vibrations of Blower/Compressors (Dynamic/Turbo Type)

The centrifugal and axial/regenerative flow blowers or compressors are widely used in fuel cells. Centrifugal blowers are configured in a way that its inlet and outlet are perpendicular. The inlet feeds air into the center of the impeller, while the outlet stays tangential to the rotation of the impeller, as shown in the image below. On the other hand, regenerative blowers have parallel inlets and outlets that are positioned perpendicular to the rotation of the impeller. Centrifugal blowers allow air to enter at the center of its rotating impeller where a number of fixed vanes act as paddles that push volumes of air to the outlet. Through centrifugal action, the air is forced to the impeller and housing where it is discharged as steady steam through the outlet. This creates a negative pressure at the center hub that sucks in more air. Unlike centrifugal blowers, regenerative blowers are fashioned differently. It consists of an impeller that spins within a housing compartment, which contains both an inboard channel and an outboard channel; thus, an alternate name is given as side channel blowers. The moment the impeller spins past the intake import, the air is drawn in and is trapped between its impeller blades. As it continues to spin, the air is pushed both inward and outward through both of the channels. This continues until the impeller stops rotating. As a result, regenerative blowers can act either as a pressure blower or a vacuum blower. As a general rule, centrifugal blowers are considered low pressure, high flow blowers, while regenerative blowers are high pressure, low flow blowers [14–18].

7.3.1 Centrifugal Type

Many blowers and compressors are centrifugal fan type, such as the compressor part in a turbocharger [19–32]. The centrifugal compressors have been used to supply air for the fuel-cell stack in the high-pressure fuel-cell system. A centrifugal compressor is generally composed of an impeller, a diffuser, and a volute. The aerodynamic noise from this kind of rotating machine can be classified by characteristics of noise, monopole, dipole, and quadruple as shown in **Figure 7.3**. In most cases of a rotating machine, the

FIGURE 7.3 The aerodynamic noise from a rotating machine.

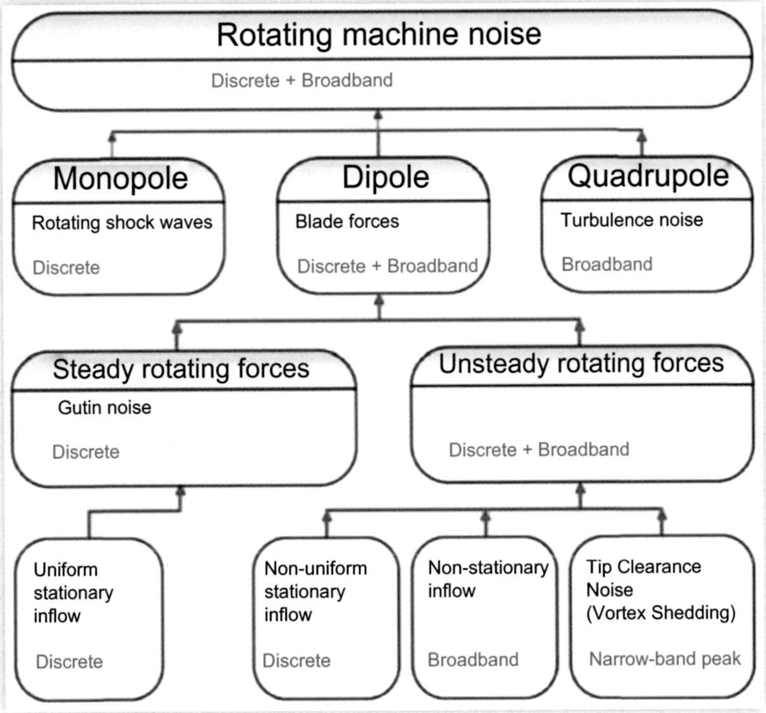

© SAE International

dipole is the dominant noise [19]. The dominant discrete tones at the blade passing frequency (BPF) Fp (Hz) are given as:

$$Fp = RPM \times Nb/60 \tag{7.1}$$

where RPM is rotor rotation speed and Nb is the number of blades. Since an aerodynamic noise wave is periodic, there exist multiple component harmonics, 2Fp, 3Fp, etc., in addition to the dominant Fp.

For the far-field sound pressure of a dipole, the absolute magnitude of the nth harmonic (based on rotational frequency) is given by

$$D_n = \frac{n\varsigma xT}{2\pi ccr^2} J_n\left(\frac{nMy}{r}\right) \tag{7.2}$$

where Ω is the rotational angular velocity, T is the dipole strength equal to the total thrust exerted by a single blade, M is the rotational Mach number $\Omega R/c$, R is the radius of the circle, and x and y are components of the distance r from the center of rotation to the observer (x is the axial component and y is the component in a direction perpendicular to the axis lying in the plane defined by the axis and the observer). For a quadrupole of strength Q, which equals the integrated mean Reynolds stress, the equivalent harmonic content is

$$Q_n = \frac{n^2\varsigma^2 x^2 Q}{2\pi c^2 r^3} J_n\left(\frac{nMy}{r}\right) \tag{7.3}$$

Mechanical noise consists of the influences from many aspects, such as bearings, defects, impact/contacts, friction, bent shaft, rotor unbalance, shaft misalignment, gears, etc. Basically, the mechanical causes of vibrations include unbalanced rotating components, damaged impellers, and non-concentric shaft sleeves, bent or warped shafts, pump and driver misalignment, pipe strain (either by design or as a result of thermal growth), inadequacy of foundations or poorly designed foundations, thermal growth of various components, especially shafts, rubbing parts, worn or loose bearings, loose parts, loose anchoring bolts, damaged parts, and others.

For example, Ha [32] studied a centrifugal compressor for FCEVs. It is a 10-kW class centrifugal compressor with an oil-free bearing system. It consists of a shaft, two airfoil journal bearings, and a pair of thrust bearings. The fuel cell system is composed of a fuel cell stack, an air processing system (APS), a fuel processing system, and a thermal management system. The rest of the system, except for the stack, is called Balance of Plant (BOP). It is a complex system that has many components connected to each other. APS has a particularly significant effect on noise, vibration, and harshness (NVH) because a compressor in the APS is one of the key components of the fuel cell system. It produces compressed air and sends it to the cathode channel of the stack. It consumes the most parasitic power of the system and it is the only rotating part with a high revolution speed in the FCEV.

Several types of compressors can be considered for the fuel cell system, such as centrifugal compressors, and screw or roots-type superchargers. A motorized centrifugal compressor has an advantage of high-pressure ratio despite its small size compared to other types of compressors. The main reason for the advantage is from the centrifugal effect due to the difference of radius between the inlet and the outlet of the impeller. The studied system has new air foil bearings (AFBs) to increase its rotational speed. A centrifugal compressor consists of aerodynamic parts, driving parts, and housing. The aerodynamic parts of the present compressor in this study have an impeller, a vaneless type diffuser, and a volute. The principle part among them is the centrifugal impeller. It can give a substantial pressure rise in a single stage due to the centrifugal effect. This effect is from its high-speed rotation and the radius difference between inlet and outlet. The impeller is driven directly by the driving parts composed of a shaft, a motor unit, and an AFB system in the current compressor.

The AFB usually is sensitive to external vibrations because it has a small load capacity and a relatively weak damping effect compared to other types of bearing. But cars are always exposed to external forces and vibrations. There are many things that can lead to damaging vibration or shock on the road, such as speed bumps, rumble strips, obstacles, pits, and holes. Therefore, the durability of auto parts from vibration is very important. This is also true for the FCEV and its parts. So, base excitation tests of the present compressor were carried out to study characteristics of the AFB on the external vibrations. The static results were verified and extended for dynamic load analysis of the AFBs system and to verify the characteristics of the compressor according to the external vibration forces.

A number of tests were carried out to investigate performance, the durability, vibration, and thermal characteristics of the compressor with airfoil bearings in a similar environment to actual driving conditions. A series of continuous drive tests, start-up-shutdown tests, and base excitation tests were performed. The applicability of AFBs and its resistance to the harsh environment of vehicles can be verified through these tests.

To reduce the overall noise level of the fan/blower package, there are several options available, but a basic understanding of the noise being generated should be discussed first [33–40]. For example, the overall blower package noise is composed of aerodynamic and mechanical noise sources.

The conventional mechanical treatments are presented as follows. If flex connectors are used, they should be lagged. When using a quiet line motor, the gear end of the blower should be treated to deaden it. All equipment should be mechanically isolated from the base. A shroud, lined with sound-absorbing foam should be placed around any rotating components not already treated. The intake/exhaust/silencer design should be modified. An enclosure should be put over the blower and motor. These treatments are rated by the level of the noise normally emanating from each of the sources given. Depending on the difference between the blower package noise level and the level required, the number of steps required to meet the specifications will vary.

7.3.2 Axial Type and Regenerative Type

The widely used axial flow fan/blowers produce noise due to the pressure fluctuation of blades [39–43]. The airborne fan/blower noise is strongly related to the aerodynamic flow field and the performance of fan/blower, so the noise control and reduction of the fan/blower must be attempted with the consideration of the interaction between aerodynamic and acoustic characteristics.

Lim et al. [44] describes the characteristics of the unsteady flow field and aeroacoustic noise of the high-pressure regenerative blower. Three-dimensional Navier-Stokes equations were introduced to analyze the internal flow of the blower and to simulate the pressure fluctuation on the impeller and the casing. Ffowcs Williams and Hawkings equation with dipole assumption was used to calculate the aeroacoustic sound generated from the regenerative blower, which includes tonal and broadband. It is noted that the dominant aeroacoustic source of the regenerative blower is located at the impeller tip and the casing outlet regions [44–51].

Noise generation in a regeneration blower is caused by the rotational speed of the impeller, which creates an unsteady pressure fluctuation on the blade and casing surfaces. It is important to find the positions of aero-acoustic noise sources to reduce the noise level of the blower. The aero-acoustic noise generated from a regenerative blower can be analyzed using the unsteady flow simulation.

Consider the two kinds of noise components. One is the discrete frequency noise at BPF and another is the broadband noise distributed over wide frequency range, which is produced due to inflow turbulence at blower inlet, turbulence within impeller and side channel and turbulent jet at blower exit.

BPF noise is produced mainly due to rotating steady fan blade thrust and blade interaction. The BPF noise component for rotating steady fan blade thrust has been analyzed by Gutin's theory where fan blades are assumed as compact moving sources.

Discrete frequency noise is produced due to the pressure difference between adjacent impeller blades rotating at BPF. In the following formula (7.4), the pressure difference is defined by $\Delta p_s / Ze$ or $2\Delta p_s / Ze$ and its pressure fluctuation can be modelled as rectangular-shaped, where Δp_s is overall blower pressure rise, and Ze, d, and r represent an effective number of impeller blades, impeller thickness, and impeller tip diameter, respectively, and θ is an angular coordinate in the direction of tangential flow. The fluid pressure rise is achieved through 1 regeneration/1 pitch of fluid at low flow capacity ($\phi < \phi_{lim}$) while it is achieved through the 1 regeneration/ 2 pitches at high flow capacity ($\phi > \phi_{lim}$). Here, ϕ is the flow coefficient as nondimensional flow capacity parameter, and ϕ_{lim} is assumed as 0.75. Under the assumption of dipole type noise radiation, classical acoustic theory [52] on the rectangular-shaped pressure fluctuation gives the following root mean square value of acoustic pressure as:

$$p_a(R,\theta) = 2\sqrt{2}\left(\frac{\Delta p_s r}{mZ\theta_c}\right) sin\left(\frac{mZd}{2r}\right)\frac{\cos(mZ\theta)}{R} \quad at\ \varphi < \varphi_{lim}$$

$$p_a(R,\theta) = 4\sqrt{2}\left(\frac{\Delta p_s r}{mZ\theta_c}\right) sin\left(\frac{mZd}{2r}\right)\frac{\cos(mZ\theta)}{R} \quad at\ \varphi > \varphi_{lim} \quad (7.4)$$

where $m = 1$ means fundamental mode, $m = 2, 3, \ldots$ mean its harmonic modes, and θc, θ, and R are side-channel extension angle, noise measuring angle, and radius.

Broadband noise is produced from three main noise sources of inflow turbulence, impeller turbulence, and exhaust turbulent jet. The present study employs a well-verified correlation model corresponding to each noise source [51, 53–55], and their noise prediction results are superimposed over the frequency range. It is noted that all the present broadband noise models are expressed in terms of blower design variables and performance parameters. Broadband noise is produced over the entire frequency range due to the turbulent boundary layer on the blade surface, inflow turbulence, and blade wake. The acoustic power spectral density function (PSDw) is expressed as

$$PSD_w(f) = \frac{dW(f)}{df} = \frac{\pi}{4}\frac{B}{\rho_0 a_0^2}\frac{f}{\left[1 - \left(\frac{r_t}{r_h}\right)^2\right]^2}\left(PSD_{F,IFT} + PSD_{F,TBL}\right) + B \times PSD_{TE} \qquad (7.5)$$

where r_t and r_h represent tip and hub radii of fan/blower, subscripts IFT, TBL, TE mean inflow turbulence, turbulent boundary layer, and trailing edge wake vortex. PSD_F spectral density function of aerodynamic force fluctuation on the blade surface.

In addition, a_0 represents the speed of sound. B is the number of fan/blower blade.

Regenerative compressors have the main characteristic of the highly three-dimensional development of the flow, which is different from axial or centrifugal compressors. However, some combined approach for axial and centrifugal compressor design can be used as an approximation. Particularly, some axial compressor approaches are used to approximate the noise features of a regenerative compressor.

Regenerative pumps and compressors are tangential flow turbomachines characterized by low specific speed, high head, and low flow rates. Such characteristics, along with the advantage of not being subject to stall or surge instability, makes them attractive for air supply devices in low-pressure fuel-cell systems [53–62].

7.4 Noise and Vibrations of Pumps/Compressors of Positive Displacement Type

The hydrogen circulation pump or compressors of positive displacement (PD) type have been used to circulate a portion of the hydrogen that did not undergo a chemical reaction, feeding this back into the fuel cell.

The reciprocating type is the original type of air compressor, in which a piston reciprocates in a cylinder to compress air. This operates at a low speed, with significant noise and vibration, however, it is at a low cost. The other types, such as screw type, scroll type, and roots types have been used in FCEVs.

A rotary-screw compressor/screw pump is a type of gas compressor that uses a rotary-type positive-displacement mechanism with two or three screws with opposing threads (e.g., one screw turns clockwise and the other counterclockwise). The screws are mounted on parallel shafts that have gears that mesh so the shafts turn together and everything stays in place. The screws turn on the shafts and drive fluid through the compressor/pump.

The gas compression process of a rotary screw is a continuous sweeping motion, so there is very little pulsation or surging of flow, as occurs with piston compressors [63].

A pump/PD compressor converts shaft energy into the velocity and pressure of a gas media. In a broader sense, it includes gases and gas mixtures by trapping a fixed amount of gas into a cavity, then compressing that cavity and discharging into the outlet pipe. A PD compressor can be further classified according to the mechanism used to move the gas, such as screw, scroll, and roots, as well as a reciprocating type.

Though each type of PD compressor has its unique shape, movement, principle, strengths, and weakness, they all have in common a suction port, a volume changing cavity, and a discharge port where a valve mechanism controls the timing of the release of gas media. Moreover, they are all cyclic in nature and possess the same cycle for the processed gas, that is, suction, compression, and discharge. Gas pulsations are a major source of system vibration, noise, and fatigue failures, taking place at the discharge side of PD type compressor, such as a screw or scroll type.

Traditionally, a serial pulsation dampener, often a reactive type silencer, is connected after the compressor, which is capable of reducing pressure pulsation. Alternative methods have also been used, such as a shunt pulsation trap using a parallel configuration, which tackles the gas pulsations before the compressor or engine discharge.

7.4.1 Noise and Vibration of Screw Compressor

Compressors are usually very noisy machines with high pressure. There are several types of compressors used in FCEVs. There are two types of noise sources: aerodynamic and mechanical.

The main source of mechanical noise is intermittent contact between the compressor rotors, especially in an oil-injected compressor. This is the result of variation in the torque transferred from the male to the female rotor. New and more efficient rotor profiles introduce a very small negative torque to the female rotor compared to that of the male rotor. This torque, generated by the pressure-induced forces acting on the female rotor, is of the same order of magnitude as that created by other means, such as contact friction and oil drag forces. Since the negative female rotor torque acts in the opposite direction to the other two, the net torque may change in sign from negative to positive within one lobe rotation cycle. This may cause instability in the female rotor rotational motion, resulting in flutter and, in the extreme case, rattling. Stošić proposed a new type of profile, called "silent" which maintains positive torque on the female rotor and prevents this kind of noise from occurring. It was suggested that another reason for noise generation is a transmission error. Transmission error occurs in the driven component of a screw pair when its instantaneous angular position differs from the theoretical angular position. This causes, earlier or later than expected, contact between rotor lobes, which generates noise. It was suggested the reasons for the existence of the transmission error might be lead mismatch, lead non-linearity, pitch errors, housing bore imperfection, bearing deflections, and the rotor deflection due to gas forces. Holmes suggests relieving the rotor's profiles to enable smoother contact between the lobes and reduce noise. The proposed noise reduction procedures resulted in a reported overall noise attenuation of 4 to 6 dBA. Theoretically, there should be no contact between the rotors in an oil-free compressor. In that case, the source of mechanical noise is the contact of the synchronizing gears. Due to the number of teeth in the gears being much greater than the number of rotor lobes, the produced noise is of a higher frequency. Apart from synchronizing gears, driving gears are present in a screw compressor. Very often, a step-up gearbox is an essential part of the machine and the gears contained in it produce noise too. The main causes of noise, described by Holmes for rotors, may be applied to noise generated by the gears [63–66].

FIGURE 7.4 (a) Screw rotor profile; (b) Motion of rotors; (c) Rotor torque.

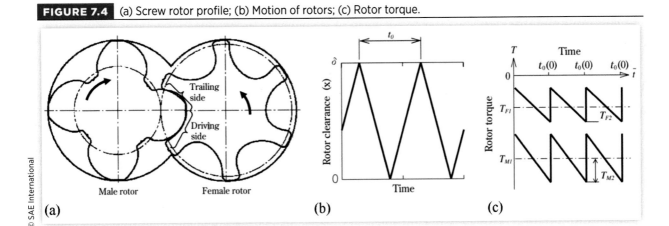

(a) (b) (c)

© SAE International

Figure 7.4a shows the rotor profiles of a pair of engaging screw rotors. The arrows indicate the respective directions of rotation. The male rotor drives the female rotor and the rotors contact each other on their driving sides. For rotors in abnormal vibration, however, the screw rotors make contact on both their driving sides and trailing sides, resulting in the motion depicted in **Figure 7.4b**. In the figure, the vertical axis (x) is the actual rotor clearance, while the horizontal axis represents time. The screw rotors have a designed clearance, δ, and $x = 0$ indicates a contact on the driving side, while $x = \delta$ indicates a contact on the trailing side. In other words, the vibration is caused by periodic impacts on the driving side and trailing side, in which alternating collisions occur at a frequency of t_0. The following summarizes the analysis of rotor behavior that periodically causes impact vibration, as shown in Figure 7.4b. The rotor motion equation of fluctuating torque acts on the screw rotors, as shown in Figure 7.4c, in which the horizontal axis represents time, t. The time $t = 0$ is a time point at which the torque changes discontinuously, and t_0 represents the variation period of the torque. The time $t = t_0$ (t_0 (0) in the figure) indicates the origin $t = 0$ of the next variation period of the torque. The torques acting on the female rotor and the male rotor are given

$$\text{TF} = \text{TF1} + \text{TF2}\left(1 - 2t/t_0\right)\left(0 \le t/t_0\right) \tag{7.6}$$

The basic noise source is the aerodynamic type, which is caused by trapping a definite volume of fluid and carrying it around the case to the outlet with a higher pressure. The pressure pulses from compressors are quite severe, and equivalent sound pressure levels can be high. The noise generated from compressors is periodic with discrete tones and harmonics present in the noise spectrum. The first-order excitation frequency can be determined by

$$F = N * \text{rpm}/60 \tag{7.7}$$

where N is the male rotor number of the screw compressor and rpm is the rotating speed.

Excessive vibrations induced by the gas pulsation or structural resonance may lead to premature failure of the system, gas leakage, and other safety problems. Gas pulsation is one of the important excitation sources of the vibrations in the compressor system. Reducing the gas pulsation within the allowable range is an effective method to control the vibration in the compressor system.

FIGURE 7.5 Suction/discharge flow and pressure pulsation in compressors.

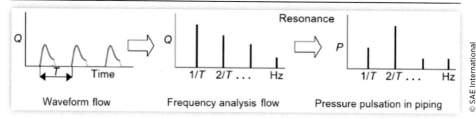

Screw compressors are sources of pulsation [67–79]. This can be explained by considering the mechanisms of the compression. The pressure of the gas is increased by the transfer of mechanical energy from the rotation of two helical rotors (female and male lobes) to the gas. At the suction side, a pocket of gas is inhaled and enclosed within a cavity between the lobes. By the design of the rotor geometry, the volume of the cavity is reduced while the gas pocket travels toward the discharge side, thus increasing the pressure. Upon opening of the pocket, the compressed gas is exhausted into the discharge piping. This occurs on an intermittent basis and thus leads to a pulsating flow. In many screw-compressor designs, four pockets per revolution (four male lobes) are transported to the discharge side. In that case, the pulsation source spectrum is dominated by the integer multiples of the pocket passing frequency (PPF). For example, at the 4th harmonic of the compressor speed (1*PPF), the 8th harmonic (2*PPF), the 12th harmonic (3*PPF), etc.

Displacement-type compressors cause large flow fluctuations because of the intermittent suction/discharge flow, as shown in **Figure 7.5**. The resulting pressure pulsations contain many harmonic components of the rotational speed.

Gas pulsations are the main source of noise generated by fluid flow in screw compressors. These are created by unsteady fluid flow through the suction and the discharge ports, which change the pressure within the suction and discharge chambers. The flow rate depends mainly on the pressure difference between the chambers and starts with the exposure and finishes with the cut-off of the suction or discharge port, as the rotors revolve past them. The amplitudes of the gas pulsations in the compressor suction and discharge chambers are very high [80, 81]. A typical frequency spectrum of pressure pulsations in the suction and discharge chambers of the test screw compressor used in this investigation is shown in **Figure 7.6**.

FIGURE 7.6 Sound pressure spectrum in suction and discharge chambers [80].

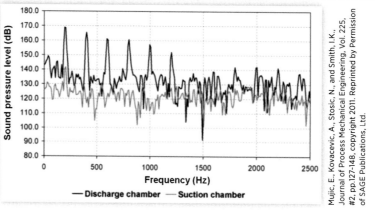

As can be seen, the gas pulsations are higher in the discharge chamber than that in the suction chamber. However, while the discharge port is completely enclosed in the housing, the suction port may be more exposed to its surroundings, which are separated from the atmosphere only by the suction filter. Therefore, despite the smaller pulsations, noise generated in the suction chamber requires similar attention to that generated in the discharge chamber.

Basic analysis has shown that the two most influential parameters affecting gas pulsations in a screw compressor discharge chamber are the pressure difference between the compressor working and discharge chambers and the discharge port area. The parameters which affect pressure difference comprise the outlet pressure, speed, built-in volume ratio, compressor clearances, sealing line length, and the number of lobes. These are determined in advance to obtain the best compressor performance. Although they influence the gas pulsations, they can hardly be varied to reduce the compressor noise, because even small variations have a large influence on the compressor performance. However, their influence on noise should not be neglected during screw compressor design. The second group of parameters like the size, shape, and position of the discharge port influence the gas pulsations and consequently compressor performance. By their variation, noise reduction can be achieved with some sacrifice of the compressor performance. However, it should be noted that optimizing these parameters can also improve compressor performance when a compressor operates in the higher pressure range.

7.4.2 Vibration and Noise in Roots Compressor

Like screw compressors, roots compressors have two types of noise sources: aerodynamic and mechanical. Since the roots compressor also uses the meshing of rotors with gear type parts, the aerodynamic noise mechanism of the roots compressor is similar to that of screw compressors.

Figure 7.7 shows a tested noise spectrum of a traditional root blower. As indicated by the peaks, the tonal whine noise is typical of roots compressors, inherent to the roots design, results from the pulses per revolution. The air induction system structure and acoustic resonators should be well designed to minimize whine.

Based on conventional helical and spur gear design theory, helical-type rotors have been adopted to replace spur-type rotors to attain quietness. The comparison of noise frequencies between helical blowers and general spur-type blowers is presented in [82]. In the low-frequency range (50-160 Hz), the noise volume is higher for spur-type rotors than for helical-type rotors. This is because the noise in the low-frequency range is not absorbed by the glass wool and leaks outside the machine chamber. The low-noise design of helical blowers significantly reduces the amount of noise output in the low-frequency band, which is the main cause of the noise. This technological ability provides a large difference in the amount of noise in the low-frequency range [82–84].

Similar to the screw blower, there are many mechanical sources having negative effects on NVH: foundation bolts being loose, which may cause vibration in the vertical direction; the couplings not cooperating well with each other; the improper support design; the connection pipes with roots blower aren't fixed well (if pipes

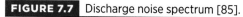

FIGURE 7.7 Discharge noise spectrum [85].

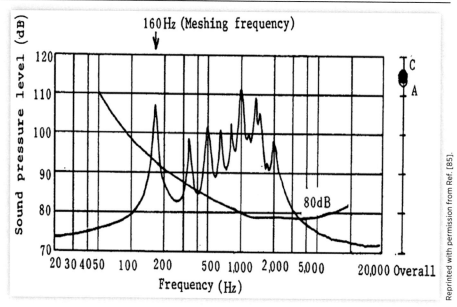

vibrate, it may cause the roots blower to vibrate); poor quality of roots blower gears; damage of roots blower bearings; impeller of roots blower lack dynamic balance; negative side affected by the flexible joints. Among these, the poor gear meshing has been a significant issue. A whistling noise is generated by tooth contact of the gears between the rotors.

In roots blower pumps, analysis of experimental data reveals that the source of the noise and vibration problem is the backlash nonlinearity due to gear teeth losing and re-establishing contact [86]. The non-smooth ordinary differential equation models were developed for the dynamics of the pump. The models include a time-dependent forcing term that arises from the imperfect, eccentric mounting of the gears. It is found that noisy solutions can coexist with silent ones, explaining why geared systems can rattle intermittently. We then consider several possible design solutions and show their implications for pump design in terms of the existence and stability of silent and noisy solutions.

A small amount of play between gears is essential to ensure that they will not jam. This means that there is always a gap between the trailing face of one tooth and the leading face of the next tooth, which is known as the backlash width. Because the gear wheels can consequently lose contact, there is a range of relative rotational displacements for which there is no restoring torque between the gears: this effect is known as free play. A tiny amount of eccentricity in the mounting of the gears introduces a forcing effect, which causes the gear teeth to repeatedly lose and re-establish contact. The key design challenge is, therefore, to change the machine design so that it is less susceptible to noisy operation driven by eccentricity.

7.5 **Noise Characteristics and NVH Refinement of FCEV**

In terms of vibration and noise, the FCEVs are very similar to EVs, but there are also some FCEV-specific phenomena, such as the noise generated by the compressor and fluid system needed for the chemical reaction between oxygen and hydrogen, and the noise associated with the airflow from the air intake and discharge.

Generating electricity requires drawing air into the fuel cell stack, and possible counter-measures to the unique noise issue associated with this intake and discharge of air are needed. Table 7.1 shows the specific NVH phenomena and frequency range in FCEV compared with similar EVs/ HEVs/internal combustion engine vehicles (ICEVs) [85].

An example is the NVH refinement development focusing on screw-type compressor in [87, 88]. As the electrochemical reaction in the fuel cell stack in principle does not generate any noise, the main NVH sources in a fuel cell-powered automobile are the auxiliary equipment of the vehicle and the air and hydrogen supply subsystems for the fuel cell.

Core components of the air supply subsystem are an air compressor (or an air pump) and an electric motor to drive it. As in many automotive fuel cell applications, NuCellSys has employed automotive proven screw-type compressor technology for its next-generation air supply subsystem.

The screw compressor technology provides several advantages, such as high compression ratios and minor back pressure dependency. However, it is very intense tonal noise and vibration profile requires extensive effort to significantly reduce the noise level emitted from the air supply subsystem and meet today's customer expectations to FCVs.

The next generation air supply system contains as core components the screw compressor and the corresponding electric auxiliary motor to drive it. Based on the power and torque spread in the compressor timing gear, it is advantageous to join the

TABLE 7.1 Specific NVH Phenomena and Frequency Range in FCEV.

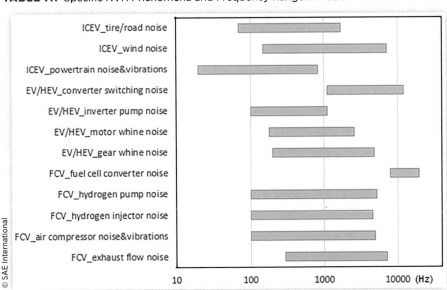

drive shaft of the motor via a flexible coupling with the male rotor of the compressor. Based on this, the first-order frequency of the next generation air supply system is defined as the speed of the auxiliary drive, which is equivalent to the speed of the coupling and the male compressor rotor.

The main noise and vibration sources are briefed as follows.

7.5.1 Imbalance of Rotating Parts

As the speed range of the auxiliary drive is 1,200 RPM (idle) to 19,800 RPM (peak load), the first-order frequency can vary in the range of 20-330 Hz as a function of the motor speed. Any vibrations with the frequency of the first-order can be traced back to imbalances of either the rotor of the auxiliary drive, the coupling, or the male compressor rotor.

The screw compressor contains two rotors that revolve at different speeds. The rotors are synchronized by the compressor timing gear. In the case of the next generation air supply system, the female rotor rotates with 3/5th of the speed of the male rotor, so any vibrations with the frequency of the 0.6th order can be traced back to an imbalance of the female compressor rotor. The 0.6th order is varying from 12-198 Hz in the given speed range.

7.5.2 Pressure Level and Peaks of the Compression Process

Among different types of air pumps, a machine having an internal compression is defined as a compressor. Screw compressors apply the internal compression by engagement of their rotor lobes [1].

The screw compressor employed in the next generation air supply system has a lobe ration of 3:5, which means that the male rotor consists of three lobes, whereas the female rotor consists of five lobes. Both rotors together create three compression chambers per revolution of the male rotor.

Each rotor lobe creates a dilatory pressure pulsation when passing the suction or discharge port of the compressor. The pulsation on the suction side is mainly caused by the high speed of the lobes passing the edges of the suction port and sealing off the compression chambers. On the discharge side, an additional effect of expansion has to be considered when the compressed air of a lobe chamber is released via the control edges of the discharge port into the discharge manifold.

In the case of the next generation air supply system, the male rotor creates three pressure peaks per revolution, whereas the female rotor creates five pulsations per revolution. Based on the inverse speed ratio between both rotors, this results in one common order, which is the third-order of the auxiliary drive speed. In the given speed range, the third-order is varying from 60-990 Hz.

7.5.3 Timing Gear Mesh Vibration

The screw compressor applied in the next generation air supply system has a self-lubricated timing gear. As the male rotor is driven, the timing gear is reducing the speed for the female rotor with a ratio of 30:50. The related teeth meshing sound is of the 30th order and varies in the range of 600-9,900 Hz as a function of the motor speed.

7.5.4 Electromagnetically Generated Noise and Vibration

The fundamental electromechanical concept of the auxiliary drive can have a significant influence on the emitted noise of the air supply system. Influenced by the number of pole pair discontinuities of the motor torque over the tor revolution have to be taken into consideration. Torque ripples can be transmitted via the coupling to the compressor, which due to relative movement of the timing gears in the magnitude of the backlash can generate a chattering noise. As the applied auxiliary drive topology proposes a six-pole rotor and a 12-tooth stator, a noise emission of the 12th order (240-3,960 Hz) and its multiples have to be taken into account.

The pole pairs of the rotor generate electromagnetic forces in the radial direction, which can result in elastic deformation of the motor housing. These forces oscillate with two times the rotational frequency [4]. Therefore, for the applied six-pole rotor, these forces oscillate with the third order (60-990 Hz).

The frequency of the power electronics circuits of the auxiliary drive can be another NVH source. As the PWM frequency in this specific application is set to 16 kHz and this is mainly above the human range of audibility, this area is not discussed here.

7.5.5 Flow Noise

Pressure pulsations due to the high velocity of the process air can be another source of the noise. To avoid issues with flow noise, the gas velocity should be limited to 30 m/s. Furthermore, the design of tubes, pipes, interfaces, and manifolds must not generate tearing edges.

7.5.6 Airborne Noise

All noise and vibration is generated in the core components of the air supply module can be emitted by the housings and be transferred via the air to the vehicle body, the driver and passengers of the vehicle, and the environment.

Pressure pulsations in the suction line can be emitted via the walls of the intake manifold to the environmental air. They can be further transferred via the process air towards the flow to the intake system and the air intake filter.

Discharge side pressure pulsations can be emitted by the outlet manifold to the environmental air. Further transmission via the process air in the tubes to the intercooler, the humidifier, and the fuel cell stack has to be considered where the pulsations can be emitted by the housings to the surrounding air.

7.5.7 Structure-Borne Noise

Structure-borne noise of the core components can be transferred via rigid joints to fastening elements, fixtures, tubes, and manifolds to other process components (e.g., intercooler, humidifier, stack) and the vehicle body. All of these parts can emit the noise and transfer it either via the air or via structure elements to the vehicle body, the driver, and passengers of the vehicle, and to the environment.

Another example is Toyota's Mirai hydrogen FCV, which is actually pretty noisy in development [89–91].

In the FCEV, as the accelerator is pressed, there are some gentle whirring and humming noises as the hydrogen pump and air compressor do their work.

Because the FCEV is propelled by a whispering electric motor and not a rumbling gasoline engine, the driver picks up sounds that might otherwise be droned out. The sounds-a distinct whirring and intermittent clicking-are unique to the Mirai's hydrogen drivetrain. They came through loud and clear during a test drive in the development. The whirring kicks in when the driver punches the pedal for quick acceleration. It is reminiscent of the motor-assist in the Prius hybrid. But in the Mirai, it is actually the hydrogen pump working overtime to flush more hydrogen through the processing stack to ramp the car up to speed. The clicking, which can grate like a noise-vibration issue, comes from the hydrogen fuel injector, which feeds the fuel from the high-pressure hydrogen tanks into the pump. The clicking speeds and slows in time with the driver's foot on the accelerator. Both sounds emanate from just under the rear floorboards where the mechanisms are housed. Engineers have worked out how to muffle the sounds.

Motors, blowers, and pumps play key roles in fuel cell designs. Their efficiency and performance are vitally important in the operation of fuel cell systems and are a critical factor in maximizing the overall efficiency of a system.

In the development of Toyota's Mirai, a lot of efforts were made for NVH refinement as elaborated as follows [89–91]. The fuel cell in an FCEV passes hydrogen and oxygen through a chemical reaction, but to do this, it needs to take in and compress air containing oxygen from the atmosphere and feed this to the fuel cell. This is the role of the air compressor.

The air compressor is a vital component, given that it must constantly provide oxygen if the FCEV is to generate electricity. On the other hand, the hydrogen circulation pump circulates a portion of the hydrogen that did not undergo a chemical reaction, feeding this back into the fuel cell.

In the development of the gas pump, the efforts are made by seeking an alternative to the scroll type.

Toyota was engaged in developing hydrogen storage alloy tanks and hydrogen gas aspiration pumps, but because the hydrogen gas aspiration pump was a reciprocating type, it did not meet modern standards for performance of noise and vibration. Engineers requested to develop a hydrogen gas aspiration pump. They changed from a reciprocating type to a scroll type and started development, also employing a scroll type compressor for the air compressor. These products were incorporated in the Toyota FCHV released at the end of 2002. Toyota continued with improvements to the scroll type, also using this on the Toyota FCHV-adv released in 2008. However, by this time, the development team had already started looking for a replacement to the scroll type, given its low compression range, which limited the amount of air (and consequently oxygen) that it could send. This was a structural problem, and they anticipated that the scroll type would not be able to respond to future demands for ever-smaller, lighter, lower cost, and low NVH products. They examined various methods and decided to adopt the "helical root type" rotor in its stead.

The reciprocating type is the original type of air compressor in which a piston reciprocates in a cylinder to compress air. This operates at a low speed, with significant noise and vibration, however, it is low cost. Scroll type refers to another method for

FIGURE 7.8 Schematic scroll type and helical root type pumps used in the Toyota Mirai.

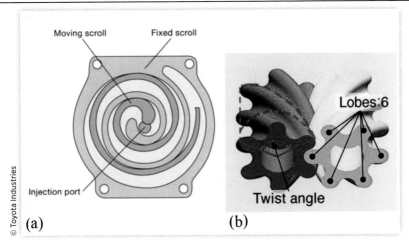

(a)　　　　　　　　　　　　(b)

compressing the air within an air compressor. This comprises a pair of spirals-one fixed and one that moves in a circular motion, thus compressing the air, as shown in **Figure 7.8a.** These are frequently used in automobiles and household air conditioning.

The helical root type rotates two root-style helical rotors, and this provides a smaller, lighter, and simpler construction, as shown in Figure 7.8b. The roots type is also called a "rotary blower," and they were known mainly for their performance as blowers rather than as compressors. Helical root type rotors are currently used in water treatment facilities and for conveying powder, such as cement. Put simply, the helical root type method is capable of smooth and efficient transport of air, water, and powder, and these have not been used much for compressing air. **Figure 7.9** shows the Helical Root Type Pump used in the Toyota Mirai.

Engineers discovered air could be compressed, depending on the number of lobes on the rotor, as well as on the twist angle. When using computer-aided design (CAD), they discovered that changing the number of lobes and the twist angle reduced the amount of air taken in. After repeated experiments and calculations, they achieved the objectives.

FIGURE 7.9 Helical root type pump used in the Toyota Mirai.

FIGURE 7.10   (a) Cross-section of air compressor; (b) sound intensity contour of the system; (c) CAE results of design with barrier.

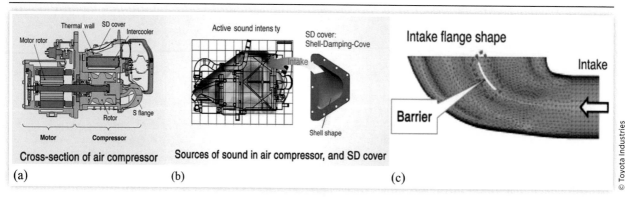

Cross-section of air compressor   Sources of sound in air compressor, and SD cover

(a)   (b)   (c)

Pinpointing the source of noise is critical. The main challenges faced in the development of air compressors are improving fuel efficiency, responsiveness, and vehicle acceleration, and reducing noise and vibration, with these last two posing the most problems. The development team thoroughly investigated the sources of the noise and determined that there were four areas in which to take measures-reviewing the shape of the housing, reducing vibration in the rotor, reducing noise in the intake and exhaust systems, and tuning the sound. The first of these was reviewing the shape of the housing. Reviewing the shape of the housing and increasing its rigidity are effective in controlling vibration, and consequently in reducing noise. **Figure 7.10a** shows the cross-section of the air compressor. Figure 7.10b shows the sound intensity contour of the system based on CAD.

The first thing they tried was eliminating waste from the housing and making it more compact. Doing so reduces the area from which sound can be emitted. They also integrated the silencer and intercooler, both of which had previously been external. Fitting the intercooler into the cylinder absorbs air intake pulsation, which is also effective in reducing housing vibration.

To reduce vibration in a cylinder by connecting one part to another was conducted by simulating fastening locations to optimally reduce vibration. They tried a variety of ways to reduce vibration and incorporated those that worked, including positioning a rib in the part of the cylinder that uses the most energy in compressing air and sandwiching a resin sheet only a few microns thick between vibration-damping steel sheets.

The rotor is a generic term for a rotating part of the compressor. This mainly refers to motors, shafts, and rotors. A silencer is a muffler for reducing the loud noise generated when air under pressure is released into the atmosphere. Intercooler air under pressure is high temperature with low density. The intercooler functions to cool the air and increase its density. It not only cools the air but also functions as a muffler. Pulsation is a regular rhythmic movement. A rib is a part added to reinforce a plate or other thin section.

An effort in the development is to reduce the shaking. When a rotor shakes, it contacts the cylinder within a certain range, and this is where the sound originates. Reducing shaking is also an effective way to reduce NVH. The rotor shaft is supported by three bearings. Shrinking that gap between support points would minimize shaking.

First, they measured the interference between the rotor and cylinder in a compressor. Then, using computer-aided engineering (CAE), they analyzed rotor shaking, and calculated where supports could be placed to minimize this. They then incorporated these changes in the machine and reevaluated. They repeated this cycle over and over. There are more effective ways to reduce shaking, which include increasing the shaft diameter or using more rigid materials, but these measures make the housing larger, increasing costs. What they needed was to minimize shaking by changing only the interior, without changing the current size, shape, or materials. If the size or shape of the housing is changed, then this will impact how the air compressor fits into the vehicle. Finding solutions given the constraints is a problem inherent to development. Another effort is to remove sound with barriers, as shown in Figure 7.10c. The air intake inlet pipeline could be the main radiation channel of aerodynamic noise created by the compressor.

They tried another approach to reduce noise and vibration-dampening sound in the intake and exhaust systems. An air compressor sucks in air from the atmosphere (intake), compresses this, and feeds it to the fuel cell (exhaust). The pulsation of the air vibrates the piping, causing sound, and this is a problem that needs to be addressed. First, in the intake system, they tried putting a barrier in the intake flange to split the air path into two. Air that has been split into two then converges again, and when these flows meet, the pulsation in each is canceled out. This reduced the amount of sound generated by a particularly jarring frequency. This was a high-pitched keening, which after time is unpleasant. Even if the volume of this is reduced, it can still cause discomfort. The problem with sound is that improvements to both its volume and its range are required. In the exhaust system, they put a barrier called a thermal wall in the exhaust piping to reduce pulsation. The aforementioned intercooler and vibration damping steel sheets within the cylinder also help to reduce exhaust noise.

The creation of a futuristic sound is attained. Yet another approach is tuning the sound. Efforts up until this point have been in reducing noise and vibration, but designers are not simply trying to achieve "a quiet FCEV." They want a sound that conveys to the driver that sense of acceleration when putting their foot down, but the sound cannot be too abstract.

Whether a sound is perceived as good or bad is subjective, but voiced sounds like "ZZZ" lack clarity, and were not an option. They wanted to create a clear sound that increased in pitch with speed. Engineers set out to create a sound taking into account both the quantitative elements of volume and register, together with the qualitative elements of sensation and emotion.

Reducing rotor shake was effective in terms of creating sound. Shaking generates a mid-range sound, and measures taken to reduce this have left a fairly linear sound. Other approaches at reducing noise and vibration all involved tuning, and instead of simply reducing NVH, they found it very difficult to leave just pleasing sounds. They run over 20 listening tests with the compressor in the vehicle, with the FCV driving on the Toyota test track. This involved driving around in the cold environment and listening to the car. In the design, the high sound-insulating body and motor propulsion at all speeds deliver outstanding quietness. After NVH refinement, the FCEV Mirai has attained excellent quietness, as shown in **Figure 7.11**.

FIGURE 7.11   Comparison of the articulation index of Mirai with other Toyota cars.

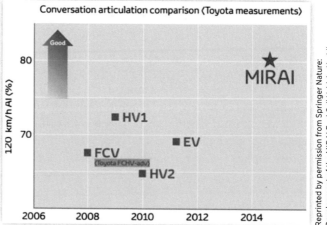

Reprinted by permission from Springer Nature: Development of the MIRAI Fuel Cell Vehicle, Yoshikazu Tanaka ©2016

References

1. Sang, J., Venturi, M., and Bocksch, R., "NVH-Challenges of Air Supply Subsystems for Automotive Fuel Cell Applications," *SAE Int. J. Engines* 1(1):258–266, 2009, doi:https://doi.org/10.4271/2008-01-0316.

2. Wilberforce, T., El-Hassan, Z., Khatib, F.N., Al Makky, A. et al., "Developments of Electric Cars and Fuel Cell Hydrogen Electric Cars," *International Journal of Hydrogen Energy* 42: 25695–25734, 2017.

3. Y. Gao and M. Ehsani, "Systematic Design of FC Powered Hybrid Vehicle Drive Trains," SAE Technical Paper 2001-01-2532, 2001, https://doi.org/10.4271/2001-01-2532.

4. T. Simmons, P. Erickson, M. Heckwolf, and V. Roan, "The Effects of Start-Up and Shutdown of a FC Transit Bus on the Drive Cycle," SAE Technical Paper 2002-01-0101, 2002, https://doi.org/10.4271/2002-01-0101.

5. D. Tran, M. Cummins, E. Stamos, J. Buelow, and C. Mohrdieck, "Development of the Jeep Commander 2 FC Hybrid Electric Vehicle," SAE Technical Paper 2001-01-2508, 2001, https://doi.org/10.4271/2001-01-2508.

6. P.J. Berlowitz and C.P. Darnell, "Fuel Choices for Fuel Cell Powered Vehicles," SAE Technical Paper 2000-01-0003, 2002, https://doi.org/10.4271/2000-01-0003.

7. D. Tran, M. Cummins, E. Stamos, J. Buelow, and C. Mohrdieck, "Development of the Jeep Commander 2 Fuel Cell Hybrid Electric Vehicle," SAE Technical Paper 2001-01-2508, 2002, https://doi.org/10.4271/2001-01-2508.

8. C.E. Thomas, B.D. James, F.D. Lomax Jr, and I.F. Kuhn Jr, "Societal Impacts of Fuel Options for Fuel Cell Vehicles," SAE Technical Paper 982496, 2002, https://doi.org/10.4271/982496.

9. S.E. Gay, J.Y. Routex, M. Ehsani, and M. Holtzapple, "Investigation of Hydrogen Carriers for Fuel Cell Based Transportation," SAE Technical Paper 2002-01-0097, 2002, https://doi.org/10.4271/2002-01-0097.

10. Automotive Engineering, Wroclaw University of Technology, http://www.ae.pwr.wroc.pl/filez/20110606094057_HEV.pdf

11. Guo, A., Chen, W., Li, Q., Liu, Z., and Que, H., "Air Flow Control Based on Optimal Oxygen Excess Ratio in Fuel Cells for Vehicles," *Journal of Modular Transport* 21, no. 2: 79–85, 2013.

12. Fans, Blowers and Pumps for Fuel Cells, http://www.ebmpapst.co.za/media/content/downloads_10/brochures/Fuel_cells_brochure_EN.pdf

13. Bang, J., Kim, H.S., Lee, D.H., and Min, K., "Study on Operating Characteristics of Fuel Cell Powered Electric Vehicle with Different Air Feeding Systems," *Journal of Mechanical Science and Technology* 22: 1602–1611, 2008.

14. Wei, K., Zuo, S., He, H., and Wang, Z., "Numerical Study on the Unsteady Behavior of a Centrifugal Compressor for the Fuel-Cell Vehicle," *Meeting of the Acoustical Society of America*, USA, 2014, 10.1121/2.0000032

15. Zuo, S. and Yan, J., "Experimental Analysis for the Interior Noise Characteristics of the Fuel Cell Car," *Vehicular Electronics and Safety, 2006. IEEE International Conference on IEEE*, Shanghai, 2006.

16. Zuo, S. and Yan, J., "Experimental Analysis for the Interior Noise Characteristics of the Fuel Cell Car," *Vehicular Electronics and Safety, 2006. ICVES 2006. IEEE International Conference on. IEEE*, Shanghai, 2006.

17. Kang, Q., Zuo, S., and Wei, K., "Study on the Aerodynamic Noise of Internal Flow of Regenerative Flow Compressors for a Fuel-Cell Car," *Proceedings of the Institution of Mechanical Engineers, Part C: Journal of Mechanical Engineering Science* 228, no. 7: 1155–1174, 2014.

18. Ha, K.K. et al., "Experimental Investigation on Aero-Acoustic Characteristics of a Centrifugal Compressor for the Fuel-Cell Vehicle," *Journal of Mechanical Science and Technology* 27, no. 11: 3287–3297, 2013.

19. Lee, Y.W. et al., "Control of Airflow Noise from Diesel Engine Turbocharger," SAE Technical Paper 2011-01-0933, 2011, https://doi.org/10.4271/2011-01-0933.

20. Pak, H., Krain, H., and Hoffmann, B., "Flow Field Analysis of a High Pressure Ratio Centrifugal Compressor," AGARD-CP-537, 1993.

21. Krain, H., Hoffmann, B., and Pak, H., "Aerodynamics of a Centrifugal Compressor Impeller with Transonic Inlet Conditions," ASME-Paper 95-GT-79, American Society of Mechanical Engineers, New York, Jun. 1995.

22. Eisenlohr, G., Dalbert, P., Krain, H., Proll, H. et al., "Analysis of the Transsonic Flow at the Inlet of a High Pressure Ratio Centrifugal Impeller," ASME Paper 98-GT-24, American Society of Mechanical Engineers Conference, Stockholm, Sweden, Jun. 1998.

23. Krain, H. and Hoffmann, B., "Flow Physics in High Pressure Ratio Centrifugal Compressors," ASME-Paper FEDSM98-4583, Summer Meeting of the American Society of Mechanical Engineers, Washington DC, USA, Jun. 1998.

24. Kameier, F. and Neise, W., "Rotating Blade Flow Instability as a Source of Noise in Axial Turbomachines," *Journal of Sound and Vibration* 203 (1997): 833–853.

25. Marz, J., Hah, C., and Neise, W., "An Experimental and Numerical Investigation into the Mechanisms Rotating Instability," *Journal of Turbomachinery—Transactions of the ASME* 124: 367–375, 2002.

26. McAlpine, A., Fisher, M.J., and Tester, B.J., "Buzz-Saw Noise: A Comparison of Measurement with Prediction," *Journal of Sound and Vibration* 290: 1202–1233, 2006.

27. Morfey, C.L. and Fisher, M.J., "Shock-Wave Radiation from a Supersonic Ducted Rotor," *The Aeronautical Journal of the Royal Aeronautical Society* 74: 579–585, 1970.

28. Hawkings, D., "Multiple Pure Tone Generation by Transonic Compressors," *Journal of Sound and Vibration* 17: 241–250, 1971.

29. Holste, F. and Neise, W., "Noise Source Identification in a Propfan Model by Means of Acoustical Near Field Measurements," *Journal of Sound and Vibration* 203: 641–665, 1997.

30. Raitor, T. and Neise, W., "Sound Generation in Centrifugal Compressors," *Journal of Sound and Vibration* 314, no. 3-5: 641–665, 2008.

31. Ffowcs Williams, J.E. and Hawkings, D.L., "Theory Relating to the Noise of Rotating Machinery," *Journal of Sound and Vibration* 10, no. 1: 10–21, 1969.

32. Ha, K., Lee, C, Kim, C, Kim, S. et al., "A Study on the Characteristics of an Oil-Free Centrifugal Compressor for Fuel Cell Vehicles," *SAE Int. J. Alt. Power.* 5(1): 164–174, 2016, https://doi.org/10.4271/2016-01-1184.

33. Füßer, R. and Weber, O., "Air Intake and Exhaust Systems in Fuel Cell Engines," SAE Technical Paper 2000-01-0381, 2000, https://doi.org/10.4271/2000-01-0381.

34. Wu, B., Matian, M., and Offer, G.J., "Hydrogen PEMFC System for Automotive Applications," *International Journal of Low-Carbon Technology* 7, no. 1: 28–37, 2012.

35. Wiartalla, A., Pischinger, S., Bornscheuer, W., Fieweger, K. et al., "Compressor Expander Units for Fuel Cell Systems," SAE Technical Paper 2000-01-0380, 2000, doi:https://doi.org/10.4271/2000-01-0380.

36. Venturi, M., Sang, J., Knoop, A., and Hornburg, G., "Air Supply System for Automotive Fuel Cell Application," SAE Technical Paper 2012-01-1225. 2012, https://doi.org/10.4271/2012-01-1225.

37. Blum, L., Meulenberg, W.A., et al., "Worldwide SOFC Technology Overview and Benchmark," *International Journal of Applied Ceramic Technology* 2, no. 6: 482–492, 2010.

38. Ha, K.K., Lee, C.H. et al., "Experimental Investigations of Effects of Centrifugal Compressor Impeller Design on Noise Characteristics," *Proceedings of ASME Turbo Expo 2014*, Diisseldorf, Germany, Jun. 16-20, 2014.

39. Venturi, M., Sang, J., Knoop, A., and Hornburg, G., "Air Supply System for Automotive Fuel Cell Application," SAE Technical Paper 2012-01-1225. 2012, doi:https://doi.org/10.4271/2012-01-1225.

40. Gangwar, H, Anderson, R., and Kim, K, "Ultra-Long Life Oil-Free Supercharger for Fuel Cell and Hybrid Vehicle Power Trains," SAE Technical Paper 2013-01-0478, 2013, doi:https://doi.org/10.4271/2013-01-0478

41. Lieblein, S., "Loss and Stall Analysis of Compressor Cascades," *ASME Journal of Basic Engineering* 81: 387–400, 1959.

42. Lieblein, S., "Incidence and Deviation Angle Correlations for Compressor Cascades," *ASME Journal of Basic Engineering* 82, no. 3: 575–584, 1960.

43. Horlock, J.H. and Lakshminarayana, B., "Secondary Flows: Theory, Experiment and Applications in Turbomachinery Aerodynamics," *Annual Review of Fluid Mechanics* 5, no. 1: 247–280, 1973.

44. Lim, T.G., Lee, S.M., Jeon, W.H., and Jang, C.M., "Characteristics of Unsteady Flow Field and Aeroacoustic Noise in a Regenerative Blower," *Journal of Mechanical Science and Technology* 29, no. 5: 2005–2012, 2015.

45. Carolus, T., Schneider, M., and Hauke, R., "Axial Flow Fan Broad-band Noise and Prediction," *Journal of Sound & Vibration* 300, no. 1-2: 50–70, 2007.

46. Lee, C., Kil, H., and Kim, J., "Low Noise Design of Regenerative Blower by Combining the FANDAS-Regen Code: Optimization Technique and Phase-Shift Cancellation Concept," *Proceedings of the 5th International Conference on Simulation and Modeling Methodologies, Technologies and Applications (SIMULTECH-2015)*, Lda, 2015, 10.5220/00055143046 90475.

47. Lee, C. and Kil, H.G., "Aero-Acoustic Performance Prediction Method of Axial Flow Fan," *3 rd International Conference on Integrity, Reliability and Failure*, Porto/Portugal, Jul. 20-24, 2009.

48. Lee, C. and Kil, H.G., "A New Through-flow Analysis Method of Axial Flow Fan with Noise Models," *Euronoise 2018 - Conference Proceedings*, Heraklion, Crete - Greece, May 27–31, 2018.

49. Lee, H.G. Kil, G.C. Kim, J.G. Kim, J.H., and Chung, K.H., "Aero-acoustic Performance Analysis Method of Regenerative Blower," *Journal of Fluid Machinery* 16, no. 2: 15–20, 2013.

50. Badami, M. and Mura, M., "Setup and Validation of a Regenerative Compressor Model Applied to Different Devices," *Energy Conversion and Management* 52, no. 5: 2157–2164, 2011.

51. Morfey, C.L., "The Acoustic Spectrum of Axial Flow Machines," *Proceedings of the British Acoustical Society* 22, no. 4: 445–466, 1972.

52. Koch, C.C. and Smith, L.H., "Loss Sources and Magnitudes in Axial-Flow Compressors," *Journal of Engineering for Power* 98, no. 3: 411–424, 1976.

53. Mugridge, B.D., "Noise Characteristics of Axial and Centrifugal Fans as Used in Industry-A Review," *Shock and Vibration Digest*, 1979, doi:10.1002/ar.1091240408

54. C. L., "Morfey Rotating Blades and Aerodynamic Sound," *Journal of Sound & Vibration* 28, no. 3: 587-617, 1973.

55. Sjolander, S.A. and Amrud, K.K., "Effects of Tip Clearance on Blade Loading in a Planar Cascade of Turbine Blade," *ASME Journal of Turbomachinery* 109: 237–245, 1987.

56. Paterson, R.W., and Amiet, R.K., "Noise and Surface Pressure Response of an Airfoil to Incident Turbulence," *Journal of Aircraft* 14 no. 4: 729–736, 2012.

57. Mugridge, B. D., and Morfey, C. L., "Sources of Noise in Axial Flow Fans," *Journal of the Acoustical Society of America* 51, no. 5A: 1411–1426, 1972.

58. Raitor, T. and Neise, W., "Sound Generation in Centrifugal Compressors," *Journal of Sound and Vibration* 314: 738–756, 2008.

59. Griffini, D., Salvadori, S., Carnevale, M., Cappelletti, A., Ottanelli, L., and Martelli, F., "On the Development of an Efficient Regenerative Compressor," *Energy Procedia* 82: 252–27, 2015.

60. Engeda, A., "Flow Analysis and Design Suggestions for Regenerative Flow Pumps," ASME Paper No. FEDSM2003-45681, 2003.

61. Engeda, A. and Raheel, M., "Theory and Design of the Regenerative Flow Compressor," *Proceedings of the International Gas Turbine Congress*, Tokyo (IGTC 2003 Tokyo TS-050), 2003.

62. Raheel, M. and Engeda, A., "(Michigan State University) Current Status, Design and Performance Trends for the Regenerative Flow Compressors and Pumps," *ASME 2002 International Mechanical Engineering Congress and Exposition*, American Society of Mechanical Engineers, no. 81: 387–400, 1959.

63. Yoshimura, S., "Impact Vibration of Screw Compressor Rotor (In the Case That Vibration Occurs at Both the Driving and Trailing Side)," *Transaction of JSME* 61, no. 586: 2216, 1995.

64. Huang, P.. Sean, X.Y., and David, H., "Gas Pulsation Control Using a Shunt Pulsation Trap," *International Compressor Engineering Conference*, USA, 2014.

65. Yoshimur, S., "Technology for Improving Reliability of Oil-flooded Screw Compressors," *Kobelco Technology Review*, Dec. 29, 2010.

66. Yoshimura, S., "Impact Vibration of Screw Compressor Rotor (In the Case That Vibration Occurs at the Driving Side)," *Transaction of JSME* 61, no. 582: 501, 1995.

67. Yoshimura, S., "Impact Vibration of Screw Compressor Rotor (3rd Report, Analysis in consideration of Inertia of Both Rotors)," *Transaction of JSME* 64, no. 617: 15, 1998.

68. Koai, K. and Soedel, W., "Gas Pulsation in Screw Compressors – Part I: Determination of Port Flow and Interpretation of Periodic Volume Source," presented at *10th International Compressor Engineering Conference*, Purdue, West Lafayette, USA, 1990.

69. Koai, K and Soedel, W, Gas Pulsation in Screw Compressors – Part I: Dynamics of Discharge System and Its Interaction with Port Flow," presented at *10th International Compressor Engineering Conference*, Purdue, West Lafayette, USA, 1990.

70. Wu, H.G., Xing, Z.W., Peng, X.Y. et al., "Simulation of Discharge Pressure Pulsation within Twin Screw Compressors," *Proceedings of the IMechE, Part A: Journal of Power and Energy* 218 : 257–264, 2004.

71. Papes, I., Degroote, J., and Vierendeels, J., "Development of a Thermodynamic Low Order Model for a Twin Screw Expander with Emphasis on Pulsations in the Inlet Pipe," *Applied Thermal Engineering* 103: 909–919, 2016.

72. Andrews, R.W. and Jones, J.D., "Noise Source Identification in Semi-Hermetic Twin-Screw Compressors," presented at *1990 International Compressor Engineering Conference*, Purdue, 1990.

73. Erol, H. and Ahmet, G., "The Noise and Vibration Characteristics of Reciprocating Compressor: Effects of Size and Profile of Discharge Port," *2000 International Compressor Engineering Conference*, Purdue, 2000.

74. Fleming, J.S., Tang, Y., and Cook, G., "The Twin Helical Screw Compressor Part 1: Development, Applications and Competitive Position," *Proceedings of the IMechE, Part C Journal of Mechanical Engineering Science* 212, no. 5: 355–367, 1998.

75. Fujiwara, A. and Sakurai, N., "Experimental Analysis of Screw Compressor Noise and Vibration," presented at the *1986 International Compressor Engineering Conference*, Purdue, 1986.

76. Gavric, L. and Cassagnet, B., "A Measurement of Gas Pulsations in Discharge and Suction Lines of Refrigerant Compressors," presented at the *2000 International Compressor Engineering Conference*, Purdue, 2000.

77. Gavric, L., "Modelling and Analysis of Excitation Mechanisms," presented at the *Short Course and Workshop on Noise and Vibration of Compressors*, Cetim, Senlis, France, Jun. 2001.

78. Holmes, C.S., "Inspection of Screw Rotors for Prediction of Compressor Performance, Reliability and Noise," *Proceedings of the 4th International Conference on Compressor and Refrigeration*, Xi'an City, China, Oct. 2003.

79. Holmes, C.S., "Transmission Error in Screw Compressors, and Methods of Their Compensation during Rotor Manufacture," *International Conference on Compressors and their Systems 2005*, London, UK, 2005.

80. Huagen, W., Ziwen, X., Xueyuan, P., and Pengcheng, S., "Simulation of Discharge Pressure Pulsation within Twin Screw Compressors," *Proceedings of the I MECH E Part A Journal of Power and Energy* 218, no. 4: 257–264, Aug. 2004.

81. Mujic, E., Kovacevic, A., Stosic, N., and Smith, I.K., "Noise Generation and Suppression in Twin Screw Compressors," *Proceedings of the Institution of Mechanical Engineers, Part E: Journal of Process Mechanical Engineering* 225, no. 2: 127–148, 2011.

82. Tryhorn, D.W., "Blower Noise and Solution: An Introduction to the A.W. Convel Roots Blower," presented at *International Compressor Engineering Conference*, West Lafayette, IN, 1976.

83. Mujić, E., Kovačević, A., Stošić, N., and Smith, I.K., "Noise Control by Suppression of Gas Pulsation in Screw Compressors," *Advances in Noise Analysis, Mitigation and Control, INTECH*, Sydney, Australia, 2016

84. Ohtani, I. and Iwamoto, T., "Reduction of Noise in Roots Blower," *Bulletin of JSME* 24, no. 189: 547–554, 1981.

85. Mason, J.F., Homer, M.E., and Wilson, R.E., "Mathematical Models of Gear Rattle in Roots Blower Vacuum Pumps," *Applied Mechanics and Materials* 5-6: 21–28, 2006.

86. Shinmaywa, Helical Rotor Blower (Roots Type), ARH-S/SP ARH-E/EP Series, 2016.

87. Society of Automotive Engineers, "Vibration, Noise and Ride Quality," Society of Automotive Engineers of Japan, 2016.

88. Connelly, T. and Hollingshead, J., "Statistical Energy Analysis of a Fuel Cell Vehicle," SAE Technical Paper 2005-01-2425, 2005, https://doi.org/10.4271/2005-01-2425.

89. "Toyota Mirai first drive review," https://www.drive.com.au/new-car-reviews/toyota-mirai-first-drive-review-20151102-gkopw8, accessed Mar. 2016.

90. Toyota Motor Corporation Reports, "Story for Mirai-for the Future," https://www.toyota-industries.com/innovation/story/story_1/chapter_2/

91. Toyota Motor Corporation Reports, "Mirai Dismantling Manual." https://www.toyota-tech.eu/HYBRID/HVDM/EN/DM32B0U_Revised_201507.pdf

8

Sound Quality of BEV/HEV/FCEV

8.1 Introduction of Sound Quality of BEV/HEV/FCEV

Sound quality is an important attribute of BEV/HEV/FCEV, even though electrified vehicles are very quiet due to not using an Internal combustion engine (ICE) at all or for part of the time [1–4].

Sound quality is the science and engineering of sounds that help to select, create, and design positive appeal to the driver of the vehicle. The study of sound quality is beyond the quantification of overall noise level. It is to investigate the content of the sound and to identify the sound that is desirable to customers. Sound quality investigation could cover all possible sounds associated with vehicles. For instance, the interior sound of the compartment has a sound quality requirement. The door-closing sound could also have the sound quality needs determined in the design phase. The engine sound has many sound-quality issues, for instance, a certain level of noise and vibration during idling may be needed to assure the driver that the vehicle is still running. Particular sound attributes may be needed to create an image associated with the vehicle, which may include the perceptions of being "sporty," "luxurious," "responsive," etc., in addition to eliminating "buzzy," "cheap," and "weak" perceptions.

The task of automotive sound quality is to identify what aspects of a sound define its quality. It has been the experience of most persons involved in noise and vibration testing that analysis of acoustic signals alone does not identify the quality of those signals. Knowing how to design the correct attributes into a vehicle sound directly impacts the appeal of the vehicle and ultimately impacts its profitability.

It is noted that there are no analysis techniques that can be used to accurately and reliably quantify the descriptive terms mentioned above without the aid of subjective testing of some kind; hence, there are needs for subjective testing and analysis for correlating and modifying the objective evaluations based on existing methods and tools [3–15]

The absence of ICE noise is a double-**edged sword** for the sound quality of BEV/HEV/FCEV. Consider wind noise as an example. The results show that BEV and ICE vehicles have equivalent wind noise loudness levels at all speeds. However, at lower speeds (50-60 mph), the BEV is judged to have more wind noise even though the sound pressure level was the same as the ICE vehicle. The difference is that in the BEV, there is no engine noise to mask the wind noise. Masking is a well-known psychoacoustic phenomenon wherein one sound can decrease the perceived level of a second, spectrally similar sound. Binaural masking occurs when the masker emanates from a different location than the signal. In a conventional vehicle, the masking effect of the ICE noise can reduce the perceived amount of wind noise. However, in the BEV, there is no engine noise and, thus, the impression is that there is increased wind noise, even though the actual loudness levels are unchanged. This effect is most prominent at speeds below 60 mph, where the engine/wind noise ratio is high, and the masking is large. This increased wind noise presence at a low speed could be quite annoying to customers accustomed to ICE vehicles. At higher vehicle speeds, the wind noise increases faster than the engine noise, masking is reduced, and the impression of BEV and conventional vehicle wind noise is similar.

Moreover, while BEV/HEV/FCEV have no engine noise, they do have powertrain noise, primarily from the planetary gears used in the transmission. During acceleration and deceleration, there is significant gear whine. If reduced to tolerable levels, this gear whine can provide audible feedback that is not unpleasant as the throttle is manipulated. At idle, BEVs make virtually no noise. Thus, any accessory noise (A/C, power steering pump, fan) will be quite prominent and may be annoying due to its strong tonal quality. While electrified vehicles will be relatively quiet, they still present significant sound quality challenges. The electric motor itself is a potential noise source, although this noise is usually high frequency and can be easily isolated by the sound package. For AC motors, the inverter is also a source of noise, which can be very annoying.

The sound quality is different for various vehicle operations, such as acceleration, deceleration, idle, and cruise. At cruise, a BEV emits little or no powertrain sound (other than the motor and inverter noise discussed above). However, during acceleration and deceleration, the planetary gearset that serves as the transmission can make a whining sound. Some subjective studies have shown that drivers of conventional vehicles find this gear whine to be particularly annoying. A BEV with a well-isolated motor produces virtually no powertrain noise at idle. The interior noise level is basically equal to that of the ambient. However, the powertrain accessories (A/C compressor, power steering pump, and vacuum pump) will produce noise. As in the discussion on low-speed wind noise, these accessory noises may be perceived as being excessive due to the lack of any engine idle noise to mask them. In a conventional ICE vehicle, there is a great deal of spectral overlap between the idle and accessory sounds because their rotational rates are very similar. This leads to a large amount of masking. The BEV/HEV/FCEV do not benefit from such masking.

Conventionally, focus on vehicle noise, vibration, and harshness (NVH) has been to reduce or eliminate the problematic noises and vibrations. Over the last decade, this focus has been extended to include the tuning of interior sound quality. While eliminating or reducing unwanted noises is still an area of focus, quality of sound perceived in the vehicle interior has become increasingly important. Interior sound quality can be used to differentiate vehicles or brands, as well as give an overall impression of the quality of the vehicle.

Vehicle sound and vibration can be evaluated by both objective and subjective methods. Subjective evaluation of sound events does not always correlate to an objective evaluation where the values have been measured using standard measurement and analysis techniques. Whether a sound is felt to be pleasant or annoying is usually dependent on the subjective criteria of the person listening.

The use of hybrid and electric powertrains in passenger vehicles brings unusual narrowband auditory stimuli into the passenger compartment. When attempting to shape and control those effects, in either a passive or active approach, it is possible that one aspect of the soundscape is addressed at the expense of evidencing other unwanted noise components. Therefore, a proper handling of the different disturbance components is needed, which would consider the intricate interrelation between those components and its cross effects on the various sound quality metrics relevant to this application. To efficiently and effectively achieve an optimized product, the sound quality should be designed with objective methods and physical tools instead of fully using a trial-and-error approach of subjective evaluations. However, the establishment of objective prediction methods and tools needs to use subjective evaluation for correlation and validation. As such, the major effort of sound quality analysis is to derive methods/models and tools of objective evaluation by using subjective evaluation and understanding the basis of sound language and sound perception.

In this chapter, we present the fundamentals of sound quality of vehicles, with special emphasis on the sound quality of BEV/HEV/FCEV.

8.2 Psychoacoustic Parameters and Sound Quality Matric, Loudness, Roughness, Sharpness, Fluctuation Strength, Etc.

For the assessment of sound quality, psychoacoustic parameters or magnitudes, such as loudness, sharpness, fluctuation strength, and roughness, have been proven useful. In many cases, a combination of these basic psychoacoustic parameters could predict sound quality ratings objectively [11–110].

It is critical to find physical metrics that show good correlation with subjective impression to predict the sound quality and find countermeasures to improve sound quality of vehicles. Many physical metrics have been proposed and some of them are shown to have good correlation with subjective perception. There are several acoustical aspects of experimentally separable ways related to the human hearing mechanism: pitch, duration, timbre, loudness, sonic texture, and spatial location. Some of these terms have a standardized definition (for instance, in the ANSI Acoustical Terminology ANSI/ASA S1.1-2013).

Pitch is perceived as how "low" or "high" a sound is and represents the cyclic, repetitive nature of the vibrations that make up sound. For simple sounds, pitch relates to the frequency of the slowest vibration in the sound (called the fundamental harmonic). In the case of complex sounds, pitch perception can vary.

The scale for pitch is mel (comes from melody), which is defined by

$$\text{Mel}(f) = 2595 * \log_{10}\left(1 + \frac{f}{700}\right) \tag{8.1}$$

Frequency and pitch correspond only in the low-frequency range, which is below 1000 Hz and pitch increases more slowly than frequency because of the effect of the logarithm function in Equation 8.1. In addition, the frequency and intensity sensitivity of the human ears affect each other.

Duration is perceived as how "long" or "short" a sound is and relates to onset and offset signals created by nerve responses to sounds. Duration is a very important dimension in human hearing, since all sounds fluctuate over time. When the duration of a sound is less than about half of a second, it affects both frequency and intensity sensitivity. The duration threshold for tonality decreases from about 60 ms at 50 Hz to approximately 15 ms at 500 Hz. Above 1000 Hz, the threshold for tonality is essentially constant and on the order of about 10 ms. Absolute intensity sensitivity threshold levels decrease when the duration of a stimuli becomes much shorter than 1 s. For durations up to 200-300 ms, a tenfold change in duration can offset an intensity change on the order of about 10 dB. Durations longer than about 1/3 s are treated by the ear as though they were infinitely long.

Timbre is perceived as the quality of different sounds and represents the preconscious allocation of a sonic identity to a sound [19, 20, 21]. Timbre is a very important factor related to sound quality. A number of different historic timbre definitions are found in the open literature. The American Standards Association specifies: "timbre is that attribute of sensation in terms of which a listener can judge that two sounds having the same loudness and pitch are dissimilar." ANSI1994A indicates "timbre, that attribute of auditory sensation with enables a listener to judge that two not identical sounds similarly presented and having the same loudness and pitch, are dissimilar." Many researches tried to decompose timbre into component attributes.

Psychoacoustic analyses have been developed to address the problems of connecting an objectively measured signal with subjectively perceived sound evaluation. Different sound features, and the perceived signature can be quantified using various developed metrics. Quantification of sound magnitudes can use the metrics of sound pressure level (dBA), loudness (stationary or nonstationary), articulation index (AI), and speech intelligibility. The modulation and roughness of sound can be quantified by fluctuation

FIGURE 8.1 Frequency range where some psychoacoustic parameters are applied to quantify customer perception

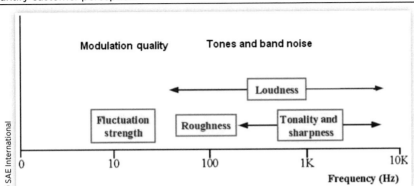

strength, roughness, or Fast Fourier Transform (FFT) of envelop function. The tonality can be characterized by tonality, pitch strength, or tone-to-noise ratio (TNR). The frequency balance can be quantified by sharpness, spectrum balance, center of gravity, and spectrum envelope metrics. The impulsiveness, transients, and intermittent sound can be quantified by using Kurtosis, skewness, crest factor, peak-to-peak, mean, and standard deviation. There are many psychoacoustic parameters previously developed and some of them are described here.

The psychoacoustic parameter loudness (N) is the subjective impression of the intensity of sound. Loudness is one of the essential factors in sound perception and the calculation of the other auditory sensations.

The fluctuation strength is a psychoacoustic analysis of the human perception of slowly varying modulation (5-20 Hz) of the signal based on a hearing model.

Roughness is the perception of amplitude or frequency-modulated tones that produce two different kinds of aural perceptions. Envelope fluctuations between 20 Hz and 300 Hz are perceived as tone modification; the tone sounds "rough." In contrast, modulation frequencies under 20 Hz are immediately detected as time modifications and result in a perception of fluctuation strength. The roughness depends on center or modulation frequency and degree of modulation, and the level dependency is very low. Sound signals whose spectral components are mainly located in the higher frequency range are heard by the human auditory apparatus as "sharp" or "shrill." Sharpness has been introduced as a unit for this sensation. Sounds that are mainly composed of spectral components at high frequencies are judged by the ear as sharp or shrill. The peak of the area below the spectrum envelope is critical to sharpness. The further the peak shifts toward the higher frequencies, the sharper the perception of the signal. **Figure 8.1** illustrates the frequency range where some of the psychoacoustic parameters are applied to quantify customer perception of vehicle sound quality.

8.2.1 Basic Psychoacoustic Parameters

The analysis using A-weighted sound levels, sound power, and noise spectra in octaves or third octaves is not sufficient to describe sound to reflect customer's perception.

The classical analysis parameters include A-weighted sound power level, power, and sound pressure levels in octaves or third octaves and follow standard norms, especially ISO 3744 and ISO 9614-2, which takes the sound power level of the source from sound pressure or intensity, respectively. Objective psychoacoustic parameters, like loudness, sharpness, and roughness and fluctuation strength can be derived by processing these recordings. Correlations can be established with the objective classical parameters.

The performance of human hearing as a sound evaluator has not been realized by available technical systems. In general, the analytical methods, such as frequency spectrum and sound pressure, do not compare with the human being as a sound receiver [3]. Hence, the conventional analytical test methods are complemented by psychoacoustic parameters for a precise and meaningful understanding. There are many standard metrics. Besides A-weighting, the parameters that influence auditory perception include the loudness, fluctuation strength, roughness sharpness, tonality, etc.

8.2.2 Loudness

Loudness represents the auditory perception character related to sound magnitude. There are several models for calculating loudness, such as the Zwicker model and Steve approach. The basic difference between these is the way they deal with masking, the reduction of sensitivity by noise in a stronger frequency band to noise in other bands. The loudness is measured in phones or sones. One sone is the loudness for pure tone sound with an amplitude of 40 dB at 1 kHz. The relationship between sone values and perceived loudness is directly proportional. Specific loudness is the distribution of loudness over the frequency axis and is measured in sone/Bark. Loudness is also expressed in loudness level, expressed in phons. Loudness level is given by the sound pressure level of a tone at 1 kHz, that is, equally loud to the tone being presented, or resulting in the same loudness sensation as the measured signal. The relation between loudness level and loudness is determined at 1 kHz and this is graphed in Figure 8.1. At sound levels greater than 40 phons, loudness doubles every 10 phons (doubles every 10 dB at 1 kHz) and doubles every three phons at lower levels.

Loudness is a sensation of the intensity and magnitude of sound experienced by the human ear. Loudness is not dependent on amplitude alone. Factors such as bandwidth, waveform, frequency, and exposure time can influence the loudness perceived by humans. Zwicker proposed a relation to objectively quantify loudness, which incorporates the factors that influence loudness.

$$N = \int_{0}^{24\,\text{Bark}} N^t \, dz \quad [\text{Sone}] \tag{8.2}$$

The Bark was proposed by Zwicker as a division of the frequency spectra, similar to third octave bands. A pure tone producing 40 dB at 1 kHz would produce 1 sone, which is the reference value for Zwicker Loudness.

8.2.3 Roughness

Roughness represents the auditory perception character related to the amplitude modulation and frequency modulation for sound with frequency modulation at middle frequency around 70 Hz. It is related to the high-frequency modulation of the sound. The unit of roughness is the asper. One asper is the roughness for a pure tone sound with an amplitude of 60 dB at 1 kHz, which is 100% modulated (modulation factor of 1) in amplitude at a modulation frequency of 70 Hz.

The evaluation of the noise for quantifying the amplitude or frequency modulation is focused between 20 Hz and 200 Hz. The index RAM refers to the degree of amplitude modulation (refer to a standard value: 1 kHz tone at 60 dB), and index RFM refers to the degree of frequency modulation (refer to a standard value: 1 kHz tone at 60 dB).

There is also no standardization for roughness. A preliminary implemented algorithm in the Artemis program calculated partial roughness from degrees of modulations of band signals and adds them to obtain a total roughness value. First, the signal is divided into 24 subbands by phase linear filters, which have a width of 2 Bark and overlap neighboring bands by 1 Bark. The bandwidth has to be wider than those used for loudness calculations to detect sufficiently large modulation frequencies inside one subband. The overlap is necessary to prevent signals at the limits of the subbands resulting in different total values than in the middle. The dependency of roughness on signal level is taken into account by doubling the partial roughness for an increase in the subband signal power of 20 dB. Total roughness is calculated as the sum of all partial roughness values as

$$R = \int_0^{24\,\text{Bark}} 0.0003 f\, \Delta f_{\text{mod}}(z)\, \Delta z \qquad (8.3)$$

8.2.4 Fluctuation Strength

Fluctuation strength is the auditory perception character related to amplitude modulation and frequency modulation for sound with frequency modulation at lower frequency around 4 Hz. The unit of fluctuation strength is the vacil. One vacil is the fluctuation strength for pure tone sound with an amplitude of 60 dB at 1 kHz, which is considered as 100% modulated (modulation factor of 1) in amplitude at a modulation frequency of 4 Hz.

The calculation of fluctuation strength takes place in a similar way to the calculation of roughness. But now, the signal is divided in 1 Bark wide, nonoverlapping bands. This is because at the low modulation frequencies which are to be detected here, the side lines of a tone modulated in amplitude lie very close beside the main line. The envelopes of the band signals are calculated and filtered with a bandpass with a center frequency of 4 Hz. The quotient of the filtered and the unfiltered envelopes is taken as a measure for the partial fluctuation strengths. An increment of the sound pressure level by 20 dB causes the fluctuation strength to rise by a factor of two, as occurs with roughness. The sum of the partial fluctuation strengths gives the fluctuation strength as

$$\text{FS} = \sum_0^{24\,\text{Bark}} \frac{0.032 \Delta L \Delta z}{\dfrac{f_{\text{mod}}(z)}{4} + \dfrac{4}{f_{\text{mod}}(z)}}, \qquad (8.4)$$

where $\Delta L = 20\log\left(\dfrac{N'(1)}{N'(99)}\right)$.

8.2.5 **Sharpness**

Sharpness characterizes auditory perception related to the spectra correlation of a sound. The calculation model of sharpness was developed by Aures and Bismarck.

The evaluation of the noise for quantifying the ratio of high-frequency level to overall level can be done by sharpness, which is given by

$$S = \frac{0.11\int_0^{24} N'zg(z)dz'}{N}$$

$$\frac{S}{\text{acum}} = 0.11\frac{\int_0^{24\,\text{Bark}} N'(z)g(z)zdz'}{\ln\left(\frac{N}{\text{sone}\times 20}+1\right)}$$

$$g(z) = e^{0.171z} \tag{8.5}$$

where

N' is the specific loudness within the critical band (Bark)

$g(z)$ is a critical band rate-dependent weighting factor that is unity between 0 Bark and 16 Bark and then increases to four at 24 Bark

z is tonality (Bark)

N is the total loudness

The unit of sharpness is acum. One acum is the sharpness for pure tone sound with an amplitude of 60 dB at 1 kHz. It is a kind of specified "center of gravity" on the frequency scale of spectral envelope, and it can be calculated from the spectral curve. Acum is the unit of sharpness. The sensation of sharpness results from high-frequency components in acoustic signals and is defined as a linear perception dimension. The sharpness of a narrowband noise of 1 kHz, bandwidth less than 150 Hz (critical bandwidth), and level of 60 dB is defined as 1 acum.

Spectral envelope defines the shape of spectrum of a sound. The shape of the spectral envelope affects the steady-state timbre features of a sound. Whether the spectral envelope is smooth or irregular, the distribution of energy over frequency is the major determination of the quality of a sound or its timbre. A sound whose spectral envelope has a lot of energy concentrated inside a narrow range of frequencies is noisier than sound with a smooth spectral envelope. In addition to the spectral envelope, a temporal envelope (the attack and decay of a nonsteady state sound) is another determinant of timbre. The main parameters of the spectral envelope are given as follows.

The spectral centroid (SC) can be thought of as the center of gravity for the frequency components of a signal and is correlated with the subjective quality of brightness. The SC is defined as

$$\text{SC}_{\text{Hz}} = \frac{\sum_{k=1}^{N=1}f[k]x[k]}{\sum_{k=1}^{N=1}x[k]} \tag{8.6}$$

where

$X[k]$ is the magnitude of frequency k
$f(k)$ is the center frequency of that signal
N is the length of the DFT
SC is the spectral centroid in Hz

The brightness can be calculated as

$$\text{brightness} = \frac{\sum_{k=1}^{Np} k a_k}{\sum_{k=1}^{Np} a_k} \tag{8.7}$$

where N_p is the number of extracted partials and a stand for their amplitude. If the partial multiplication k is replaced by the frequency of the partial, the brightness is expressed in Hertz. Generally, sounds with dark qualities tend to have more low-frequency components and those with a brighter sound are dominated by higher frequencies.

The spectral flatness measure (SFM) describes the flatness properties of the spectrum of a signal. A high spectral flatness (approaching 1.0 for white noise) indicates that the spectrum has nearly the same power all over the spectral bands. It typically sounds similar to white noise, and the shape of the spectrum is relatively flat and smooth. A low spectral flatness (approaching 0.0 for a pure tone) indicates that the spectral power is concentrated in a relatively small number of spectral bands. It sounds like a mixture of sine signals. The SFM is defined as the ratio between the geometric mean (G_m) and the arithmetic mean (A_m)

$$\text{SFM}_{dB} = 10\log\left(\frac{Gm}{Am}\right) = 10\log_{10}\left(\frac{\left(\prod_{k=0}^{k=N-1}|X[k]|\right)^{1/N}}{\frac{1}{N}\sum_{k=0}^{N-1}|X[k]|}\right) \tag{8.8}$$

where $x[k]$ represents the magnitude of frequency component N.

The roll-off is defined as the frequency boundary where 85% of the total power spectrum energy resides. It is commonly referred to as skew of the spectral shape and is frequently used in differentiating percussive and highly transient sounds (which exhibit higher frequency components) from more constant sounds, such as vowels.

$$\sum_{k=0}^{R} X[k] = 0.85 \sum_{k=0}^{N-1} X[k] \tag{8.9}$$

where R is the frequency roll-off point with 85 % of the total energy.

Kurtosis is a measure of impulsiveness of the sound signal. Basically, it sums up all time sample level differences from the signal mean value. The method represents the impulses in the sound and a high kurtosis value normally reflects poor sound quality.

$$\text{kurtosis} = \frac{\mu_4}{\sigma_4} - 3 \tag{8.10}$$

where

μ_4 is the fourth moment about the mean
σ is the standard deviation

8.2.6 Tonality

Tonality is a measure of the proportion of tonal components in the spectrum of a signal and allows a distinction between tones and noises. Tones consist mainly of tonal components, which appear in the spectrum as pronounced peaks.

The hearing threshold of pure tone varies with frequency. The ear is far more sensitive in the middle (1000-5000 Hz) of its perceived frequency range than in both the low and high frequency ranges. The absolute threshold curve shows a distinct valley around 3000-4000 Hz and a peak around 8000-9000 Hz, which means sounds in these frequency ranges are more detectable.

The human hearing system is thought to contain a bank of band-pass filters with overlapping pass bands. The psychoacoustics term "critical band" refers to the frequency bandwidth of the human auditory filters. Generally, the critical band is the band of frequencies within which a second tone will interfere with the perception of the ordinary tone by masking effect. The Bark scale, which is proposed by Zwicker, ranges from 1 to 24 and corresponds to the first 24 critical bands of human hearing response. The relationship between frequency and Bark is shown as follow:

$$\text{Bark} = 13\arctan(0.00076f) + 3.5\arctan\left(\left(\frac{f}{7500}\right)^2\right) \tag{8.11}$$

The width of the critical band is

$$\Delta f_c = 25 + 75 \times \left[1 + 1.4 \times (f_0/1000)^2\right]^{0.69} \tag{8.12}$$

$$f_0 = \sqrt{f_1 \times f_2}$$

$$\Delta f_c = f_2 - f_1$$

where

Δf_c is the width of the critical band
f_0 is the center frequency
f_1 is the lower band edge frequency
f_2 is the upper band edge frequency

Another concept associated with the human auditory filter is the equivalent rectangular bandwidth (ERB). The ERB represents the relationship among the auditory filter, frequency, and the critical bandwidth. The relationship between frequency and ERB is described as

$$\text{ERB} = 24.7 \times \left[(4.37 \times f_c) + 1\right] \tag{8.13}$$

where f_c is the filter center frequency in kHz.

The results of two tones are also suitable to multiple tone situations. Multiple tones have some other special features: sounds with widely spaced frequencies are more detectable, frequency components at the edge of a complex sound are more easily heard than inner ones, and sounds with more frequency components are more detectable.

Bark is the unit of the tonality scale. The scale corresponds to segments of the basilar membrane (inner ear) of equal length. Critical bands are segments of equal length on the tonality scale. The audible frequency range has been divided into 24 critical bands on a scale of 0 to 24 Bark by Zwicker. Noise and broadband noises are considered as having no or little tonality.

The detailed approaches to quantify tonality will be presented in the next section dedicated to motor sound tonality characterization.

8.2.7 Psychoacoustic Annoyance

This method is based on the calculation of psychoacoustic parameters, such as loudness, roughness, sharpness, and fluctuation strength. These four parameters are used to form another parameter that is used to evaluate the annoyance known as psychoacoustic annoyance (PA) or objective noise index (ONI). On the other hand, another typical method is based on listening tests in combination with the application of questionnaires. Different kinds of questionnaires and different scales of evaluation were used in the existing studies as subjective evaluation, which is then correlated with the above objective evaluation.

Regulations or international standards are not available yet for sound quality. There are no clear subjective indications as in the objective part. Some psychoacoustic parameters defined by Zwicker have been used for the objective part of the sound quality, which indicates how the sound quality can be evaluated with a unique value that combines the influence of loudness, sharpness, roughness, and fluctuation strength. This is done through the PA or ONI, to objectively quantify subjective evaluation.

$$\text{ONI} = N\left(1 + \sqrt{\left[g_1(S)\right]^2 + \left[g_2(R, F)\right]^2 + \cdots}\right) \tag{8.14}$$

in which

N is the loudness
S is sharpness in acum
R is roughness in asper
F is fluctuation strength in vacil
g_1, g_2 are specific functions

$$g_1 = \left(\frac{S}{\text{acum}} - 1.75\right).0.25.\log\left(\frac{N}{\text{sone}} + 10\right) \text{ for } S > 1.75 \text{ acum}$$

$$g_2 = \frac{2.18}{\left(\dfrac{N}{\text{sone}}\right)^{0.4}}\left(0.4.\frac{F}{\text{vacil}} + 0.6.\frac{R}{\text{asper}}\right) \tag{8.15}$$

in which

S is sharpness (acum)
F is the fluctuation strength (vacil)
R is the roughness (asper)

Zwicker and Fastl also described a formula For calculation of sensory pleasantness developed by Aures. In this formula, sensory pleasantness was calculated from the parameters of fluctuation strength, roughness, tonality, and sharpness.

A major difference between annoyance and pleasantness is that loudness has a greater influence on annoyance than on pleasantness. Therefore, characterization of annoyance should be used rather than a characterization of pleasantness. According to Aures, the tonality increases the pleasantness, while Kryter and Pearson indicated that tonal sounds are more annoying than noises under the same circumstances. Tonality was not included in this formula For PA, which presupposes no pure tones in the sound, and it was assumed small because of the spectral width of the impulsive closing sounds.

8.2.8 Sensory Pleasantness

Sensory pleasantness is a compound metric that is influenced by loudness, roughness, sharpness, and tonality. Sensory pleasantness, P, is calculated by

$$\frac{P}{P_0} = e^{-\left(0.023\left(\frac{N}{N_0}\right)\right)^2} e^{-0.7\left(\frac{R}{R_0}\right)} e^{-1.08\left(\frac{S}{S_0}\right)} \left(1.24 - e^{-2.43\left(\frac{T}{T_0}\right)}\right) \tag{8.16}$$

where the subscript 0 indicates reference values for each of the elementary perceptions.

8.2.9 Articulation Index

Articulation index (AI) is a measure of the intelligibility of speech or language. The AI is determined by understanding spoken words correctly with the scale ranging from 0 for a completely unintelligible value to 1 for a fully intelligible value.

The AI can be used to estimate sentence intelligibility. The higher AI is desirable inside the vehicle, so as to allow background noise not to affect the conversation of passengers or cell phone usage. The AI indicates the extent to which noise reduces intelligibility of speech. Intelligibility depends on both the level and frequency of background noise. Language and the audible range of human hearing can be represented as a spectrum region of sound pressure (whispering to shouting) and frequency (200 Hz to 6300 Hz).

AI is usually quantified as a function of velocity. The interior noise of a vehicle is proportional to speed. As the Revolutions Per Minute (RPM) increases, the noise changes in such a way that AI continues to decrease (for instance, from 95% to 35%).

The AI analysis includes testing to what extent the spectral components of the background noise (third-octave level curve) are important frequency components for intelligibility (limited by the speech area); the third-octave level curve dissects the speech area. This results in a breakdown of the AI (the table of fixed values assigned to the area). When the noise spectrum is in the lower section of the speech area, communication is only slightly disturbed and AI is of high value (close to 1). When the noise spectrum is in the upper section of the speech area, communication is made more difficult and the AI is of low value (close to 0).

Assume the spectrum of speech is represented as $H(f)$ in 1/3 octave range. If the background noise level is higher than this speech level, the hearing is not clear. When the noise level is higher than the speech by 12 dB, the speech is not entirely intelligible. The corresponding background noise is defined as an upper limit noise, $UL(f)$, and the relationship between $UL(f)$ and $H(f)$ is

$$UL(f) = H(f) + 12\,\text{dB} \tag{8.17}$$

Similarly, if the noise level is lower than speech by 30 dB, the speech is totally intelligible. The corresponding background noise is defined as a lower limit noise, $LL(f)$, and the relationship between $LL(f)$ and $H(f)$ is

$$LL(f) = H(f) - 30\,\text{dB} \qquad (8.18)$$

The above-defined difference between the upper and lower noise limits is valid for all frequencies, or 30 dB for all. However, the speech sound is frequency-dependent. Hence, a weighting function, $W(f)$, is needed where the speech sound with a frequency outside the range of 200 Hz to 6300 Hz is ignored. AI can be represented as

$$\text{AI} = \sum W(f)D(f)/30 \qquad (8.19)$$

where $D(f)$ is defined as follows. When the noise exceeds the upper limit, or $N(f) > UL(f)$, speech is entirely unintelligible and AI = 0, then

$$D(f) = 0 \qquad (8.20)$$

When the noise falls between the upper and lower limits, $LL(f) < N(f) < UL(f)$, speech is partially intelligible and

$$D(\text{f}) = UL(f) - N(f) \qquad (8.21)$$

When the noise is lower than the lower limit, $N(f) < LL(f)$, speech is entirely intelligible and

$$D(f) = 30 \qquad (8.22)$$

As an example, two group noises, 1 and 2, are given in terms of their level, $D(f)$ and AI. The corresponding AI for the two noises are, respectively, 49.3% and 33.6%.

The sound pressure level of the second noise is higher than the first one, thus the first noise has a higher AI.

8.2.10 Speech Interference Level

Speech interference level (SIL) is another sound quality index. SIL is calculated by averaging octave levels in the frequency bands considered to be relevant for speech intelligibility. The value is used to quantify the interference of the measured background noise on speech intelligibility.

The following different SILs are common.

- SIL-3: Average of octave levels at 1 kHz, 2 kHz, and 4 kHz
- SIL-4: Average of octave levels at 500 Hz, 1 kHz, 2 kHz, and 4 kHz
- P-SIL ("Preferred SIL"): Average of octave levels 500 Hz, 1 kHz, and 2 kHz

In contrast to the more commonly used energetic averaging, levels are averaged directly in the dB domain. SIL is a metric indicating the ability of a noise background to interfere with the understanding of speech. The speech signal has frequency components from about 200 Hz to 5000 Hz, and it varies in time with the natural rise and fall of the voice.

The SIL of noise indicates the extent to which that noise reduces the intelligibility of speech. Intelligibility depends on the level and frequency of background noise and on the speech spectrum itself. The SIL index is calculated according to ANSI S3.5-1997 "Methods for Calculation of the Speech Intelligibility Index." This standard is a major revision of ANSI S3.5-1969 "Methods for Calculation of the Articulation Index." The SIL is calculated using two spectra: the noise spectrum and the speech spectrum. The details of the calculation depend on the method selected in the band properties (octave, 3rd octave, or critical bands).

8.2.11 Other Psychoacoustic Indexes

In the automotive sound quality community, engineers have tried to identify what aspects of a sound define its quality. It has been the experience that analysis of acoustic signals alone does not identify the quality of those signals. Individuals usually use descriptive words, such as buzz, cheap, luxurious, weak, etc., to describe the defining attributes in sounds. Knowing how to design correct attributes into a vehicle sound directly impacts the appeal of the vehicle and ultimately impacts the profitability of a vehicle line. No instruments or analysis techniques have been able to quantify the terms mentioned above without the aid of subjective testing. The correlation between the physiological acoustic matrix and the subjective evaluation has been explored by regression analysis. Some empirical formulations have been proposed to quantify favorable sound quality, which requires the control of dominant order, tonality component, linearity, and lower noise level. Consider that the index depends on vehicle acceleration performance, frequency range, combination of relevant engine orders (CEO), and the ratio of engine orders to background noise (REO), the sportive sound quality can be given by

$$Sportness = f\left(REO, \ Tonality, \frac{\Delta rpm}{\Delta t}, \cdots \right) \tag{8.23}$$

This requires the firing order and harmonic component to be salient in contrast to lower background noise. The powerful sound quality is described by

$$Powerfulness = f\left(CEO, \ REO, \ Tonality, \frac{\Delta rpm}{\Delta t}, \cdots \right) \tag{8.24}$$

This requires the better combination of relevant engine orders of better linearity. The order components are higher than nonorder components in noise. The luxury sound quality is described by

$$Luxury = f\left(AI, \ Tonality, \ Less \ roughness, \cdots \right) \tag{8.25}$$

This requires better speech intelligibility and less noise roughness.

It has been proposed to use linear combinations of the elementary perceptions for describing the sound quality of representative sound or auditory stimuli. The metrics are combined in a formulation with the following general form:

$$Y = w^T F \tag{8.26}$$

where

$F \in R^0$ is a vector composed by sound quality metrics

$W \in R^0$ is a vector that weights the relevance of each metric in a given auditory scenario

From the point of view of optimization, this kind of formulation owns three interesting advantages over sensory pleasantness, namely, (i) the possibility of incorporating other relevant metrics in a specific auditory condition, rather than merely loudness, roughness, sharpness, and tonality; (ii) the advantage of incorporating decision-maker preferences by using weights; and (iii) the possibility of minimizing/maximizing all or some of the associated elementary perceptions. On the other hand, the implementation of weights should be carried out carefully, as this possibility is indeed a means of constraining the optimization toward a single result.

The vector optimization could be implemented by several approaches, such as multiobjective evolutionary algorithms, for example, applied to an electric motor driven powertrain, as perceived in the passenger compartment of a hybrid vehicle. Some criteria concerning the identification and optimization of the responsible narrowband components for the psychoacoustic perceptions could be developed.

Finally, active sound quality control could be designed and implemented through computer simulations for tackling a synthesized, stationary, powertrain-induced noise, hence, demonstrating the viability of accomplishing the desired sound quality targets devised during the vector optimization stage.

The use of hybrid and electric powertrains in passenger vehicles brings unusual narrowband auditory stimuli into the passenger compartment. When attempting to shape and control those effects, in either a passive or active approach, it is possible that one aspect of the soundscape is addressed at the expense of evidencing other unwanted noise components. Therefore, a proper handling of the different disturbance components is needed, which would consider the intricate interrelation between those components and its cross effects on the various sound quality metrics relevant to this application.

More general sound quality index could be developed for reliable objective assessment of the sound quality of BEV/HEV/FCEV. The sound quality index is based on the determination of weighting factors for several selected psychoacoustic quantities. The weighting factors have been obtained by the correlation and subsequent statistical processing of the objective binaural measurement and the questionnaire survey with many respondents. Point scores obtained from the respondents were subsequently statistically correlated with measured psychoacoustic parameters by the multicriteria method, namely, the weighted sum method (Pearson). By the application of the obtained weighting factors to the measured psychoacoustic parameters, it is possible to objectively determine the sound quality index of BEV/HEV/FCEV without the questionnaire survey.

To quantify the sound quality, objective indexes have been set up by applying acoustic and psychoacoustic metrics, and by correlating with the correspondent subjective evaluations through correlation analyses between subjective judgments on sound quality and objective parameters. Regression analyses were applied to develop models of perceived sound quality and annoyance.

8.2.12 Subjective Evaluation and Correlation

Automotive companies have tried hard to improve vehicle sound quality. The existing research provides a set of guidelines intended to be used as a reference for the practicing automotive sound quality engineer with the potential for application to the field of general consumer product sound quality.

For the assessment and design of sound quality, which depends on subjective impressions, psychoacoustic parameters or a combination of the basic psychoacoustic parameters are used or developed to predict sound quality ratings objectively. It is critical to find physical metrics, which show good correlation with subjective impressions to predict the sound quality and find countermeasures to improve sound quality of vehicles. Many physical metrics have been proposed and some of them are shown to have good correlation with subjective perception.

However, due to the complex relationship between the psychoacoustic parameters and human hearing, there is still no approach that is completely recognized by the public and directly used to map the objective evaluation parameters to subjective sensations. Therefore, based on some vehicle interior noises or reference sound in design, an investigation on correlations between the subjective and objective evaluation parameters of sound quality has been investigated.

After the objective psychoacoustic parameters are derived, the subjective part of the sound quality can be established through a set of surveys to a representative sample of customers who listen to the binaural recordings. Those data allow statistics to be derived and to establish correlations with classical and sound quality parameters.

The process of a jury panel study usually consists of the following steps: select jury members representing customers; get the jury trained in the meaning of the questions and in the scaling of responses to allow them to respond to sounds in a structured way; record a group of sounds; and conduct listening by the jury with headphones or loudspeakers, preferably binaural listening. Jury testing is a simple subjective testing done with a group of persons, rather than one person at a time. Subjective testing can be done with a single person or many people at the same time; both cases have their own set of benefits and caveats.

For subjective evaluation, there are several methods applied: scale rating, paired comparison, semantic differential, magnitude estimation, and their combinations.

Conducting subjective jury evaluations can be useful for understanding the importance of various sound metrics in selecting varied reference sound designs. Customer preference alone, based on valid jury evaluation results can provide sufficient information for sound/system selection. The reference sound or passenger binaural head acoustic data can be acquired for different system designs/acoustic options when the system or vehicle is conducted on dynamometer in anechoic chamber or road with different speeds.

For example, the following is a typical practice used [22]. Specific topics are covered by standardized procedures, including specifying the listening environment, subjects, sample (sound) preparation, test preparation and delivery, jury evaluation methods, analysis methods, and subjective to objective correlation.

The listening environment specification includes room acoustics/frequency characteristics, ambient noise control, the listening room decoration/lighting, air circulation, temperature, and humidity control.

Subjects indicate any person who takes part in the evaluation of sounds in a listening study, its selections, subject type, listening experience, product experience, demographics, number of subjects, simple evaluation tasks, complex evaluation tasks, recruiting subjects, company employees, customers subject training, simple evaluation tasks, and complex evaluation tasks.

Sample preparation includes good recording/measurement practice, level setting and calibration, recording practices, recording range, measurement variation, recording variations sample (sound) selection sample editing and preparation, extraneous noise equalizing loudness, sample length, binaural recording equalization free-field equalization (FF), diffuse-field equalization (DF), independent of direction equalization (ID), other recording issues.

Test preparation and delivery covers sound presentation (play) order scaling task presentation of samples, pacing or timing, paced juries sample size, test length, sound reproduction, method headphones data collection environment, and subject instructions.

Jury evaluation methods include rank order, response (rating) scales, paired comparison method, semantic differential, and magnitude estimation.

Analysis methods include distribution analysis, mean, median, mode, measures of variability, range, interquartile range, variance and standard deviation, measures of shape, skewness, kurtosis, test for normality, graphical techniques, scatter plots, quantile-quantile and normal probability plots, histograms, confidence intervals, testing and comparing sample means, one/two sample t-test, comparing equality of means for k samples, an extensive treatment of analysis of variance (ANOVA), Fisher's Least significant difference (LSD), linear regression analysis, residual analysis, and factor analysis,

The logic behind performing subjective to objective correlation centers on the concept that one can possibly replace subjective testing with mere objective characterizations of the stimuli. By doing this, one can reduce subjective testing that is costly from a time, equipment, facilities, and general logistics standpoint. If one can reliably replace subjective testing and the costs involved with objective characterizations of the stimuli, which are usually less costly, then that gives a significant impetus to finding strong correlation between the two views of the stimuli. The eventual goal, of course, is to guide the establishment of design and testing specifications for the product that will guarantee the product is the one that the marketplace desires.

Moreover, increasingly more complicated and powerful correlation methods have been proposed and applied as follows.

Scatterplot method. If the sound event is strongly correlated to a single, physical dimension of the stimuli, a simple scatterplot will often yield significant insight into relationships between the subjective responses and the physical stimuli. The process merely involves plotting the subjective response for a stimulus against some scalar on the vertical axis versus some objective measure of the stimuli on the horizontal axis. Most popular spreadsheet and statistical software packages provide means for generating scatterplots.

Linear regression method. Linear regression takes scatterplots one level of information higher by providing a mathematical relationship between the subjective and objective characterizations of the stimuli. The mathematical relationship in this case is

$$y = B_0 + B_1 x + \varepsilon \qquad (8.27)$$

where
    x is the objective characterization of the stimuli or regressor
    y is the response
    B_0 is the intercept
    B_1 is the slope of the straight line relationship between y and x

The error term ε is considered random with mean of zero and an unknown variance. Least squares estimation is used to find the values of the coefficients on the regressors. One looks for the objective characterization, whether physical or psychophysical, that gives the strongest correlation and hence, the higher coefficient of determination, R^2.

Multiple linear regression method. Going one step further than single variable linear regression, one can perform multiple linear regression where the straight-line relationship is now between the subjective response data and some linear combination of scalar, objective characterizations of the stimuli. In this case, the model regression equation is

$$y = B_0 + B_1 x_1 + B_2 x_2 + \cdots + B_n x_n \qquad (8.28)$$

Again, the goal is to find the relationship that gives the strongest relationship between the regressors and the response, yielding the highest R^2. Issues that need to be addressed when performing multiple linear regression are feasibility of using multiple regressions in setting test and design specifications, and collinearity of regressors.

Nonlinear regression method. Relationships between subjective responses and objective characterizations are not limited to the linear type. They can be nonlinear relationships, nonlinear combinations of the objective characterizations that correlate to the subjective response data. The mathematics involved here can be more complicated than for linear regression.

Neural network method. Another method to derive relationships between subjective response and objective characterizations of a set of stimuli is the use of neural networks. This method is relatively untapped in the field of automotive sound quality but does offer some exciting potential in the areas of pattern recognition. Neural networks are advertised as an attempt to model the neural structures in the human brain to perform artificial learning in a system. Much like the mysteries involved with how the brain organizes information, neural networks operate as a black box with known input and output-but little usable information about the sequence and logic of events that happen inside. Typically, these networks require a great deal of input/output data (training) to produce reliable results. Moreover, the grey system method has also been used.

Putting one's instincts to work is also critical. From the perspective of a person that is at least partly responsible for automotive sound quality, this person will most likely have some knowledge and suspicions about the relationships that might exist between particular sound events and the mathematical transformations of those same signals. Performing a proper subjective to objective correlation will give one the chance to test and challenge those suspicions. The process involved is technically the input and output data must make sense. One will need to defend any derived relationship when using it to establish design and test specifications. There are always chance relationships which will, from time to time, appear in correlation. This arises usually from too small of a number of stimuli and/or subjective evaluations being used.

It is very easy to fit straight lines to a small number of data points. One also needs to guard against using multiple regressions that use dimensions that are highly colinear. For example, if a multiple regression shows that a 3-regressor equation gives a high R^2

value, but those regressors are all colinear (i.e., strongly correlated), which is to say they are all telling nearly the same thing about the stimuli, the regression is probably no more useful for setting test and design specifications than a single dimension regression using three regressors. The only information proven by the 3-regressor correlation was that each regressor gives a slightly different view of the stimuli. No correlation that is performed should be thought of as proof of causality between the stimuli and the subjective response.

The pair comparison method has been widely applied [23, 24]. The typical approach is known as least squares estimation and works on the premise of the model regression equation. The files were presented to the jurors for subjective evaluation using specific methods, such as the paired comparison test. The jury studies were run in the jury room using a specific procedure or jury evaluation software package. The number of jurors used in the study could be tens of people. A paired comparison of preference evaluation was conducted in which subjects were asked to choose which sound they preferred for an interior acoustic package. The details of the measurement conditions were discussed with each juror. However, the test was blind in that the jurors were not aware of which sound corresponded to which design. The jurors were presented with pairs of sounds and were asked to select which they preferred. This process was repeated for every possible pair and presented in random order with all pairs repeated. This type of testing was used to remove presentation order error and minimize bias. Subjects were asked to select Sound A or Sound B and were not given the option to select "Neither." Many jury experts agree that ties should be allowed for naïve subjects, while expert subjects should be forced to choose. The jurors, mostly engineers and technicians, were considered to be expert subjects and were thus forced to make a choice. Two major quality indicators, repeatability and consistency, were used in the paired comparison evaluations to determine if the evaluation was a good one. All paired comparisons were presented to the evaluators twice to calculate the repeatability of their responses and all responses were used in the study. Consistency is obtained by evaluating all triads (a, b, c), i.e., if a is preferred to b and b is preferred to c, then the triad is consistent if a is preferred to c and inconsistent if c is preferred to a. The following thresholds were used for juror acceptance: 60% consistency and 60% repeatability. Consistency measured above 70% for all jurors while 7 of the 24 jurors were rejected, all for repeatability. Jurors did present some issues of fatigue as well as the inability to choose a preferred sound in some pairs. The repetition of pairs likely played a role in the fatigue of the jurors and thus was a factor in producing repeatability errors [80–83].

However, the pair comparison method does not provide an indication of the magnitude of the winning performance margin. The other methods, such as bipolar semantic differential scales, also provide information on the preferred psychoacoustic metric, while yielding a degree of magnitude for the winner. Bipolar semantic differential scales are thus very useful and are used in industry for subjective NVH testing [24–27]. The semantic differential method with a reference noise [27–30] is widely used for subjective evaluation of vehicle noises, in which a reference noise as an anchor is introduced into the evaluations for improving the accuracy of the jury tests, and the words and scaling scores in the jury tests are templated.

8.3 Sound Quality Index for High-Frequency Tonal Noise: Prominence Ratio and Tone-to-Noise Ratio

It is well known that prominent tones contribute to unpleasantness and that this is particularly true for high-frequency tones. Some metrics have been developed for evaluating the prominence of each individual tone, such as TNR or prominence ratio (PR), as defined in standards (e.g., DIN 45681-2002 or ISO S1.13-1995). Some more complex metrics, such as Aures' tonality aims at evaluating the tone of a sound in which several tones can be detected. The knowledge of the relation between tone and the sound quality have been improved [3, 20–35].

In BEV/HEV/FCEV, there have been many needs for the identification and evaluation of prominent discrete tones. One of the most important issues is interior whine noise, which is frequency-related tonal noise. It is well-known from literature that noise containing tonal components is more annoying than noise without tonal components.

The below is an example of the sound quality evaluation of interior noise of EV vs. ICE vehicles in acceleration WOT operation. As shown in **Figure 8.2** and **Table 8.1**, the SPL, sharpness, AI, PR, and TNR are compared for the two vehicles SIEMENS.

FIGURE 8.2   SPL and sharpness comparison of two vehicles

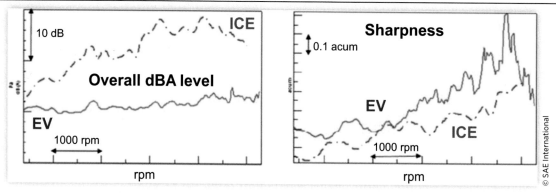

© SAE International

TABLE 8.1   Comparison of sound indicators for two vehicles

Articulation index	ICE 56%
	EV 73%
Prominence ratio	EV order 4xnp
	9.27 dB
	>9dB threshold
Tone-to-noise ratio	EV order 4xnp
	11.03 dB
	>8 dB threshold

© SAE International

Overall, the representative sound quality metrics of TNR and PR are suitable for SQ quantification and target setting.

In general, the audible threshold or the prominence of a tonal component is decided by the relationships between the tonal component level and the surrounding band noise level, which is masking the tonal component. The frequency bandwidth is called the critical band that is centered at the frequency of the tone [3].

In the standards, such as the ISO 7779, ECMA 74, and ANSI S1.13, the judgment methods of the certain prominent discrete tone in information technology devices are presented as the TNR and PR.

The TNR is defined as the decibel value of the ratio of the power of the tonal component and other noise component in the critical band. According to ECMA-74, a tone is classified as prominent if the TNR is higher than 8 dB for tone frequencies of 1 kHz and higher. For frequencies below 1 kHz, this threshold value is increased by 2.5 dB per octave.

In case that multiple peaks exist in the same critical bandwidth or the noise level next to the critical bandwidth is considerable, the TNR value tends to show bigger or smaller.

The PR is defined as the decibel value of the ratio of the critical bandwidth power, including the tonal component and the critical bandwidth power on both sides. When using the PR method according to ECMA-74, a tone is classified as prominent if the difference between the level of the critical band centered on the tone and the average level of the adjacent critical bands is equal to or greater than 9 dB for tone frequencies of 1 kHz and higher. For frequencies below 1 kHz, this threshold value is increased by 3 dB per octave.

The relevant standards have detailed TNR and PR procedures for the identification and evaluation of prominent discrete tones.

The discrete tones and tonality are treated separately. The tonality can arise not only from discrete tones but also from other phenomena and conjunctions of phenomena.

The advanced psychoacoustic tonality calculation method procedure based on the hearing model has also been developed.

Discrete tones occurring at any frequency within the one-third octave bands having center frequencies from 100 Hz to 10,000 Hz can be evaluated by the procedures (i.e., discrete tones between 89.1 Hz and 11,220 Hz).

The TNR method may prove to be more accurate than the PR method for multiple tones in adjacent critical bands, for example, when strong harmonics exist. The PR method can be more effective for multiple discrete tones within the same critical band and is more readily automated to handle such cases. The TNR may underestimate prominence of discrete tones that overlap with elevated or sloped noise spectra. The PR may be applicable to tonalities like discrete tones and noise with narrow or sloped spectra, to the degree that critical bands adjacent to the tonalities of interest are free of such tonalities.

Based on the psychoacoustical principle, a discrete tone which occurs together with broad-band noise is partially masked by that part of the noise contained in a relatively narrow frequency band, called the critical band, that is centered at the frequency of the tone. Noise at frequencies outside the critical band does not contribute significantly to the masking effect. The width of a critical band is a function of frequency. In general, a discrete tone is just audible in the presence of noise when the sound pressure level of the tone is about 4 dB, depending on frequency below the sound pressure level of the masking noise contained in the critical band centered around the tone. This is sometimes referred to as the threshold of detectability. For the purposes of this text, a discrete tone is classified as prominent when using the TNR method if the sound pressure level of the tone exceeds the sound pressure level of the masking noise in the critical band by 8 dB for discrete tone frequencies of 1000 Hz and higher, and by a greater amount for discrete tones at lower frequencies. This corresponds, in general, to a discrete tone being prominent when it is more than 10 dB to 14 dB above the threshold of detectability. When using the PR method, a discrete tone is classified as prominent if the difference between the level of the critical band centered on the tone and the average level of the adjacent critical bands is equal to or greater than 9 dB for tone frequencies of 1000 Hz and higher, and by a greater amount for tones at lower frequencies.

8.3.1 Tone-to-Noise Ratio Method

The sound pressure level of the discrete tone, L_t, is determined from the FFT spectrum in the narrow band that "defines" the tone. The width of this frequency band, Δf_t, in Hz, is equal to the number of discrete data points ("the number of spectral lines") included in the band, times the resolution bandwidth ("line spacing").

The sound pressure level of the masking noise, Ln, is taken as the value determined using the following two-step procedure.

The first step is to compute the total sound pressure level in the critical band. The width of the critical band is determined from the width of the critical band,

$$\Delta f_c = 25 + 75 \times \left[1 + 1.4 \times \left(f_0 / 1000 \right)^2 \right]^{0.69} \tag{8.29}$$

The critical band is modelled as an ideal rectangular filter with center frequency f_0, lower band-edge frequency f_1, and upper band-edge frequency f_2, where

$$f_2 - f_1 = \Delta f_c \tag{8.30}$$

For $89\text{Hz} \leq f_0 \leq 500\text{Hz}$, the critical band approximates a constant-bandwidth filter, and the band-edge frequencies are computed as

$$f_1 = f_0 - \frac{\Delta f_c}{2} \qquad f_2 = f_0 + \frac{\Delta f_c}{2} \tag{8.31}$$

For $500\text{Hz} < f_0 \leq 11200\text{Hz}$, the critical band approximates a constant-percentage bandwidth filter

$$f_0 = \sqrt{f_1 \times f_2} \qquad f_1 = -\frac{\Delta f_c}{2} + \frac{\sqrt{\left(\Delta f_c \right)^2 + 4 f_0^2}}{2} \qquad f_2 = f_1 + \Delta f_c \tag{8.32}$$

with f_0 set equal to the frequency of the discrete tone under investigation, f_t, and with lower band-edge frequency, f_1, and upper band-edge frequency, f_2, as given in Equation 8.32.

From the FFT spectrum, the total sound pressure level of the critical band, L_{tot}, is computed. Depending on the instrumentation used, this may be performed on the FFT analyzer itself using band cursors, on an external computer using appropriate software, or by some other means. In any event, the width of the frequency band used to compute this value, Δf_{tot}, in Hz, is equal to the number of discrete FFT data points included in the band times the resolution bandwidth.

The second step is to calculate the masking noise sound pressure level of the masking noise, L_n. The TNR, ΔL_T, in decibels, is calculated from one of the following formulas:

$$\Delta L_T = L_t - L_n \, dB \tag{8.33}$$

$$L_n = 10\lg\left(10^{0.1L_{tot}} - 10^{0.1L_t}\right) + 10\lg\left(\frac{\Delta f_c}{\Delta f_{tot} - \Delta f_t}\right) dB$$

where sound pressure level of the masking noise, L_n, the sound pressure level of the discrete tone, L_t, the total sound pressure level of the critical band, L_{tot}, with f_0 set equal to the frequency of the discrete tone under investigation, f_t, and with lower band-edge frequency f_1 and upper band-edge frequency f_2.

Prominent discrete tones criteria were set for the TNR method. A discrete tone is classified as prominent in accordance with the TNR method if one of the following conditions is met.

$$\Delta L_T \geq 8.0 + 8.33 \times \lg\left(\frac{1000}{f_t}\right) dB \quad \text{for } 89.1\,Hz \leq f_t \leq 1000\,Hz$$

$$\Delta L_T \geq 8.0 \, dB \, for \, 1000\,Hz \leq f_t \leq 11200\,Hz \tag{8.34}$$

The multiple tones in a critical band could be treated as follows. The noise emitted by a machine may contain multiple tones, and several of these may fall within a single critical band. If one or more discrete tones are audible, the procedure above is followed for each tone, with the following differences. The discrete tone with the highest amplitude in the critical band is identified as the primary tone, and its frequency is denoted as f_p. For the critical band centered on this primary tone, the discrete tone with the second-highest level is identified as the secondary tone and its frequency denoted as f_s.

If the secondary tone is sufficiently close in frequency to the primary tone, then the two are considered to be perceived as a single discrete tone and the prominence is determined by combining their mean-square sound pressures (or sound pressure levels). Two discrete tones may be considered sufficiently close or "approximate" if their spacing $\Delta f_{sp} = |f_s - f_p|$ is less than the proximity spacing, Δf_{prox}, in Hz, defined by

$$\Delta f_{prox} = 21 \times 10^{1.2 \times \left[\lg\left(f_p/212\right)\right]^{1.8}} Hz, \quad \text{for } 89\,Hz \leq f_p \leq 1000\,Hz \tag{8.35}$$

Figure 8.3 shows how a single tone in a critical band is analyzed using the TNR method. Figure 8.3 shows how the tone-to-noise ratio method is used when multiple tones exist in a critical band.

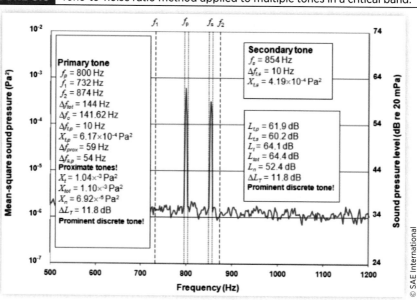

FIGURE 8.3 Tone-to-noise ratio method applied to multiple tones in a critical band.

8.3.2 Prominence Ratio Method

To quantify this index, the measurement using FFT analyser could be done first. Then, the level of the middle critical band I determined.

The sound pressure level of the middle critical band, L_M is defined as the sound pressure contained in the critical band centered on the discrete tone under investigation. The width of the middle critical band, Δf_M, and the lower and upper band-edge frequencies, $f_{1,M}$ and $f_{2,M}$ are determined from the relationships in the following equation with f_0 set equal to the frequency of the discrete tone under investigation, f_t. The band-edge frequencies then become,

For $f_t \leq 500$ Hz

$$f_{1,M} = f_t - \frac{\Delta f_M}{2} \quad \text{and} \quad f_{2,M} = f_t + \frac{\Delta f_M}{2} \tag{8.36}$$

For $f_t > 500$ Hz

$$f_{1,M} = -\frac{\Delta f_M}{2} + \frac{\sqrt{(\Delta f_M)^2 + 4 f_t^2}}{2} \quad \text{and} \quad f_{2,M} = f_{1,M} + \Delta f_M \tag{8.37}$$

The level of the lower critical band is determined as follows. The sound pressure level of the lower critical band, LL, is defined as the total sound pressure contained in the critical band immediately below, and contiguous with, the middle critical band. $f_2 = f_1 + \Delta f_c$ govern this lower critical band, with center frequency $f_{0,L}$, bandwidth Δf_L, and lower and upper band-edge frequencies $f_{1,L}$ and $f_{2,L}$, respectively. Since this lower critical band must be contiguous with the middle critical band, it follows that $f_{2,L} = f_{1,M}$. However, because $f_{0,L}$ is not known a priori, Equations 8.36 and 8.37 cannot be used

TABLE 8.2 Parameters for calculation of $f_{1,L}$.

Frequency range	$C_{L,0}$		$C_{L,2}$
Hz	Hz	$C_{L,1}$	Hz^{-1}
$89.1 \leq f_t \leq 171.4$	20.0	0.0	0.0
$171.4 < f_t \leq 1600$	-149.5	1001	-6.90×10^{-5}
$11200\,\text{Hz} > f_t > 1600$	6.8	0.806	-8.20×10^{-6}

TABLE 8.3 Parameters for calculation of $f_{2,U}$.

Frequency range	$C_{U,0}$		$C_{U,2}$
Hz	Hz	$C_{U,1}$	Hz^{-1}
$89.1 \leq f_t \leq 1600$	149.5	1.035	7.70×10^{-5}
$11200 > f_t > 1600$	3.3	1.215	2.16×10^{-5}

directly to determine the value of $f_{1,L}$, and an iterative method of solution would ordinarily has to be used. For the purposes of this annex, the value of $f_{1,L}$ shall be computed from Equation 8.38 (which has been derived from an iterative solution using curve fitting).

$$f_{1,L} = C_{L,0} + C_{L,1}f_t + C_{L,2}f_t^2 \tag{8.38}$$

where

f_t is the frequency of discrete tone under investigation

$C_{L,0}, C_{L,1}, C_{L,2}$ are constants given in **Table 8.2**

The level of the upper critical band is determined as follows. The sound pressure level of the upper critical band, LU, is defined as the total sound pressure level contained in the critical band immediately above and contiguous with the middle critical band.

$$f_{2,U} = C_{U,0} + C_{u,1}f_t + C_{U,2}f_t^2 \tag{8.39}$$

where f_t is the frequency of discrete tone under investigation. $C_{U,0}, C_{U,1}, C_{U,2}$ are constants given in **Table 8.3**.

The PR is then determined as follows. The PR, ΔL_p, in decibels, is calculated as (for discrete tone frequencies greater than 171.4 Hz)

$$\Delta L_p = 10\lg\left(10^{0.1L_M}\right) - 10\lg\left[\left(10^{0.1L_L} + 10^{0.1L_U}\right) \times 0.5\right]\text{dB for } f_t > 171.4\text{Hz} \tag{8.40}$$

Prominent discrete tone criterion for the PR method is presented as follows. A discrete tone is classified as prominent in accordance with the PR method if

$$\Delta L_p \geq 9.0 + 10\lg\left(\frac{1000}{f_t}\right)\text{dB for } 89.1\,\text{Hz} \leq f_t < 1000\,\text{Hz}$$

$$\Delta L_p \geq 9.0\text{dB for } 11200\,\text{Hz} > f_t > 1000\,\text{Hz} \tag{8.41}$$

FIGURE 8.4 Illustration of the PR method for prominent discrete tone identification.

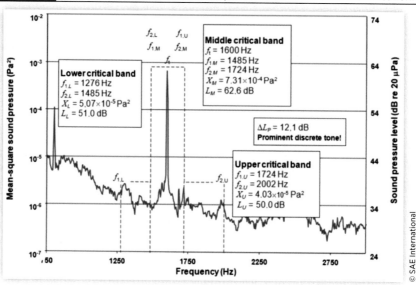

© SAE International

The following is an example of PR method. The PR method is illustrated graphically in **Figure 8.4**. The PR was calculated in accordance with Equation 8.41 and was found to be $\Delta L_p = 12.1$ dB for the 1600 Hz discrete tone. Because the result is more than 9 dB, which is the PR criterion at 1600 Hz, the discrete tone is classified as prominent.

Figure 8.5 shows the criteria for prominence for both TNR and PR as a function of frequency.

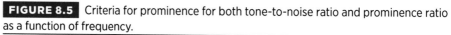

FIGURE 8.5 Criteria for prominence for both tone-to-noise ratio and prominence ratio as a function of frequency.

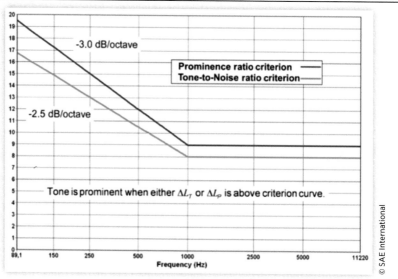

© SAE International

8.3.3 **New Approaches**

For many years, in vehicle and other product noise assessments, tonality measurement procedures, such as the TNR, PR, and DIN 45681 Tonality, have been available to quantify the audibility of prominent tones. Especially through the recent past, as product sound pressure levels have become lower, disagreements between perceptions and measurements have increased across a wide range of product categories, including automotive, information technology, and residential products. One factor is that tonality perceptions are caused by spectrally elevated noise bands of various widths and slopes, as well as by pure tones, and usually escape measure in extant tools. Near-superpositions of discrete tones and elevated narrow noise bands are increasingly found in low-level technical sounds. Existing pure-tone methodologies tend to misrecognize an elevated noise band as general masking, lowering the audibility of a tone in the measured vicinity, whereas perceptually they add. To address such issues, the new psychoacoustically based tonality model has been developed, which evaluates the nonlinear and time-dependent loudness of both tonal and broadband components, separating them via the autocorrelation function (ACF) and giving their spectral relationships. Based on a hearing model of Sottek, the model has been validated by many listening tests. Its background and current state have been widely reported with particular attention to automotive situations, such as perceptions inside electric and hybrid vehicles in stationary or low-speed operation, and IC sound design criteria balancing desired order tonalities and order-caused roughness [102, 105–123].

The new approach could be more promising than PR, TNR, DIN 45681 Tonality, or the old Aures-Terhardt Tonality method.

TNR and PR are mandated by ECMA-74 (1) to quantify the tonality of identified discrete tones, but do not respond well or even at all to tonalities caused by narrow bands of noise or nonpure tones, and thus are particularly useless with many frequently encountered tonalities. The very important topic of hard disk drive cover plate tonalities is an example. The latter involve combinations of elevated noise bands due to structural resonances and pure rotating-mechanism tones.

The psychoacoustic parameter tonality was introduced to quantify the perception of tonal content. However, existing methods for tonality calculation show problems when applied to technical sounds. In the new approach to tonality calculation based on a hearing model, the calculation of tonality is therein performed upon the basis of the partial loudness of the tonal content.

The new perceptually accurate tonality assessment method based on a hearing model of Sottek was developed, which evaluates the nonlinear and time-dependent loudness of both tonal and broadband components, separating them via the ACF. This new perception-model-based procedure, suitable for identifying and ranking tonalities from any source, is proposed for the next edition of ECMA-74 as an alternative to TNR and PR.

A hearing model approach to calculate psychoacoustic parameters is a perception-model-based procedure for determining the specific loudness of a sound. This result can be used as a basis for further psychoacoustic analyses. The described psychoacoustic hearing model transforms calibrated sound pressure data into psychoacoustic loudness, from which subsequent calculations can be made.

Figure 8.6 displays the basic hearing model structure for calculating specific loudness as the basis for determining other psychoacoustic sensations. Subsequently, the different signal processing blocks of the hearing model are briefly explained. The input signal is a discrete time signal containing sound pressure values.

The preprocessing consists of filtering the input signal with outer and middle ear transfer functions. The transfer function of the filter is chosen such that the filtering together with the loudness threshold leads to a loudness estimation emulating the equal-loudness contours from 20 to 90 phon and the lower threshold of hearing. The filter is optimized on the equal loudness contours of ISO 226:2003 for frequencies higher than 1 kHz. For lower frequencies, the equal loudness contours of ISO 226:1987 are chosen as target because there is a large uncertainty of the experimental data at low frequencies and recent studies showed that the less noise-sensitive curves of ISO 226:2003 do underestimate low-frequency noise issues. Thus, the more sensitive curves are preferred.

The identification and evaluation of prominent tonalities using a psychoacoustic tonality calculation method is briefed as follows. The next section describes a perception-model-based procedure for determining whether or not noise emissions contain prominent tonalities, and if present, their strengths: the psychoacoustic tonality calculation method.

Prominent perceived tonalities arise from a variety of causes including, but not limited to, prominent discrete tones: discrete tones, nonpure tones, narrow elevated noise bands, combinations of tones and narrow elevated noise bands, band-edges of various slopes terminating elevated noise bands of various bandwidths, and combinations of these.

The discrete tone tonality methods may be considered hybrids, as they apply a psychoacoustic concept (the critical bandwidth), but operate with conventional sound pressure level information rather than perception-based level information (loudness)

FIGURE 8.6 Basic hearing model structure.

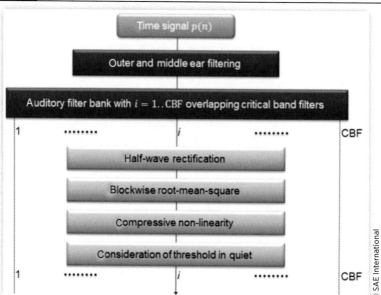

as input data. Thus, they do not inherently consider loudness-based perceptual tonal masking variation and the threshold of hearing, though an auxiliary procedure for hearing threshold compensation has been described.

The existing research results show a strong correlation between tonality perception and the partial loudness of tonal sound components. Therefore, the new hearing model approach to tonality on the basis of the perceived loudness of tonal content has been developed. The new model evaluates the nonlinear and time dependent specific loudness of both tonal and broadband components, which are separated using the ACF. This model has been validated by many sound situations and listening tests.

Moreover, the sliding ACF is used as a processing block in the hearing model for the calculation of roughness and fluctuation strength and later for other psychoacoustic quantities, like tonality and loudness. It was proposed to use the ACF of the band-pass signals to separate tonal content from noise. The ACF of white Gaussian noise is characterized by a Dirac impulse. Any broadband noise signal has at least a nonperiodic ACF with high values at low lags, whereas the ACF of periodic signals also shows a periodic structure. Thus, the loudness of the tonal component can be estimated by analyzing the ACF at a certain range with respect to the lag, and also the loudness of the remaining (noisy) part.

Figure 8.7 shows the schematic of the calculation of tonality based on the scaled ACFs, but with frequency-dependent analysis window borders.

FIGURE 8.7   Calculation of tonality based on the scaled ACFs with frequency-dependent analysis window borders.

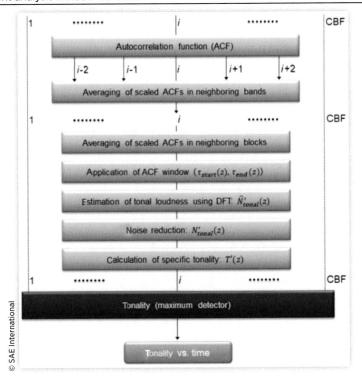

© SAE International

8.3.4 Application of Sound Quality Metric of Tonality for BEV/HEV/FCEV

The sound quality metric of tonality has been used for many engineering systems [3, 4].

The vehicle interior noise of HEV is composed of road noise, wind noise, ICE noise, battery fan noise, electric machine noise, and other unusual sounds. The noise from electric machines, such as motors and generators, manifests in the form of whine noise, i.e., tonal noise (typically in the 400 Hz-2000 Hz range). The tonal nature of the whine noise from the electric machines can be annoying to the customer. The tonal noise issues can play a larger role on BEV/HEV/FCEV due to the lack of sufficient "normal" masking noise from the ICE under the all electric range (AER) and regeneration modes of operation.

Figure 8.8 shows an example of interior noise measurements made on an electric vehicle operating in AER mode [102]. Specifically, the vehicle was tested by first ramping up the vehicle speed, holding a constant vehicle cruise speed, and then conducting a braking operation. It was found that there is significant whine noise present under both the vehicle speed ramp-up phase and the braking phase. The absence of masking background noise from the lack of ICE operation underscores the need to minimize the levels of whine noise shares in the interior noise of the electric vehicle in AER mode.

As shown in Figure 8.8, the electric machine whine order is prominent during vehicle acceleration and braking (due to regenerative braking). In addition, the whine noise order amplitudes change during vehicle acceleration and braking due to differences in the loading of the electric machines.

The lack of ICE noise can render the presence of electric whine noise in EV and ReEV in AER mode very audible and hence, objectionable. To understand potential vehicle NVH issues related to electrical whine noise, it necessary to understand the influence of masking levels (from other noise sources) that might make the whine noise

FIGURE 8.8   Interior noise of an electric vehicle in AER mode [102].

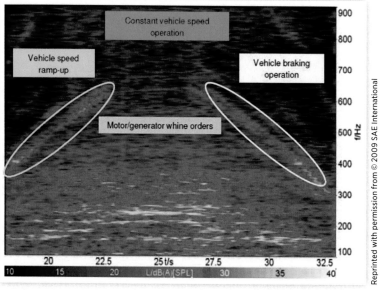

more or less objectionable to the driver. Specifically, listening studies have shown that it is important to evaluate the whine orders relative to masking noise bands composed of noise from other sources (e.g., ICE noise, road noise, wind noise) in a given frequency band. The whine order drops below the masking bands only for vehicle speeds above 45 mph. For vehicle speeds above 45 mph, relatively higher levels of road and wind noise shares effectively masks the electric whine noise under the tested conditions. However, the whine noise on the tested vehicle was clearly objectionable for vehicle speeds lower than 45 mph. The tonal character of the electric whine and its rather wide frequency range (corresponding to the vehicle speed range) makes the NVH development of ReEV and EV challenging.

It is important to understand the relative contributions from various structure-borne and airborne transfer paths, so as to develop appropriate NVH countermeasures to meet the refinement needs of the vehicle [18]. It is important to integrate sound quality design in the design specifications for new product developments and include it in the engineering design process. By doing this, automotive manufactures can reduce the later work of troubleshooting and troublesome adjustments, which will be time consuming and expensive. Some basic design principles based on the summed psychoacoustic results have been proposed. Pure tone, located in the 1000-8000 Hz range, especially in the 3000-4000 Hz range, is usually more detectable. Therefore, countermeasures to shift the frequency component location, like changing the transmission gear tooth numbers, fan blade numbers and so on, can be applied.

The in-vehicle whine issues are usually multiple-tone situations and many design principles have been developed, just to name a few [100–103].

1. Both the lowest and highest frequencies dominate the perceptions of the complex whine. It is better to avoid the location of the lowest and highest frequency components to be in the sensitive frequency range.

2. The more frequency components in the complex whine, the more detectable it is. For the vehicle cabin, the engine timing gears, belt, electrical motors, and transmission gears are all sources of whine noise. It is inevitable that these parts work simultaneously because of their working principle. Designers may modify their frequency components to be the same or to be harmonics of each other. This is because for a steady complex tone with many harmonics, it is usually perceived with a single pitch, which is corresponding to fundamental frequency and the individual harmonics are not separately perceived.

3. When there is a beat, it is more detectable even when the beat is in the inner of the complex whine. Therefore, all the frequency components should avoid the beat and roughness.

4. SC is related well with human hearing response.

The distribution of tonal components in many hybrid/electric vehicles was studied [119, 120]. The number of prominent orders, their maximum levels, and frequency separation were analyzed for the most critical driving conditions. The study is based upon measurements made on 13 electrified cars on the market. Table 8.4 shows the description of the test objects used in the study. Two established tonal metrics, PR and TNR, have been

TABLE 8.4 Description of the test objects used in the study [119, 120].

Test object	Slow speed	Full acceleration	Type	E-motor max power/torque	E-motor position	ICE max power / torque
Car 1	×	×	PHEV	2 * 60 kW/332 Nm	Front and rear	121 hp/190 Nm
Car 2	×	×	PEV	80 kW/254 Nm	Front	×
Car3	×		Mild hybrid	15 kW/160 Nm	Front	306 hp/370 Nm
Car4	×	×	PEV	285 kW/440 Nm	Rear	×
Car5	×		HEV	27 kW/200 Nm	Rear	163 hp
Car6	×		RE	111 kW/370 Nm	Front	80 hp
Car7	×	×	PEV	49 kW/200 Nm	Rear	×
Car8		×	PEV	82 kW/220 Nm	Front	×
Car9		×	PEV	82 kW/220 Nm	Front	×
Car10		×		89 kW/250 Nm	Front	×
Car11	×		HEV	183 kW/317 Nm +50 kW/139 Nm	Front and rear	249 hp/317 Nm
Car12	×		HEV	33 kW/211 Nm	Front	211 hp/350 Nm
Car13	×		PHEV	49 kW/200 Nm	Rear	215 hp/440 Nm

used for the data analysis and have also been evaluated and compared. Examples of when the two metrics provide significantly different results are examined and recommendations of when to use either one are given.

The main specific acoustic signature of the electric powertrain comprises orders (frequency components related to the rotational speed) due to primarily electromagnetic forces, tones, and/or orders due to DC/AC converter Pulse Width Modulation (PWM) and finally gear mesh order(s) from the reducer. All those phenomena yield tonal and high frequency sound characteristics, which can be perceived as annoying if being prominent. The prominence of tonal components is preferably described using psychoacoustic metrics in contrast to sound pressure levels.

Many research studies explored the interior tonal noise content of today's hybrid/electric cars [102, 119–123]. More specifically, the number of prominent orders, their maximum levels, and frequency separation were analyzed. This was done for the driving conditions identified as generally most critical, for example, when the emergences of tones are most evident.

It also evaluates and compares TNR and PR which were used for quantifying maximum levels and occurring frequencies. In some cases, those two metrics provide differences. The uncertainties associated especially with PR are that it does not take multiple tones located in the same critical band into account. In practice, multiple orders can be close in terms of frequency, hence present in any of the critical bands used to calculate PR or TNR.

In the study, two driving cases were identified where the presence of the tones from the electric powertrain (orders related to the electromagnetic forces, inverter PWM tones, and gear whine from the gearbox of the electric motor) are generally prominent. The first driving case was from stand still up to about 20-30 kph at a very low acceleration rate. With increasing torque, the order excitation is increased; thus, recordings were made while driving in a moderately leaning uphill road with smooth asphalt. This slow speed driving was varied from car to car in terms of acceleration rate and stop speed. The purpose was to record the most critical case for each car, rather than making a comparison between the cars for a strictly defined driving case. The second driving case was a maximum acceleration from standstill to typically 100 kph performed on a flat test track with smooth asphalt. In the case of hybrid electric vehicles, as powerful acceleration as possible while having the combustion engine turned off was carried out. During this driving mode, maximum available torque from the electric motor is offered, while sweeping a very wide frequency range in terms of electromagnetic and gear mesh excitation. Tones were not always prominent in both driving cases for all cars. In the test, 10 cars were recorded in the slow speed condition, while 7 of the 13 cars were recorded in the full load acceleration case. Recordings were made with Head Acoustics portable four-channel unit SQuadriga II. In this device, the microphones are integrated in a headset to have a faithful sound reproduction. For the analysis, spectrograms of TNR vs. time and PR vs. time were extracted with the software Artemis 12.01.400 by Head Acoustics. An overlap of 50 % was used and resampling and spectrum sizes were altered for the different driving conditions to achieve a frequency resolution as close to and never less than 1 % of the tone's frequency. For the TNR analyses, Discrete Fourier Transform (DFT) transformation was selected. For the PR analyses, "Show tones only" and "Comp. threshold of hearing" were selected.

The maximum TNR and PR levels are collected based on existing research [128] that has shown that instead of judging the average of an event, subjects tend to judge peaks and also the end of the event. This "peak-end rule" has also been confirmed for loudness judgment of sounds.

The first analysis aimed to find how the different car models were clustered with respect to maximum TNR and PR levels, and for which frequencies this occurred. **Figures 8.9-8.10** display the results for the slow speed driving and the full load acceleration condition, respectively.

FIGURE 8.9 Maximum TNR and PR levels as a function of frequency for the slow speed driving case. Type of metric (TNR or PR) is shown in case of deviation in terms of level and/or corresponding frequency between the two.

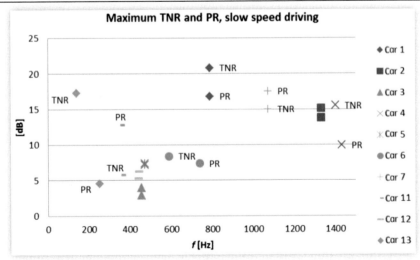

FIGURE 8.10 Maximum TNR and PR levels as a function of frequency for the full load acceleration driving case. Type of metric (TNR or PR) is shown in case of deviation in terms of level and/or corresponding frequency between the two.

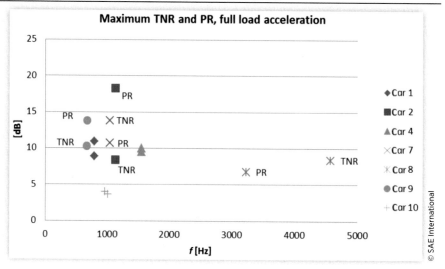

For the slow speed case, the maximum TNR and PR levels and the corresponding frequencies vary considerably between the cars. Levels cover up a range from about 3 to 20 dB and the occurring frequencies range from about 140 to above 1400 Hz. The occurring frequencies for the samples of cars are more clustered in the full load acceleration case, with 6 out of the 7 cars having their maximum level between 660 and 1600 Hz. Still, levels vary between 4 and 18 dB.

For 6 out of the 10 records in the slow-speed case and 5 out of the 7 records in the full load acceleration case, the TNR and PR levels are similar (less difference than 3 dB). The average record difference is in total 3.4 dB (standard deviation is 3.6 dB), but one outlier shows a difference up to 13 dB. The occurring frequencies for maximum TNR and PR are the same or very similar for a majority of the records.

It shows that the level deviations depend on the metric.

So why do the maximum TNR and PR levels sometimes vary substantially, e.g. as in the slow speed case where the maximum TNR level is 17 dB (at 141 Hz) while maximum PR is below 5 dB for car 13? The main reason for the large deviation between the two metrics for this case is that the sound pressure level in the lower critical band is very high. The power of the middle critical band containing the tone is not considerably higher than the average of the power of the lower and upper critical band for the frequency range where the tonal power is high. Since TNR is based on narrower bandwidths, it is not affected by the low-frequency noise as drastically as PR. In addition, the lower critical band is truncated when the tone's frequency is below 171.4 Hz, which introduces uncertainties. Interestingly, according to the criteria in ECMA-74, a tone located at 150 Hz is prominent if it exceeds 15 dB (TNR) or 17 dB (PR). Another phenomenon that can yield large variations between TNR and PR was present for car 2 for the full load acceleration case. In this case the maximum levels were due to a very high order, meaning that the tone's frequency changes very rapidly. During the time block, this particular order has high sound pressure level for all frequencies in the middle critical band, which explains the rather low TNR level. For PR however, the effect is opposite. The power of the middle band becomes large in comparison to the more remote (in terms of frequency) lower and upper critical bands; hence, the PR level is very high.

In this study, a number of hybrid/electric vehicles were investigated with respect to the tonal phenomena originating from the electric powertrain. Two customer-relevant driving conditions where the emergence of orders usually is severe, were identified: slow take-off from zero up to 20-30 kph uphill and full load acceleration from zero to about 100 kph. The maximum tonal prominence, quantified by TNR and PR, occurred between 140 and 1400 Hz for slow speed driving (10 out of the 13 cars were recorded in this condition) with levels ranging from about 3 to 20 dB. For the full load acceleration case, 6 out of the 7 cars available for this condition had maximum level ranging from 4 to 18 dB occurring between 660 and 1600 Hz. The maxima originated from electromagnetic orders or gear whine. PWM tones were more subdued, although one car was noted for a TNR level of 9.5 dB at 6.5 kHz during full load acceleration due to the inverter's switch frequency. Further, analyses were carried out to investigate the number of harmonic and nonharmonic prominent orders present. The measured cars exhibit between one and four prominent orders. The most common case is that one single order dominates, especially in the slow speed condition. Here, 8 (TNR) or 9 (PR) out of the 10 cars had either one single prominent order or had a composition of orders where the most dominant was more than 5 dB above the second most dominant in terms of maximum level. With a few exceptions, a majority of the dominant orders were harmonic in relation to each other. The gear whine order originating from the reducer is one component that is expected to show up as a nonharmonic component.

In many research studies, the whine indexes evaluating the whine sound masked by the background have been developed and used as a sound metric. This metric was employed for the development of an annoyance index and detectability index for BEV/HEV/FCEV [26, 124–139].

8.4 Sound Quality Development of BEV/HEV/FCEV

This section presents the sound quality development techniques in BEV/HEV/FCEV. Any process to address sound or vibration quality issues should always start from the voice of the customer to understand what features are objectionable and what are desirable. Once the features are understood, they need to be quantified.

As shown in **Figure 8.11**, the two steps can be performed in parallel for vehicle sound/vibration quality target development [140].

For sound quality assessment, there are two main methods to measure subjective response to a noise or vibration issue: real-time and off-line. The real-time method is referred to as the "what-do-you-think" method. The engineer/customer experiences the

FIGURE 8.11   Sound/vibration quality target development process.

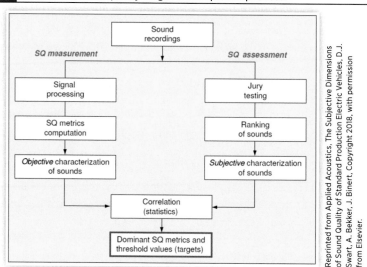

noise/vibration by walking around the item running in a lab (or driving, in case of a vehicle) and expresses an opinion, typically in relation to a previously tested sample. This method is by far the most preferred because it is easy, it provides instant feedback, the engineer/customer has a real feel for the product, and it facilitates discussion among engineers. This method may also be the only one possible in cases in which the concern cannot be reproduced by artificial devices, such as loudspeakers or shakers. The off-line methods are those in which the sound or vibration of multiple samples have been recorded and are reproduced to either one or more people at the same time in a laboratory environment. Naturally, the precision of the recording and the fidelity of the reproduction play a major role for this type of task and require an up-front investment of equipment and a facility.

The off-line methods are preferred in a benchmarking situation, when there is a need to compare the S/V quality of competitor samples, or when a target needs to be established. Either objective requires robustness of procedure and data accuracy since they will lead to specific countermeasures to improve the product and/or achieve the target and we all know that these will be judged on the "1/dB" scale. This means that the subjective experiment must be as controlled as possible to minimize the risk of biases in the result and misleading information.

The signature of electric and hybrid vehicles, in general, is quite different from vehicles powered by ICEs, but so are customers' expectations, since the degree of "green-ness" of the vehicle weighs the fuel efficiency/fuel independence more heavily than look and feel.

From a sound quality standpoint, there are two main design challenges: interior noise, which needs to provide an image of quality and "cool," and exterior noise, first to ensure safety and next to be used for brand recognition.

Cars with ICE usually have the following attributes. Engine sound was the car's voice, powerfully conveying excitement and emotion as part of the brand image. In-cabin engine sound underpinned the dynamic driving experience by giving instant feedback about how a car is operating, which enhanced the connection between driver and vehicle. For decades, the industry has worked to engineer and optimize engine sound quality [134, 143]. Whereas, electric vehicles have the following attributes. They deliver strong and responsive vehicle performance, but naturally lack the acoustic feedback of ICE. They are great for in-cabin comfort and environmental noise pollution, except for pedestrians risks. They uncover previously unnoticed noises from nonpower unit sources (road, wind, ancillaries). They are terrible for the dynamic driving experience, excitement, emotion, and the brand image, severely damaging the acceptance and appeal of electric cars. Customer expectations for electric car sound are not yet fully defined-the challenge is to tap into the existing "language" of car sound, while retaining an authentic link to the power unit.

Sound quality refinement of ICEs has been an ongoing activity for many decades among car manufacturers [119–121]. Sound quality of electric traction motors replacing the ICE in hybrid/electric vehicles is a mostly unexplored area. The signature sound from automotive e-motors, for example, typically permanent magnet synchronous motors, is completely different from ICEs. Besides being generally quieter, they are characterized by multiple high-frequency tonal components originating from harmonics of the electromagnetic force waves acting on the stator housing. High-frequency tones appearing in a broad band mix of sound have been found to be perceived as annoying in many different contexts. It has been shown that perceived unpleasantness increased with increasing level and frequency for pure tones ranging from 100 to 3000 Hz appearing in a uniformly masking noise. For automotive applications, Lennström et. al. found that an increase in levels of individual high-frequency components (>1 kHz) resulted in higher rating in perceived annoyance, sharpness, aggressiveness, and powerfulness, and also lowered the impression of overall sound quality satisfaction. PR was found to be an appropriate metric for quantifying the relative levels of the tones compared to the adjacent broad band random noise.

The sound quality metric PR has been used for target setting of EVs. Once the accepted PR levels for all potentially occurring e-motor harmonics are decided for an upcoming vehicle, the corresponding maximum allowed sound pressure level for the respective harmonics need to be derived. Those will probably differ from the prototype vehicle, since road and wind noise loads are expected to be different and need to be estimated. If the harmonics are expected to be airborne (typically those harmonics that originate from radial electromagnetic forces), the allowed radiated sound power from the stator housing can be determined if the acoustic transfer functions can be estimated. Similarly, for the structure-borne shares, the noise transfer functions at body attachment points relating sound pressure to force are required to determine the maximum allowed dynamic forces. As for the loads, the transfer functions of the future vehicle will be different compared to the prototype and the delta needs to be estimated. A schematic presentation of how the complete vehicle and component targets for a future vehicle can be derived from the PR requirement from a prototype vehicle is presented in [120].

In the investigation, the following is a typical test setup. Motor bay sound measurements were recorded with the use of a half-inch prepolarized microphone. The microphone was secured underneath the hood of the vehicle and mounted in close proximity to the electric motor and inverter [120]. Care was taken to position the microphone such that the exterior wind noise was minimized. In addition, the motor bay microphone was wrapped in foam to isolate it from structure-borne vibration. The interior vehicle sound was measured using a binaural SQuadriga headset from Head Acoustics. The measurements were recorded on the SQuadriga portable data acquisition system.

Constant speed tests of 60 kph and 80 kph were maintained for all vehicles. wide open throttle (WOT) drives were conducted on all full electric vehicles, therefore excluding the Hybrid Porsche Panamera, as the electric mode did not allow for the maximum acceleration of this vehicle.

The subjective evaluation was conducted as follows. A subjective evaluation was conducted by a jury of 31 members in a half anechoic chamber. The subjective responses of jury members were evaluated for two electric vehicle sound signatures (recorded in the motor bay), as well as three enhanced sound signatures. The study evaluated jury responses through twelve bi-polar semantic differential pairs, as listed in **Figure 8.12**. These bi-polar semantic pairs were subsequently correlated with the calculated objective metrics.

The objective evaluation is given as follows. Sound pressure level and loudness analyses were performed on the measurements of the constant speed drives at 60 and 80 kph. First, the SPL and loudness were compared to establish if any changes or variation

FIGURE 8.12 Second environments for subjective evaluation [139].

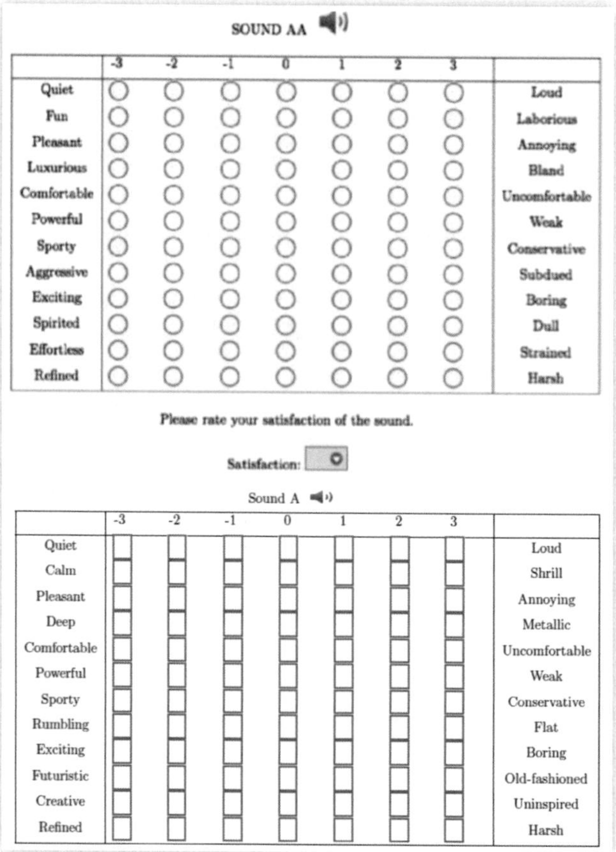

SOUND AA

	-3	-2	-1	0	1	2	3	
Quiet	○	○	○	○	○	○	○	Loud
Fun	○	○	○	○	○	○	○	Laborious
Pleasant	○	○	○	○	○	○	○	Annoying
Luxurious	○	○	○	○	○	○	○	Bland
Comfortable	○	○	○	○	○	○	○	Uncomfortable
Powerful	○	○	○	○	○	○	○	Weak
Sporty	○	○	○	○	○	○	○	Conservative
Aggressive	○	○	○	○	○	○	○	Subdued
Exciting	○	○	○	○	○	○	○	Boring
Spirited	○	○	○	○	○	○	○	Dull
Effortless	○	○	○	○	○	○	○	Strained
Refined	○	○	○	○	○	○	○	Harsh

Please rate your satisfaction of the sound.

Satisfaction:

Sound A

	-3	-2	-1	0	1	2	3	
Quiet	□	□	□	□	□	□	□	Loud
Calm	□	□	□	□	□	□	□	Shrill
Pleasant	□	□	□	□	□	□	□	Annoying
Deep	□	□	□	□	□	□	□	Metallic
Comfortable	□	□	□	□	□	□	□	Uncomfortable
Powerful	□	□	□	□	□	□	□	Weak
Sporty	□	□	□	□	□	□	□	Conservative
Rumbling	□	□	□	□	□	□	□	Flat
Exciting	□	□	□	□	□	□	□	Boring
Futuristic	□	□	□	□	□	□	□	Old-fashioned
Creative	□	□	□	□	□	□	□	Uninspired
Refined	□	□	□	□	□	□	□	Harsh

could be detected between the two methods. In addition, the specific loudness was compared to the third octave and analysis to determine if differences could be observed.

Transient psychoacoustic metrics, such as loudness, sharpness, and roughness versus time, were calculated for the interior and motor bay sound signatures of the pure electric vehicles. The transient metrics of fluctuation strength and SIL were calculated, in addition to the above-mentioned metrics for the BMW and Renault motor bay sound signatures. Furthermore, three enhanced sound signatures were generated and subjected to the described analysis. The correlation is given as follows.

The objective results from different vehicles and the three enhanced sound signature concepts were used to determine a correlation between objective metric scores and subjective responses from the semantic bi-polar test. A statistical analysis software package was used to perform a Spearman correlation test between the subjective and objective attributes of the stimuli. The subjective scores comprised averaged subjective semantic values, which were calculated for every semantic pair for the different stimuli. The averaged semantic values were correlated against several different single value methods that represented the transient objective metric results. These single values included the average, median, maximum, root mean square (RMS), and integration values.

In statistics, Spearman's rank correlation coefficient or Spearman's rho, named after Charles Spearman and often denoted by the Greek letter ρ is a nonparametric measure of rank correlation (statistical dependence between the rankings of two variables). It assesses how well the relationship between two variables can be described using a monotonic function.

The automotive industry is currently exploring the global sound sphere to identify a pleasant, safe, and unique electric vehicle signature sound. Drive-train acoustics contribute to the performance benchmark of vehicles in the marketplace. Electric vehicle sound signatures differ vastly from those of ICEs. Questions arise as to how these signature sounds relate to consumer experiences, and how the positive attributes of these sounds can be extracted and enhanced.

The method to evaluate the perception of noise annoyance is normally through listening tests and surveys. This questionnaire consisted of two questions related to noise annoyance based on the questions recommended by ICBEN. Likewise, two answer scales are proposed for these two questions: a 5-point verbal scale and 0-10 numeric scale (11-point scale). These scales have been used because a slight difference of perception between a numerical scale and a verbal scale could exist [133].

In a conventional in-room paired comparison test, all possible combinations of a set of sounds are presented as pairs and the juror is required to choose between the sounds in each pair using a subjective criterion, such as "Which sound is the most powerful?" [134–136].

Jury testing was selected as the preferred methodology to perform subjective evaluations due to the shortage of electric vehicles and the unavailability of a vehicle for in-car evaluations [26, 28, 33–35, 133, 137–165]. The pool of words was obtained from a subjective evaluation. The subjective evaluations of noise produced by electric vehicles were investigated through jury testing and a subjective evaluation form. The form utilized word association and a bi-polar semantic differential scale to evaluate the subjective response of the jury to a variety of sound stimuli.

The second testing environment was a bi-polar semantic differential evaluation, which was used to determine the difference in perception of the interior and under-hood sounds from an electric vehicle [144, 166]. In addition, the general perception of the sound signature of an EV compared to an ICE vehicle was investigated. The two EV sounds were accompanied by the interior sounds from a two ICE vehicle sound. These two ICE sounds are different in sound character to provide variation in the data. A computer-generated sound was also evaluated in an attempt to find a link between the ICE and EV sounds. The sound in question was generated by means of frequency and amplitude modulation in combination with or de-filtering of the lower motor orders. Finally, the juror was required to assess the satisfaction of each sound clip in the second environment to establish a link between subjective sound metrics and perceived consumer satisfaction.

The second evaluation was designed using 12 bi-polar semantic differential pairs, as detailed [133, 138–140]. The study selected several semantics from similar sound sources, such as washing machines, trains, and aircrafts. Suitable semantics for electric vehicles were chosen based on the results from this study. Several sound stimuli were evaluated using the bi-polar semantics pairs in Figure 8.12. The bipolar semantics are separated by a seven-point scale. The bipolar semantic evaluation provides information regarding the subjective sound character of each sound, as perceived by the jury. A satisfaction rating from 0 to 10 was also added to determine the correlation between the semantics and overall juror sound satisfaction [121, 133, 138–140].

Upon completion, the form was summarized, and it was decided to conduct all the tests at one specific location with the same equipment. The subjective evaluation tests were conducted in a silent room.

The word association completed in the first environment was analyzed to determine the most frequently selected word, or the mode of the dataset. It was found that 70.6% of the jurors preferred sound clip B, while 23.5% preferred sound clip D. Interestingly, these stimuli correspond to the interior and under-hood recorded sounds of a specific vehicle. The different modes for the entire data set, as well as the preferred sounds, are summarized and the results show that the words "powerful," "deep," and "rumbling" are associated with electric vehicle sounds. The strong character of these words suggests that the aspects relating to vehicle power are significant in electric vehicles. The data also shows that the words "noisy" and "shrill" are used to describe sounds similar to that of an electric vehicle. The words "pleasant" and "quiet" were not selected once by any of the jurors, illustrating that these words are not commonly associated with the sound of electric vehicles. In the analysis, the bi-polar semantics data was averaged across all participating jurors. The data was averaged for each semantic characteristic corresponding to the specific sound stimulus. The averaged data was then graphed using polar plots, which can plot multiple axes on a single graph. The polar plots are graphed, and the resulting graph can be seen as a type of sound map with reference to the specific semantics used. These sound maps can then be compared to determine the characteristics that drive the perception of the superiority of one sound over another. The resulting polar plots from the bi-polar semantic evaluation are illustrated in **Figure 8.13**. The EV sound signature appears to be more comfortable than that of the benchmark vehicle, Mercedes, which could result in a greater satisfaction rating, but further analysis is required to find the principal components that can account for this result.

FIGURE 8.13   EV and Mercedes interior sound comparison of semantic characteristics.

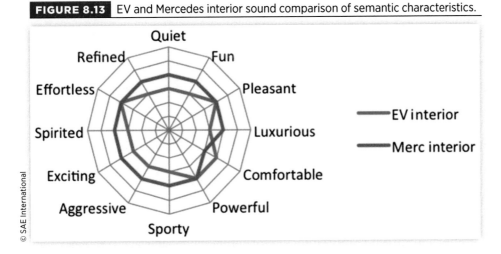

© SAE International

Polar plots are useful because they map out the sound character of each sound according to the semantic differential pairs. The extracted jury data could be further analyzed using various statistical methods, such as factor analysis, principal component analysis, and cluster analysis. These methods were used to determine the semantics that provide the most variance, as well as establish a correlation between the subjective semantics and the juror satisfaction ratings [28, 30, 33–35, 140–169].

The investigation of vehicle sound quality is usually governed by two approaches: a subjective evaluation and an objective evaluation. The subjective evaluation approach is where jurors evaluate sound quality through physical test drives or listening room evaluations. Some of the advantages of the subjective evaluation approach is that the researcher is presented with a relevant response with regards to true perception of the stimuli in question. Disadvantages include that it is time consuming, costly, requires a significant number of participants to ensure validity and reliability, and has natural limitations on the number of stimuli that can be evaluated due to jury fatigue. In contrast, the objective evaluation approach utilizes an analytical and calculated method to determine the quantification of sound quality through analytical models, objective sound metrics, and computer software. The advantages of the objective evaluation is that it is fast, efficient, and has virtually no limitations in terms of sample size or iterations per evaluation. The disadvantages of the objective approach include that the interpretation of the analyses with respect to the true perceived experience can be complex or detached. For example, if the calculated roughness for Car A is found to exceed Car B by X asper, how does this value relate to the perceived difference in roughness as experienced by a person?

Furthermore, does the specific psychoacoustic metric, for example, Zwicker Loudness, only explain the perceived sensation of loudness, or could other characteristics also influence the specific subjective sensation? These are typical questions that sound quality engineers face. It is, therefore, necessary to attempt to reconcile the gap between the interpretations of subjective and objective approaches, such that sound quality can be assessed in a fast and efficient manner, which also relates to actual consumer experiences. This work investigates the sound experience and identifies the links between subjective sound experiences and objective metrics that govern and describe electric vehicle sound signatures, by means of jury testing, psychoacoustic software, and statistical analyses. Subjective experiences were evaluated through a bi-polar semantic evaluation in a listening room. Spearman rank correlations and factor analyses were exercised to evaluate and understand the subjective sound space. Similarly, various time-varying or transient psycho-acoustic metrics were calculated using the Head Acoustics ArtemiS Suite. Finally, a multiple linear regression analysis was performed to establish a proposed consumer satisfaction model that links the calculated objective metrics to the subjective experiences.

The comparisons of subjective semantics vs. objective metrics have been widely conducted. The subjective semantics and objective metrics are both used to investigate the attributes of sound character and quality. The existing research regarding the relationship between the subjective semantics and objective metrics for electric vehicles is sparse. Some research [119, 120, 121] investigated the relationship between perceived annoyance and the psychoacoustic metric of PR. Another study investigated the relationship between unpleasantness and several psychoacoustic metrics [166]. However, no studies have been found that illustrate the relationship between multiple subjective semantics and psychoacoustic metrics for electric vehicles. To this end, a Spearman correlation was determined between the subjective semantics and objective metrics, as presented in **Table 8.5**. The metric group indicates the psychoacoustic metric with the strongest positive (+) or negative (-) correlation with multiple SVRs as a whole.

The column for best SVR denotes the SVR technique of the metric group with the strongest correlation, followed by the Spearman rank correlation (R). The last column indicates an additional metric group which also correlated strongly. The psychoacoustic metrics that correlate best with the subjective semantics include sharpness, loudness, and impulsiveness. Furthermore, it can be seen that maximum sharpness and 95th percentile loudness (N5) results in the strongest correlations for SVR techniques. Impulsiveness is best represented by either the maximum or RMS single values. Interestingly the "sporty" semantic did not correlate in a significant manner with any of the psychoacoustic metrics and is likely attributed to the lack of noteworthy sportiness in the electric vehicle sound character.

TABLE 8.5 Correlations between subjective semantics and objective metrics for electric vehicles.

Semantic	Metric Group	Best SVR	R	Other Metrics
Calm	-Sharpness	Max	0.720	-SIL
Comfortable	-Sharpness	Max	0.709	+SIL
Pleasant	-Sharpness	Max	0.645	-SIL
Quiet	-Loudness	N_5	0.952	-Sharpness
Refined	-Loudness	N_5	0.855	-Impulsiveness
Powerful	+Impulsiveness	Max	0.900	-SII
Rumble	+Impulsiveness	Max	0.945	-SII
Deep	+Impulsiveness	RMS	0.855	+ Loudness
Futuristic	-Impulsiveness	RMS	0.852	-Loudness
Creative	-Loudness	Mean	0.925	+Roughness
Exciting	-SIL	Median	0.662	none

The significant terms are the semantics "calm," "comfortable," and "pleasant" that correlate inversely with the sharpness psychoacoustic metric, or proportionally with the "shrill," "uncomfortable," and "annoying" bi-polar counterparts.

The "calm," "comfortable," and "pleasant" semantics were previously found to have the strongest correlation with perceived satisfaction, thus signifying that a reduction in maximum sharpness could lead to an increase in perceived satisfaction.

To investigate and validate this kind of claim, a set of multiple linear regression analyses have been performed. The cross-correlation coefficients of the objective quantities and the subjective quantities shown in Table 8.5 are all greater than 0.5, indicating that the psychoacoustical parameters are correlated with the subjective evaluations. The physical metrics related to the unpleasant and pleasant impressions associated with EV sounds could be determined based on the correlation of the results of the psychological evaluations to those of measurements of the physical metrics [28, 30, 33–35, 137, 154–169].

References

1. Bray, W.R., "Using the Relative Approach for Direct Measurement of Patterns in Noise Situations," *Sound & Vibration Magazine*, Sept. 2004, 22–27.

2. Bray, W.R., Blommer, M., and Lake, S., "Sound Quality, *2005 Workshop, SAE Noise & Vibration Conference*," Traverse City, 2005. SAE International, Warrendale, PA.

3. Fastl H, Zwicker E. *Psychoacoustics: Facts and Models*. Berlin, Germany: Springer, 2006.

4. William M. Hartmann, *Signals, Sound, and Sensation*, Woodbury, NY: American Institute of Physics, 2004.

5. Otto, N.C., Simpson, R., and Wiederhold, J., "Electric Vehicle Sound Quality," SAE Technical Paper 1999-01-1694, 1999, https://doi.org/10.4271/1999-01-1694.

6. Alt, N.W., Wiehagen, N., and Schlitzer, M.W., "Interior Noise Simulation for Improved Vehicle Sound," SAE Technical Paper 2001-01-1539, 2001, https://doi.org/10.4271/2001-01-1539.

7. ISO2631-1, "Mechanical Vibration and Shock - Evaluation of Human Exposure to Whole-Body Vibration," International Stand ard ISO 2631-1532B:1975, International Organization for Stand ardization, Geneva, Switzerland

8. ISO532B, Acoustics-Method for Calculating Loudness Level," International Stand ard ISO 532B:1975, International Organization for Stand ardization, Geneva, Switzerland .

9. Stucklschwaiger, W., de Mendonca, A., and Alves dos Santos, M., "The Creation of a Car Interior Noise Quality Index for the Evaluation of Rattle Phenomena," SAE Technical Paper 972018, 1997, https://doi.org/10.4271/972018.

10. Mosquera-Sanchez, J.A., Villalba, J., Janssens, K., and de Oliveira, L.P.R., "A Multi Objective Sound Quality Optimization of Electric Motor Noise in Hybrid Vehicles," in *Proceedings of ISMA2014*, Leuven, Belgium.

11. Gelfand , S. A. *Hearing: An Introduction to Psychological and Physiological Acoustics*, 4th ed. (New York: Marcel Dekker, 2004).

12. "Reference threshold of hearing under free-field and diffuse-field listening conditions", Reference zero for the calibration of audiometric equipment Part 7, ISO 389-7, 2005.

13. Moore BCJ, *An Introduction to the Psychology of Hearing* (Bingley: Emerald, 2012).

14. Plomp, R., Levelt, W.J.M., "Tonal Consonance and Critical Band width," *J. Acoust. Soc. Am.* 38: 548, 1965.

15. Buus S., Schorer E., Florentine M., Zwicker E., "Decision Rules in Detection of Simple and Complex Tones," *J. Acoust. Soc. Am.* 80: 1646-1657, 1986.

16. Green D M, Mason C R, "Auditory Profile Analysis: Frequency, Phase, and Weber's Law," *J. Acoust. Soc. Am.* 77: 1155, 1985.

17. Moore BCJ, "Effects of Level and Frequency on the Audibility of Partials in Inharmonic Complex Tones," *J. Acoust. Soc. Am.* 120, no. 2: 934, 2006.

18. Guo, D., Shi, Q., and Yi, P., "In-Vehicle Whine Perception Based on Psychoacoustics and Some Design Principles," SAE Technical Paper 2015-01-2338, 2015, https://doi.org/10.4271/2015-01-2338.

19. Beauchamp J., "Synthesis by Spectral Amplitude and "Brightness" Matching of Analyzed Musical Instrument Tones," *J. Acoust. Eng. Soc.* 30, no. 1: 143, 1982.

20. McAdams S., Winsberg S., Donnadieu S., de Soete G.et al, "Perceptual Scaling of Synthesized Musical Timbres: Common Dimensions, Specificities, and Latent Subject Classes," *Psychological Research* 58: 177-192, 1992.

21. Patsouras Christine, Fastl Hugo, "Psychoacoustic Evaluation of Tonal Components in View of Sound Quality," *Acoustical Science and Technology, AST.* 23, No. 2: 23–27, 2003.

22. Ford Motor Company, "Guidelines for Jury Evaluations of Automotive Sounds Norm Otto and Scott Amman," Amman, Detroit, MI, 2001.

23. Schneider, M., Wilhelm, M., and Alt, N., "Development of Vehicle Sound Quality - Targets and Methods," SAE Technical Paper 951283, 1995, https://doi.org/doi:10.4271/951283.2.

24. Otto, N., Amman, S., Eaton, C., and Lake, S., "Guidelines for Jury Evaluations of Automotive Sounds," SAE Technical Paper 1999-01-1822, 1999, https://doi.org/10.4271/1999-01-1822.

25. Kuwano, S. and Namba, S., "Dimensions of Sound Quality and Their Measurements," *Proceedings of the 17th International Congress on Acoustics*, Rome, Italy, 2001, 3–4.

26. Jennings, P.A., Dunne, G., Williams, R. and Giudice, S., "Tools and Techniques for Understand ing the Fundamentals of Automotive Sound Quality," *Proceedings of the Institution of Mechanical Engineers, Part D: Journal of Automobile Engineering* 224, no. 10: 1263–1278, 2010.

27. Von Gosler, J. and Van Niekerk, J.L.: "Sound Quality Metrics to Assess Road Noise in Light Commercial Vehicle," *R & D Journal of the South African Institution of Mechanical Engineering* 24, no. 1: 19–25, 2008.

28. Fastl, H., "The Psychoacoustics of Sound Quality Evaluation," *Acustica* 83: 754-764, 1997.

29. Norm, O., A. Scott and E. Cris. "Guidelines for Jury Evaluations of Automotive Sounds," SAE Technical Paper 1999-01-1822, 1999, https://doi.org/10.4271/1999-01-1822. Odden, M., Heinrichs, R. and Linow, A., "Sound Evaluation of Interior Vehicle Using an Efficient Psychoacoustic Method," Proceed Euro-Noise 2: 631-642, 1998.

30. Trapenskas, D., "Sound Quality Assessment Using Binaural Technology," Ph.D. thesis, Lulea, University of Technology, 2002.

31. Parizet, E., "Paired Comparison Listening Tests and Circular Error Rates," *Acta Acustica united with Acustica* 88: 594-598, 2002, https://hal.archives-ouvertes.fr/hal-00849430.

32. Gonzalez, M., de Diego, F. M., G. Piero, and J.J. Garcia-Bonito, "Sound Quality of Low-Frequency and Car Engine Noises after Active Noise Control," *Journal of Sound and Vibration* 265: 663-679, 2003.

33. Leite, R.P., Paul, S., and Gerges, S.N.Y., "A Sound Quality-Based Investigation of the HVAC System Noise of an Automobile Model," *Applied Acoustics* 70: 636-645, 2009.

34. Cho, W.-H., Ih, J.-G., Shin, S.-H., and Kim, J.-W., "Quality Evaluation of Car Window Motors Using Sound Quality Metrics," *Int. J. Automotive Technology* 12: 443, 2011.

35. Brizon, C.J.S. and Medeiros, E.B., "Combining Subjective and Objective Assessments to Improve Acoustic Comfort Evaluation of Motor Cars," *Applied Acoustics* 73: 913–920, 2012.

36. Adams, M. and Ponseele, P., "Sound Quality Equivalent Modeling for Virtual Car Sound Synthesis," SAE Technical Paper 2001-01-1540, 2001, https://doi.org/10.4271/2001-01-1540.

37. Ajovalasit, M. and Giacomin J., "Effect of Automobile Operating Condition on the Subjective Equivalence of Steering Wheel Vibration and Sound," *Int. J. Vehicle Noise and Vibration* 3: 197, 2007.

38. Amman, S., Mouch, T., Meier, R., and Gu, P., "Sound and Vibration Perceptual Contributions during Vehicle Transient and Steady-State Road Inputs", *Int. J. Vehicle Noise and Vibration* 3: 157, 2007.

39. Blommer, M., Amman, S., Abhyankar, S., and Dedecker, B., "Sound Quality Metric Development for Wind Buffeting and Gusting Noise," SAE Technical Paper 2003-01-1509, 2003, https://doi.org/10.4271/2003-01-1509.

40. Deblauwe, F. and Ponseele, V., "A Sound Quality System for Engineers," SAE Technical Paper 2001-01-3834, 2001, https://doi.org/10.4271/2001-01-3834.

41. Fridrich, R.J., "Pitch Intervals: Linking Sound Quality Engineering and Musical Acoustics," SAE Technical Paper 2003-01-1503, 2003, https://doi.org/10.4271/2003-01-1503.

42. Hueser, M.G., Govindswamy, K., Wolff, K., and Stienen, R., "Sound Quality and Engine Performance Optimization Development Utilizing Air-to-Air Simulation and Interior Noise Synthesis," SAE Technical Paper 2003-01-1652, 2003, https://doi.org/10.4271/2003-01-1652.

43. Hoshino, H and Katoh, H., "SEvaluation of Wind Noise in Passenger Car Compartment in Consideration of Auditory Masking and Sound Localization," SAE Technical Paper 1999-01-1125, 1999, https://doi.org/10.4271/1999-01-1125.

44. Hoshino, H. and Kato, H, "An Objective Evaluation Method of Wind Noise in a Car Based on a Model of Subjective Evaluation Process," *Japanese Society of Automotive Engineers, Annual Congress* 12: 9–12, 2000.

45. Lee, M.R. and McCarthy, M., "Exhaust System Design for Sound Quality", SAE Technical Paper 2003-01-1645, 2003, https://doi.org/10.4271/2003-01-1645.

46. Otto, N.C., Simpson, R., and Wiederhold, J., "Electric Vehicle Sound Quality," SAE Technical Paper 1999-01-1694, 1999, https://doi.org/10.4271/1999-01-1694

47. Mansfield, N.J., Ashley, J. and Rimell, A.N., "Changes in Subjective Ratings of Impulsive Steering Wheel Vibration due to Changes in Noise Level: A Cross-Modal interaction", *Int. J. Vehicle Noise and Vibration* 3: 185–196, 2007.

48. Qatu, M.S., Abdelhamid, M.K., Pang, J., and Sheng, G., "Overview of Automotive Noise and Vibration," *Int. J. Vehicle Noise and Vibration* 5: 1 2009.

49. Quinn D.C. and Hofe, R.V., "Engineering Vehicle Sound Quality," SAE Technical Paper 972063, 1997, https://doi.org/10.4271/972063.

50. Radavich, P.M. and Selamet, A., "Approximating Engine Tailpipe Orifice Noise Sound Quality Using a Surge Tank and In-Duct Measurements," SAE Technical Paper 2003-01-1641, 2003, https://doi.org/10.4271/2003-01-1641.

51. Schneider, M., Wilhelm, M. and Alt, N., "Development of Vehicle Sound Quality – Targets and Methods," SAE Technical Paper 951283, 1995, https://doi.org/10.4271/951283.

52. Shkreli, V. and Vand enbrink, K.A., "The Use of Subjective Jury Evaluations for Interior Acoustic Packaging," SAE Technical Paper 2003-01-1506, 2003, https://doi.org/10.4271/2003-01-1506.

53. Terazawa, N., Kozawa Y., and Shuku, T., "Objective Evaluation of Exciting Engine Sound in Passenger Compartment during Acceleration", SAE Technical Paper 2000-01-0177, 2000, https://doi.org/10.4271/2000-01-0177.

54. Noumura, K. and Yoshida, J., "Perception Modeling and Quantification of Sound Quality in Cabin," SAE Technical Paper 2003-01-1514, 2003, https://doi.org/10.4271/2003-01-1514.

55. Pietila, G., Jay, G., and Frank, E., "Evaluation of Different Vehicle Noise Reduction Test Methods for Tire Sound Quality Synthesis,|" SAE Technical Paper 2007-01-2252, 2007, https://doi.org/10.4271/2007-01-2252.

56. Gossler, J., "NVH Benchmarking during Vehicle Development Using Sound Quality Metrics," University of Stellenbosch, South Africa, 2007.

57. Allman-Warda, M., Williamsb, R., Dunnec, G., and Jennings, P., "The Evaluation of Vehicle Sound Quality Using an NVH Simulator," in *Proceedings of The 33rd International Congress and Exposition on Noise Control Engineering*, Prague-Czech Republic, Aug. 20-25,SS 2004.

58. HEAD Acoustics, "Binaural Measurement, Analysis and Playback," HEAD acoustics Application Note, http://www.head-acoustics.de/eng/nvh_hsu_III_3.htm accessed Apr. 2016.

59. Cerrato Jay, G., "The Sound Quality Engineering Process - Applications," presented at the SAE Vehicle Noise Control Engineering Academy, Sept. 2005.

60. Penne, F., "Shaping the Sound of the Next-Generation BMW," in *ISMA 2004 International Conference on Noise and Vibration Engineering*, Katholieke Universiteit Leuven, Sept. 20-22, 2004.

61. Churchill, C., Maluski, S., Cox, T., "Simplified Sound Quality Assessment for UK Manufacturers," in *Inter-Noise 2004*, Prague, Czech Republic, Aug. 22-25, 2004.

62. Cerrato Jay, G., "Sound Quality and Jury Techniques," in *Brüel & Kjær Sound and Vibration Conference*, Novi, MI, Sept. 2006.

63. Cerrato Jay, G. and Lowery, D., "Sound Quality Evaluation of Compressors," in *2002 Sixteenth International Compressor Engineering Conference at Purdue*, West Lafayette, IN, Jul. 2002.

64. Allman-Ward, M.et al., "The Evaluation of Vehicle Sound Quality Using a NVH Simulator," in *Inter-Noise 2004*, Prague, Czech Republic, Aug. 22-25, 2004.

65. Roussarie V.et al., "What's so Hot about Sound? - Influence of HVAC Sound on Thermal Comfort," in *Inter-Noise 2005*, Riode Janeiro, Brazil, Aug. 7-10, 2005.

66. Begault, D., "Overview of Spatial Hearing Part I and Part II," in *3-D Sound For Virtual Reality and Multimedia* (Academic Press, 1994).

67. Cerrato-Jayet al., "Implementation of Sound Quality Measurements in Component Rating Tests," in *Sound Quality Symposium at Inter-Noise 2002*, Dearborn, MI, Aug. 22, 2002.

68. Moravec, M., Ižaríkov, G., Lipta, P., Badida, M., and Badidová, A., "Development of Psychoacoustic Model Based on the Correlation of the Subjective and Objective Sound Quality Assessment of Automatic Washing Machines," *Applied Acoustics* 140: 178-182 Nov. 2018.

69. Voland ri, G., Di Puccio, F., Forte, P., Mattei, L., "Psychoacoustic Analysis of Power Windows Sounds: Correlation between Subjective and Objective Evaluations," *Applied Acoustics* 134: 160-170, May. 2018.

70. AES20-1996, "AES Recommended Practice for Professional Audio - Subjective Evaluation of Loudspeakers," Audio Engineering Society Stand ard, New York, 1996.

71. ANSI S3.1-1991, "Maximum Permissible Ambient Noise Levels for Audiometric Test Rooms," American National Stand ard.

72. ANSI S12.2-1995, "Criteria for Evaluating Room Noise," *American National Stand ard.*

73. Bech, S., "Planning of a Listening Test – Choice of Rating Scale and Test Procedure," in *Symp. on Perception of Reproduced Sound*, Denmark, 1987.

74. Otto, N., "Listening Test Methods for Automotive Sound Quality," in *Proceedings of the Audio Engineering Soc.,* New York, 1997.

75. Bech, S., "Selection and Training of Subjects for Listening Tests on Sound Reproducing Equipment," *J. Audio Eng. Soc.* 40: 590, 1992.

76. Bech, S., "Training of Subjects for Auditory Experiments," *Acta Acustica* 1: 89, 1993.

77. David, H., *The Method of Paired Comparisons*, Oxford University Press, 1988. Stevens, S., Psychophysics. (New York: John Wiley & Sons, 1975).

78. Levitt, H., "Transformed Up-Down Methods in Psychoacoustics," *J. Acoust. Soc. Am.* 49: 467, 1971.

79. Kousgaard, N., "The Application of Binary Paired Comparisons to Listening Tests," in *Symp. on Perception of Reproduced Sound*, Denmark, 1987.

80. Otto, N. and Wakefield, G., "The Design of Automotive Acoustic Environments: Using Subjective Methods to Improve Engine Sound Quality," in *Proceedings of the Human Factors Society*, Atlanta, 1992.

81. Otto, N. and Feng, B., "Wind Noise Sound Quality," in *Proceedings of SAE Noise and Vibration Conference*, Traverse City, MI, 1995.

82. Otto, N. and Feng, B., "Automotive Sound Quality in the 1990s," in *Third Int. Congress on Air- and Structure-Borne Sound and Vibration*, Montreal, 1994.

83. Amman, S. and Otto, N., "Sound Quality Analysis of Vehicle Windshield Wiper Systems," in *Proceedings 1993 SAE NVH Conference*, Traverse City, MI, Paper 931345, 1993.

84. Meilgaard, M., Civille, G., and Carr, B., *Sensory Evaluation Techniques* (Boca Raton, FL: CRC Press, 1991).

85. Malhotra, N., *Marketing Research – An Applied Orientation* (Englewood Cliffs, NJ:, Prentice Hall, 1993).

86. Bisping, R., Giehl, S., and Vogt, M., "A Stand ardized Scale for the Assessment of Car Interior Sound Quality," in Proceedings 1997 SAE NVH Conference, Traverse City, MI, Paper 971976, 843-847.

87. Chambers, J., Cleveland , W., Kleiner, B., and Tukey, P., *Graphical Methods for Data Analysis* (New York: Chapman and Hall, 1983).

88. Blommer, M., Amman, S., and Otto, N., "The Effect of Powertrain Sound on Perceived Vehicle Performance," in *Proceedings 1997 SAE NVH Conference*, Traverse City, MI, Paper 971983, 891-896.

89. Montgomery, D. and Peck, E., *Introduction to Linear Regression Analysis* Hoboken, NJ, John Wiley.

90. Kim, J. and Mueller, C., *Introduction to Factor Analysis - What It Is and How To Do It* (Beverly Hills, CA: Sage Publications, 1978).

91. Bisping, R., "Emotional Effect of Car Interior Sounds: Pleasantness and Power and Their Relation to Acoustic Key Features," in *Proceedings 1995 SAE NVH Conference*, Traverse City, MI, Paper 951284, 1203-1209.

92. Murata, H.et al., "Sound Quality Evaluation of Passenger Vehicle Interior Noise," in *Proceedings 1993 SAE NVH Conference*, Traverse City, MI, Paper 931347, 675-681.

93. Takao, H.et al., "Quantification of Subjective Unpleasantness Using Roughness Level," in *Proceedings 1993 SAE NVH Conference*, Traverse City, MI, Paper 931332, 561-570.

94. N. Otto and G. Wakefield, "A Subjective Evaluation and Analysis of Automotive Starter Sounds," *Noise Control Engineering Journal* 94, no. 3: 377-382, 1993.

95. Champagne, A. and Amman, S., "Vehicle Closure Sound Quality," in *Proceedings 1995 SAE NVH Conference*, Traverse City, MI, Paper 951370, 1109-1114.

96. Staffeldt, H., "Correlation between Subjective and Objective Data for Quality Loudspeakers," *Audio Engineering Society Journal* 22, no. 6: 103–115, 1974.

97. Kruskal, J. and Wish, M., *Multidimensional Scaling* (Beverly Hills, CA: Sage Publications, 1978).

98. Nemura, T., Adachi, N., and Suzuki, K., "Research in Regard to Sensory Characteristics Measuring for the Impulse Noise of the Engine Valve System," in *Proceedings 1991 SAE Int. Congress and Exp.*, Detroit, MI Paper 910620.

99. Schneider, M., Wilhelm, M., and Alt, N., "Development of Vehicle Sound Quality - Targets and Methods," SAE Technical Paper 951283, 1995, https://doi.org/10.4271/951283.2. Otto, N., Amman, S., Eaton, C., and Lake, S., "Guidelines for Jury Evaluations of Automotive Sounds," SAE Technical Paper 1999-01-1822, 1999, https://doi.org/10.4271/1999-01-1822.

100. Toepken, S., Verhey, J., Weber, R., "Perceptual Space, Pleasantness and Periodicity of Multi-Tone Sounds," *J. Acoust. Soc. Am.* 138: 288-298, 2015.

101. Parizet, E., Bolmont, A., and Fingerhuth, S., "Subjective Evaluation of Tonalness and Relation between Tonalness and Unpleasantness," *Proc. Internoise 2009*, Ottawa.

102. Govindswamy, Kiran and Wellmann, T., and Eisele, G., "Aspects of NVH Integration in Hybrid Vehicles," SAE Technical Paper 2009-01-2085, 2009, https://doi.org/2009-01-2085.

103. ANSI S1.13-2005, "Measuring Sound Pressure Levels in Airy Annex A Identification and Evaluation of Prominent Discrete Tones," ANSI S12.10, ISO 7779, ECMA 74.

104. Stand ard ECMA-74, "2018 Measurement of Airborne Noise emitted by Information Technology and Telecommunications Equipment."

105. Sottek, R. and Bray, W., "Application of a New Perceptually-Accurate Tonality Assessment Method," SAE NVH, 2015,

106. Sottek, R., and Bray, W., "Application of a New Perceptually-Accurate Tonality \ Assessment Method," *SAE International Journal of Passenger Cars - Electronic and Electrical Systems* 8, no. 2: 462-469, Jun. 2015, https://doi.org/10.4271/2015-01-2282.

107. Bray, W.R., "Methods for Automating Prominent Tone Evaluation and for Considering Variations with Time or Other Reference Quantities," *The Journal of the Acoustical Society of America* 123: 3685, 2008, https://doi.org/10.1121/1.2935056.

108. Sottek, R., "Progress in Calculating Tonality of Technical Sounds," Internoise 2014, Jan. 2014.

109. Sottek, R,Kamp, F., and Fiebig, A., "A New Hearing Model Approach to Tonality," in *Proc. Internoise*, Innsbruck, 2013, 3.

110. Bray, W. and Caspary, G, "Automating Prominent Tone Evaluations and Accounting for Time-Varying Conditions," in *Sound Quality Symposium, SQS 2008*, Detroit, 2008.

111. Sottek, R., "A Hearing Model Approach to Time-Varying Loudness," *Acta Acustica united with Acustica* 102, no. 4: 725-744, 2016.

112. Sottek, R., "Loudness Models Applied to Technical Sounds," in *Proc. Noise-Con. 2010*, Baltimore, 2010.

113. Sottek, R., Kamp, F., and Fiebig, A., "Perception of Loudness and Roughness of Low-Frequency Sounds," in *Proc. Inter-Noise 2012*, *New York*, 2012.

114. Sottek, R., "Improvements in Calculating the Loudness of Time Varying Sounds," in *Proc. Inter-Noise 2014*, Melbourne, 2014.

115. Sottek, R., Kamp, F., and Fiebig, A., "A New Hearing Model Approach to Tonality," in *Proc. Inter-Noise 2013*, Innsbruck, 2013.

116. Sottek, R., "Progress in Calculating Tonality of Technical Sounds," in *Proc. Inter-Noise 2014*, Melbourne, 2014.

117. Sottek, R., "Calculating Tonality of IT Product Sounds Using a Psychoacoustically-Based Model," in *Proc. Inter-Noise 2015*, San Francisco, 2015.

118. HEAD, "Acoustics GmbH: Using the new psychoacoustic tonality analyses Tonality (Hearing Model)," Application Note, 2018.

119. Lennström, D. and Nykänen, A., "Interior Sound of Toda's Electric Cars: Tonal Content, Levels and Frequency Distribution," SAE Technical Paper 2015-01-2367, 2015, https://doi.org/10.4271/2015-01-2367.

120. Lennström, D., Lindbom, T., and Nykänen, A., "Prominence of Tones in Electric Vehicle Interior Noise," in *Proceedings of InterNoise*, Innsbruck, Austria, 2013.

121. Lennström, D., Ågren, A., and Nykänen, A., "Sound Quality Evaluation of Electric Cars: Preferences and Influence of the Test Environment," in *Proceedings of the Aachen Acoustics Colloquium*, Aachen, Germany, 2011, 95–100.

122. Bézat, M.-C., Richard, F., Roussarie, V., "Acoustics of Hybrid Vehicles and Emergences of Whining Noises," in *Proceedings of SIA Conf. on NVH of Hybrid and Electric Vehicles*, 2010.

123. Sarrazin, M., Gillijns, S., Anthonis, J., Janssens, K., and van der Auweraer, H., "NVH analysis of a 3 phase 128 SR motor drive for HEV Applications," in *EVS27 International Battery, Hybrid and Fuel Cell Electric Vehicle Symposium*, Barcelona, Spain, 2013.

124. Kerkmann, J., Schulte-Fortkamp, B., and Fiebig, A., "*Acceptance of Synthetic Driving Noises in Electric Vehicles*," in *Proceedings of Forum Acusticum*, Krakow, Poland , 2014.

125. Petiot, J.-F., Kristensen, B.G., and Maier, A.M., "How Should an Electric Vehicle Sound? User and Expert Perception," in *Proceedings of the ASME 2013 International Design Engineering Technical Conferences & Computers and Information in Engineering Conference IDETC/CIE 2013*, Portland, OR, 2013.

126. Sukowski, H., Kühler, R., van der Par, S., and Weber, R., "Perceived Quality of the Interior Sounds in Electric and Conventional Motor Vehicles," in *Proceedings of InterNoise*, Innsbruck, Austria, 2013.

127. Kahneman, D., Fredrickson, B.L., Schreiber, C.A., and Redelmeier, D.A., "When More Pain Is Preferred to Less: Adding a Better End," *Psychological Science* 4: 401-405, 1993.

128. Västfjäll, D., Gulbol, M.-A., and Kleiner, M., "When More Noise Is Preferred than Less: Exploring the "Peak-End Rule" in Retrospective Judgments of Sound Quality," in *Proceedings of InterNoise*, Dearborn, MI, USA, 2002.

129. Lee, S.K., Lee, S.M., Shin, T., and Han, M., "Objective Evaluation of the Sound Quality of the Warning Sound of Electric Vehicles with a Consideration of the Masking Effect: Annoyance and Detectability," *International Journal of Automotive Technology* 18, no. 4: 699–705, (2017).

130. Jay, G.C., "Sound/Vibration Quality Engineering, Part 1 - Introduction and the SVQ Engineering Process," *Sound and Vibration, Jul. 2007.*

131. Otto, N., Simpson, R., and Wiederhold, J., "Electric Vehicle Sound Quality," SAE Technical Paper 1999-01-1694, 1999, https://doi.org/10.4271/1999-01-1694.

132. Maunder, M. and Munday, B., "Sound Quality Enhancement in Electric Vehicles, an Augmented Approach," in *IQPC EV Noise & Vibration Conference* Berlin, Germany.

133. Swart, D.J. and Bekker, A., "The Subjective Evaluation of Interior Noise Produced by Electric Vehicles," in *Proceedings of the 9th South African Conference on Computational and Applied Mechanics (SACAM 2014)*, 432, South African Association for Theoretical and Applied Mechanics (SAAM), Jan 2014, ISBN 9781634397162.

134. Otto, N., Amman, S., Eaton, C., and Lake, S., "Guidelines for Jury Evaluations of Automotive Sounds," SAE Paper 1999-01-1822, in *Noise and Vibration Conference*, Traverse City, MI, May 1999.

135. Baker, S., Jennings, P., Dunne, G., and Williams, R., "Improving the Effectiveness of Paired Comparison Tests for Automotive Sound Quality," in *Proceedings of the 11th International Congress on Sound and Vibration*, 2004.

136. Fry, J., Jennings, P., Williams, R., and Dunne, G., "Understand ing How Customers Make Their Decisions on Product Sound Quality," in *Proceedings of the 33rd International Congress and Exposition on Noise Control Engineering*, paper n. 170. Prague Czech.

137. Kuwano, S., Namba, S., Schick, A., Hoege, H., Fastl, H., Filippou, T., Florentine M., and Muesch, H., "The Timbre and Annoyance of Auditory Warning Signals in Different Countries," in *Proc. Internoise 2000*, Nice, France, 2000.

138. Swart, D.J., Bekker, A., and Bienert, J., "The Comparison and Analysis of Stand ard Production Electric Vehicle Drive-Train Noise," *The International Journal of Vehicle Noise and Vibration*: 260–276, 2016.

139. Swart, D.J., Bekker, A., and Bienert, J., "The Subjective Dimensions of Sound Quality of Stand ard Production Electric Vehicles," *Applied Acoustics* 129: 354-364, 2018, ISSN 1872910X.

140. Swart, D.J. and Bekker, A., "Interior and Motorbay Sound Quality Evaluation of Full Electric and Hybrid-Electric Vehicles Based on Psychoacoustics," in *Proceedings of the 45th International Congress and Exposition of Noise Control Engineering*, Hamburg, Germany, 2016, 5400-5410.

141. Stellenbosch University Library Web Servive, https://library.sun.ac.za/en-za/Research/oa/Pages/SUNScholar.aspx.

142. Cerrato, G., "Automotive sound quality-powertrain, road and wind noise," *Sound & Vibration* 43, no. 4: 16-24, 2009.

143. Otto, N., Amman, S., Eaton, C., and Lake, S., "Guidelines for jury evaluations of automotive sounds," *Sound and Vibration* 35, no. 4: 24–47, 2001.

144. Sottek, R., "Loudness models applied to technical sounds," *The Journal of the Acoustical Society of America* 127, no. 3: 1880–1880, 2010.

145. HEAD Acoustics GmbH, "Psychoacoustic Analyses I," available at: https://www.head-acoustics.de/downloads/eng/application_notes/OrderAnalysis_e.pdf [November 19, 2017], 2016, Application Note-12/16.

146. Genuit, K. and Sottek, R., "Application of a New Hearing Model for Determining the Sound Quality of Sound Events," Seoul National University, Seoul, 1995.

147. Sottek, R. and Genuit, K., "Perception of Roughness of Time-Variant Sounds," *The Journal of the Acoustical Society of America* 133, no. 5: 3598, 2013, ISSN 00014966.

148. Bucak, T., Bazijanac, E., and Juricic, B., "Correlation between SIL and SII in a Light Aircraft Cabin during Flight," in *Proceedings of the 14th International Congress on Sound and Vibration (ICSV14)*, Cairns, Australia, 2007.

149. ANSI, A.: S3.5-1997, "Methods for the Calculation of the Speech Intelligibility Index," New York: American National Stand ards Institute, vol. 19: 90–119, 1997.

150. Sottek, R., "Progress in Calculating Tonality of Technical Sounds," in *Proceedings of the 43rd International Congress on Noise Control Engineering*, Melbourne, Australia, 2014.

151. ECMA-74, "Tone-to-Noise Ratio Method," ECMA International 9th Edition D.7, 2005, Geneva, Switzerland .

152. Tang, K.-T., "Mathematical Methods for Engineers and Scientists 3," in *Fourier Analysis, Partial Differential Equations and Variational Methods (v. 3)* (Berlin Heidelberg: Springer, 2007), ISBN 3540446958.

153. Vaseghi, S.V.: *Multimedia Signal Processing: Theory and Applications in Speech, Music and Communications.* (UK: John Wiley & Sons, 2007). ISBN 9780470062012.

154. HEADAcoustics GmbH, "Order Analysis," available at: https://www.headacoustics.de/downloads/eng/application_notes/OrderAnalysis.pdf accessed Aug. 2015.

155. Cocron, P., Bühler, F., Franke, T., Neumann, I., and Krems, J.F., "The Silence of Electric Vehicles-Blessing or Curse," in *Proceedings of the 90th Annual Meeting of the Transportation Research Board*, Washington, DC, 2011.

156. Mosquera-Sanchez, J.A., Villalba, J., Janssens, K., and de Oliveira, L.P.R., "A Multi Objective Sound Quality Optimization of Electric Motor Noise in Hybrid Vehicles." *Proceedings of ISMA2014*, Leuven, Belgium.

157. Gonzalez, A., Ferrer, M., de Diego, M., Piero, G., and Garcia-Bonito, J. J., "Sound Quality of Low-Frequency and Car Engine Noises after Active Noise Control," *Journal of Sound and Vibration* 265: 663–679, 2003.

158. Aures, W., "A Procedure for Calculating Auditory Roughness," *Acustica* 58: 268–281, 1985.

159. Brand l, F.K. and Biermayer, W., "A New Tool for the Onboard Objective Assessment of Vehicle Interior Noise Quality," SAE Technical Paper 1999-01-1695, 1999, https://doi.org/10.4271/1999-01-1695.

160. Daniel, P. and Weber, P., "Psychoacoustical roughness: Implementation of an optimized model," *Acustica* 83: 113-123, 2003.

161. Guski, R., "Psychological Methods for Evaluating Sound Quality and Assessing Acoustic Information," *Acta Acustica* 83: 765-774, 1997.

162. Haderlein, T., Bocklet, T., Maier, A., Noth, E., Knipfer, C. and Stelzle, F., "Objective vs. Subjective Evaluation of Speakers with and without Complete Dentures," *Lect. Notes Comput. Sci.* 5729: 170-177, 2009.

163. ISO 532B, "Method for Calculating Loudness Level," International Organization for Stand ardization, Geneva, 1975. Lee, H.H., Kim, S.J., and Lee, S.K., "Design of New Sound Metric and Its Application for Quantification of an Axle Gear Whine Sound by Utilizing Artificial Neural Network," *J. Mech. Sci. Techn.* 23: 1182-1193 2009.

164. Murata, H., Tanaka, H., Takada, H., and Ohsasa, Y., "Sound Quality Evaluation of Passenger Vehicle Interior Noise," SAE Technical Paper 931347, 1993, https://doi.org/10.4271/931347.

165. Norm, O., Scott, A., and Cris, E., "Guidelines for Jury Evaluations of Automotive Sounds," SAE Technical Paper 1999-01-1822, 1999, https://doi.org/10.4271/1999-01-1822.

166. Matuszewski, M. and Parizet, E., "Sound Quality inside Electric Vehicles," DIEMS, Università.

167. Odden, M., Heinrichs, R., and Linow, A., "Sound Evaluation of Interior Vehicle Using an Efficient Psychoacoustic Method," *Proceed Euro-Noise* 2: 631-642, 1998.

168. Wang, Y.S., "Sound Quality Estimation for Nonstationary Vehicle Noises Based on Discrete Wavelet Transform," *J. Sound Vib.* 324: 1124-1140, 2009.

169. Zwicker, E., Deuter, K., and Peisl, W., "Loudness Meters Based on ISO 532B with Large Dynamic Range," *Proc. Inter Noise* 85, no. 2: 1119-1122, 1985.

Case Study: NVH Study of HEV Powertrain

This chapter presents a case study of vibration analysis of a series-parallel hybrid powertrain system [1]. The powertrain system of a series-parallel hybrid vehicle contains multiple excitation sources such as engine, motor, and generator. To reduce the noise and vibration of powertrain is quite difficult in its multiple operational and switching modes. This chapter presents an example of noise, vibration, and harshness (NVH) of a series-parallel hybrid powertrain system that contains the engine, motor, and planetary gear subsystems. This case study considered a typical working condition that is based on the power control strategy and established the torsional vibration mechanical model of the hybrid powertrain system. The inherent characteristics and transient vibration response due to electric mode, hybrid mode, and parking charging mode were studied, and it was demonstrated that the repetitive frequency of the powertrain system under the three working modes is the same, which is only related to inertia and meshing stiffness of planetary gear system. The non-repetitive frequency and corresponding vibration modes associated with the electric mode and parking charging mode are close. The transient response of body acceleration has the same frequency component as that of the interference excitation torque. The body vibration caused by torque ripple increases substantially in the mode of starter driving the engine and in the mode of regenerative braking at low engine speed. The angular acceleration of the planetary gear system significantly increases due to the torque fluctuations caused by the switches of the working mode.

9.1 **VH Development/Metrics of HEV Powertrain**

The research of hybrid electric vehicles (HEVs) has become even significant in the development of automobile industry, and customers pay more attention to the NVH performance of HEVs powertrain system. Compared to conventional internal combustion (IC) engine vehicles, HEVs are equipped with new assemblies and components such as drive motors, starter motors, and planetary gear systems to achieve pure electric, hybrid, parking, regenerative braking, and other working conditions. Accordingly, new vibration problems occur as a result of changes in the powertrain system. First of all, in addition to the IC engine, HEVs add motor as an excitation source, and the torque fluctuation of the motor will affect the vibration response of the powertrain system. Secondly, the transmission path and inherent characteristics of the powertrain system under different working modes vary, which can lead to resonance problems. Finally, switches in working conditions and modes can lead to severe transient impact. The torsional vibration of the powertrain system and the longitudinal vibration of the car body are difficult problems to handle. The NVH refinement of the hybrid powertrain system has been critical in the comfort and stability of HEVs.

There have been many researches on the NVH performance of the powertrain system. However, there is very limited research on the torsional vibration characteristics of the series-parallel hybrid powertrain system [2-4]. The existing work mainly focuses on vibration response characteristics under fixed input and the modeling of the hybrid powertrain system is relatively simple. The study on the influence of transient characteristics of the powertrain system caused by different working modes and its switches has been not thorough. Tang [5] used the given cycle input to study the transient response of the hybrid system; Ma [6] used the fixed motor speed input to study the transient response of the hybrid system; Zhang [7] and Yu [8] used multibody dynamics software to study the transient response of the hybrid system. Tang [9] used the three-mass model to study the transient response of the hybrid system, the accuracy of the model is limited. The existing work mentioned above does not take the influence on transient characteristics of the powertrain system caused by different working modes and its switches into account. However, it is an important factor in the transient impact of the powertrain system, and the NVH performance of the powertrain system can be evaluated comprehensively under typical transient conditions. Therefore, it is of great theoretical and practical significance to study the vibration problem of the hybrid powertrain system under typical working conditions and modes.

Aiming at the series-parallel hybrid powertrain system that contains engine, motor, and planetary gear subsystems, the following study considered a typical working condition that is based on the power control strategy and established the torsional vibration mechanical model of the hybrid powertrain system. The inherent characteristics and transient vibration response of the electric mode, hybrid mode, and parking charging mode were studied, and the main factors influencing the torsional vibration behavior of hybrid powertrain system were illustrated, which can be used for the refinement of NVH of the series-parallel hybrid powertrain system.

9.2 Structure and Working Modes of Hybrid Powertrain System

Figure 9.1 shows the composition of a hybrid powertrain system. The engine, electric motor, and generator provide motive force to the differential part of reducer via a power split device and the planetary gear system. The motive power drives the wheel and the vehicle via the drive shaft. The engine is connected to the planet carrier. The generator is connected to the sun gear and the motor is connected to the ring gear. The motive force of the engine is separated to drive the wheels and to drive the generator. The generated electrical power is used directly to drive the motor or is converted by an inverter into direct current to be stored in a high-voltage battery. The hybrid powertrain system consists of five types of working conditions:

(1) *Slow starting and low-speed condition*: Under this condition, the system is working in pure electric mode. In this mode, the planet carrier is fixed. The engine is shut down and the generator idles. The vehicle is driven only by the motor.

(2) *Normal driving condition*: Under this condition, the system is working in hybrid mode. In this mode, the basic components of the planetary gear system are not fixed. Part of the power provided by the engine is directly used to drive the vehicle, and the other is used to generate electricity by a generator to drive the motor.

(3) *Rapid acceleration and climbing condition*: Under this condition, the system is also working in hybrid mode. In this mode, the basic components of the planetary gear system are not fixed. The vehicle needs more power under this

FIGURE 9.1   Construction of a hybrid powertrain system.

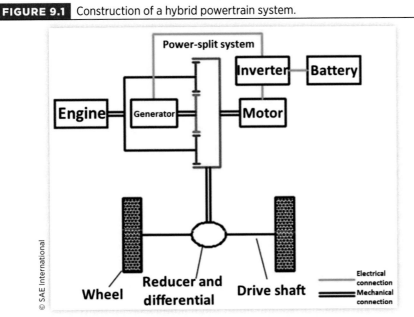

© SAE International

working condition, so the motor gets power from both the generator and the battery.

(4) *Deceleration and braking condition*: Under this condition, the system is working in regenerative braking mode. In this mode, the engine stops working. If the vehicle speed is high, the engine will idle, and basic components of the planetary gear system are not fixed; if the vehicle speed is low, the engine will shut down and the planet carrier will be fixed. After pressing the brake pedal, the electric motor is motored by the ring gear and it generates electricity as a generator, charging the battery and generating braking torque at the same time.

(5) *Parking charging condition*: Under this condition, the system is working in parking charging mode. In this mode, the ring gear is fixed. When the vehicle stops while driving and the power switch is not turned off, the engine will still work in a predetermined period of time before it shuts down, if the parameters such as water temperature, battery temperature, and electric load state are not within the specified range. When the vehicle stops and the power switch is turned off, the engine and the generator will start to generate electricity if the battery management system detects any items that do not meet the specified conditions.

According to the working conditions, it can be found that the main working modes of the hybrid powertrain system include pure electric mode, hybrid mode, regenerative braking mode, and parking charging mode. Based on transmission system structure, in pure electric mode, the engine shuts down and the carrier is fixed. In hybrid mode, basic components of the planetary gear system are not fixed, which means that planetary carrier, sun gear, and ring gear are all movable. In regenerative braking mode, there are two different cases in different working conditions: one is carrier fixed and the other is basic components not fixed. In parking charging mode, the vehicle is stationary. The ring gear is fixed. The reduction/differential assembly and tires are still. Therefore, there are only three different transmission structures: carrier fixed, basic components not fixed, and the ring gear fixed. These respectively correspond to three typical working modes: pure electric mode, hybrid mode, and parking charging mode.

9.3 System Modeling of HEV Powertrain

According to Figure 9.1, a hybrid powertrain system model is mainly composed of engine, motor, generator, planetary gear system, reducer/differential assembly, axle, tire, and body sub-models. The dynamics model structure is shown in **Figure 9.2**. It can be seen that the engine is connected to the planet carrier. The motor is connected to the ring gear and the generator is connected to the sun gear. The different transmission structures of the planetary gear system correspond to different working modes. The carrier fixed

FIGURE 9.2 Dynamics model of a hybrid powertrain system.

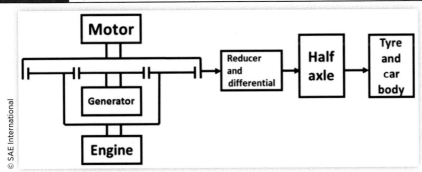

is in pure electric mode. The basic components of the planetary gear system not fixed are in hybrid powertrain mode. The ring gear fixed is in the parking charging mode. To elaborate the torsional vibration characteristics of the system, it is necessary to simplify the dynamic model with limited degrees of freedom. The system is simplified by the lumped parameter method.

In this study, a four-cylinder four-stroke engine is used. By taking into account the equivalent moment of inertia of the crank connecting rod mechanism, the moment of inertia of the flywheel and the torsional stiffness and damping between each other, the torsional dynamic model of the engine crankshaft system is shown in **Figure 9.3**. The kinetic equation is given in Equation (9.1).

In the system, J_{bi} is the equivalent moment of inertia of the crank connecting rod mechanisms. J_f is the moment of inertia of the flywheel. θ_{bi} is the torsional vibration angular displacement of crankshafts and θ_f is the torsional vibration angular displacement of the flywheel. k_i is the torsional stiffness of the shafts. c_i is the inner damping coefficient of the shafts. c_{0i} is the external damping coefficient acting on the inertia elements. T_i is the engine excitation torque acting on crankshafts and T_f is the engine excitation torque acting on the flywheel. $i = 1, 2, 3, 4$.

FIGURE 9.3 Dynamics model of crankshaft system.

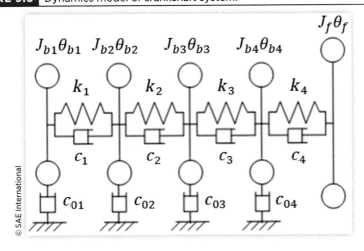

$$
\begin{cases}
J_{b1}\ddot{\theta}_{b1} + k_1(\theta_{b1} - \theta_{b2}) + c_1(\dot{\theta}_{b1} - \dot{\theta}_{b2}) + c_{01}\dot{\theta}_{b1} = T_1 \\[6pt]
J_{b2}\ddot{\theta}_{b2} + k_1(\theta_{b2} - \theta_{b1}) + k_2(\theta_{b2} - \theta_{b3}) + c_1(\dot{\theta}_{b2} - \dot{\theta}_{b1}) \\[6pt]
+ c_2(\dot{\theta}_{b2} - \dot{\theta}_{b3}) + c_{02}\dot{\theta}_{b2} = T_2 \\[6pt]
J_{b3}\ddot{\theta}_{b3} + k_2(\theta_{b3} - \theta_{b2}) + k_3(\theta_{b3} - \theta_{b4}) + c_2(\dot{\theta}_{b3} - \dot{\theta}_{b2}) \\[6pt]
+ c_3(\dot{\theta}_{b3} - \dot{\theta}_{b4}) + c_{03}\dot{\theta}_{b3} = T_3 \\[6pt]
J_{b4}\ddot{\theta}_{b4} + k_3(\theta_{b4} - \theta_{b3}) + k_4(\theta_{b4} - \theta_f) + c_3(\dot{\theta}_{b4} - \dot{\theta}_{b3}) \\[6pt]
+ c_4(\dot{\theta}_{b4} - \dot{\theta}_f) + c_{04}\dot{\theta}_{b4} = T_4 \\[6pt]
J_f\ddot{\theta}_f + k_4(\theta_f - \theta_{b4}) + c_4(\dot{\theta}_f - \dot{\theta}_{b4}) = T_f
\end{cases}
\tag{9.1}
$$

The engine excitation mainly includes the cylinder gas pressure-induced torque variation and the reciprocating moment of inertia from the crank connecting rod mechanism. In fact, the interference torque value caused by a reciprocating moment of inertia is small and cannot cause resonance. So, it is always ignored and only the interference torque caused by the cylinder gas pressure is considered. Therefore, the torque formula for each single cylinder is:

$$
M_g = \frac{\pi D^2}{4} P_g \frac{\sin(\alpha + \beta)}{\cos \beta} R
\tag{9.2}
$$

where

   P_g is the cylinder gas pressure acting on the unit area of the piston
   D is the cylinder diameter
   R is the crank radius
   α is the crank angle based on the top dead center (TDC)
   β is the swing angle of the connecting rod

As the motor rotor rotates, the electromagnetic frequency is high, so electromagnetic vibration is ignored, and only mechanical vibration is considered. The rotation inertia of the motor rotor is large, therefore, the rotor's torsional deformation can be ignored. The moment of inertia of the motor shaft is small and consider it concentrated on the rotor. But its torsional stiffness is equivalent to a torsion spring. As the material of the motor shaft is metal, its damping can be ignored. Therefore, the dynamics equation of the motor is given by:

$$
J_M\ddot{\theta}_M + k\left(\theta_{Mf} - \theta_{Mr}\right) = T_{M\_ref} + T_{rip}
\tag{9.3}
$$

where

   J_M is the moment of inertia of the motor rotor
   k is the equivalent torsional stiffness of the motor shaft
   $\theta_M, \theta_{Mf}, \theta_{Mr}$ are respectively the torsion displacements of the motor rotor, the front
       end of the motor shaft and the rear end of the motor shaft
   T_{M_ref} is the target torque of the motor
   T_{rip} is harmonic torque

The fundamental frequency of the motor current is $f = pn/60 = 2n/15$. The pole pairs of the motor are 8 and n is the motor speed. The fluctuation frequency of motor torque is six times the fundamental frequency. Therefore, the harmonic component of the motor torque is given by:

$$T_{rip} = T_{mag}\sin(12\pi ft + \varphi) = T_{mag}\sin(6\omega_{Me}t + \varphi) \tag{9.4}$$

where

φ is the initial phase of the harmonic torque

w_{Me} is the angular velocity of the motor

$w_{Me} = 2\pi pn/60$

T_{mag} is the fluctuation amplitude of the harmonic component

Set the fluctuation rate of the motor torque to 4% [1], then $T_{mag} = 0.02T_{M\_ref}$.

The main assumptions of the planetary gear system model include: first, planet gears are evenly distributed along the planet carrier and the gear parameters of them are the same. The gyro effect and the impact of the centripetal force are ignored. Second, the static transmission error of the gear meshing and the time variation of the meshing stiffness are ignored. The meshing stiffness is considered as constant. Third, the gear side clearance is not considered. According to these assumptions, the dynamics model of the planetary gear system is shown in **Figure 9.4**. C, r, and S are respectively the symbol of the carrier, the ring gear, and the sun gear. $k_j (j = C, r, S)$ is the torsional stiffness between the component j and the external connection. $\theta_j (j = C, r, S, 1, 2, ..., N)$ is the twist angle displacement of the planet carrier, the ring gear, the sun gear, and the planet gears. k_{rn} is the meshing stiffness between the planet gear n and the inner ring gear; and k_{sn} is the meshing stiffness between the planet gear n and the sun gear. C_{rn} is the meshing damping between the planet gear n and the ring gear; and C_{sn} is the meshing damping between the planet gear n and the sun gear. ψ_n is the inertia phase of the planet gear n.

FIGURE 9.4   Schematic of torsional vibration model of the planetary gear system.

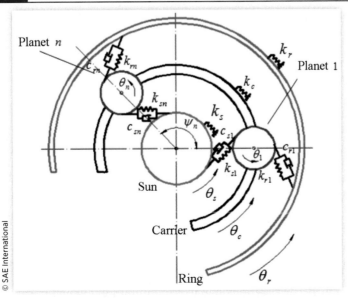

Therefore, the differential equation of torsional vibration for each component is given as follows:

Planet carrier:

$$\left(J_c + \sum_{n=1}^{N} J_n + N m_p r_c^2 \right) \ddot{\theta}_c + k_c \theta_c - r_c \sum_{n=1}^{N} \left(k_{rn} \delta_{rn} + c_{rn} \dot{\delta}_{rn} \right) \cos \alpha_r$$

$$- r_c \sum_{n=1}^{N} \left(k_{sn} \delta_{sn} + c_{sn} \dot{\delta}_{sn} \right) \cos \alpha_s = T_{p1} \tag{9.5}$$

Ring gear:

$$J_r \ddot{\theta}_r + k_r \theta_r + r_r \sum_{n=1}^{N} \left(k_{rn} \delta_{rn} + c_{rn} \dot{\delta}_{rn} \right) = T_{p2} \tag{9.6}$$

Sun gear:

$$J_s \ddot{\theta}_s + k_s \theta_s + r_s \sum_{n=1}^{N} \left(k_{sn} \delta_{sn} + c_{sn} \dot{\delta}_{sn} \right) = T_{p3} \tag{9.7}$$

Planet gear:

$$J_n \ddot{\theta}_n - r_n k_{rn} \delta_{rn} - r_n c_{rn} \dot{\delta}_{rn} + r_n k_{sn} \delta_{sn} + r_n c_{sn} \dot{\delta}_{sn} = T_{p4} \tag{9.8}$$

$$\begin{cases} \delta_{sn} = r_s \theta_s + r_n \theta_n - r_c \theta_c \cos \alpha_s \\ \delta_{rn} = r_r \theta_r - r_n \theta_n - r_c \theta_c \cos \alpha_r \end{cases} \tag{9.9}$$

where

δ_{rn} is the displacement between the planet gear n and the ring gear in the direction of the meshing line

δ_{sn} is the displacement between the planet gear n and the sun gear in the direction of the meshing line

Subscripts p refers to planet gear

N is the number of the planet gears

m_p is the mass of the planet gear

α_r is the meshing angle of internal gearing

α_s is the meshing angle of the external gearing

$J_i (j = C, r, S, 1, 2, ..., N)$ is the moment inertia of component i

r_c is the radius of the carrier

r_r is the radius of the base circle of the ring gear

r_s is the radius of the base circle of the sun gear

$r_j (j = C, r, S, 1, 2, ..., N)$ is the radius of the base circle of each planet gear

For a planetary gear system, when the system is in equilibrium, torque balance and power balance should be satisfied. It means that the algebraic sum of the torque of all the basic components is zero. And the algebraic sum of the transmitted power is also zero. Therefore, when the system is in equilibrium, 72% of the engine torque is transmitted directly to the driveshaft and 28% to the generator.

The connection stiffness and meshing stiffness between the gears in the reducer/differential part is very large and it can be approximated as rigid, regardless of the static transmission error and gear side clearance. According to the principle of kinetic energy equalization and transmission ratio [10], the inner ring gear, the reducer, and the differential can be equivalent to an equivalent moment of inertia J_r' and the corresponding twist angle is θ_r. The angle of the system is $\theta_r' = \theta_r / i$ after passing through the reducer/differential part.

The simplification of the axle is similar to that of the motor shaft. It is equivalent to two torsion springs, without considering of damping. The kinetic equation is given as follows:

$$T_{ax} = K_{ax}\left(\theta_{axf} - \theta_{axr}\right)$$ (9.10)

where

T_{ax} is the torque acting on the axle

K_{ax} is the equivalent torsional stiffness of the axle

θ_{axf} and θ_{axr} are respectively the torsional angular displacement of the front end of the axle and rear end of the axle

Based on the classic "brush" model, the tire mechanics model is established as shown in **Figure 9.5**. Based on the structure of the radial tire, the tire assembly is divided into two parts: one is the moment of inertia of the hub; the other is the moment of inertia of the rest of the wheel J_b. The torsional vibration model of the tire is established as given in Equation (9.11).

$$\begin{cases} J_a\ddot{\theta}_a + C_\theta\left(\dot{\theta}_a - \dot{\theta}_b\right) + K_\theta\left(\theta_a - \theta_b\right) = M_f \\ J_b\ddot{\theta}_b + C_\theta\left(\dot{\theta}_b - \dot{\theta}_a\right) + K_\theta\left(\theta_b - \theta_a\right) = -R_d F_{bx} \end{cases}$$ (9.11)

FIGURE 9.5   Tire model.

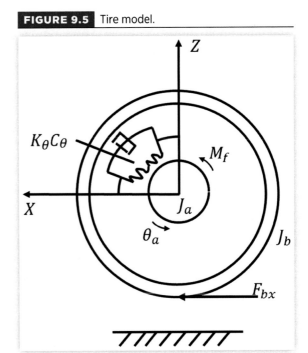

where

K_θ is the torsional stiffness of the tire

C_θ is the torsional damping of the tire

R_d is the rolling radius of the tire

M_f is the torque acting on the hub

F_{bx} is the tangential force of the ground to the tire, which is associated with the normal force of the vehicle by the ground adhesion coefficient and the slip rate [11]

θ_a is the angular displacement of the hub

θ_b is the angular displacement of the wheel ring

The key parameters in the model are given in **Table 9.1**.

The operation of the hybrid powertrain system includes three different working modes: pure electric mode, hybrid power mode, and parking charging mode. The constraints of the planetary gear system in different working modes are different. Therefore, based on the established torsional vibration model, the appropriate boundary conditions also need to be determined to analyze the transient torsional vibration

Reprinted with permission from "Vibration Analysis of Series-Parallel Hybrid Powertrain System Considering Typical Operating Conditions", The Journal of Tongji University (Natural Science Edition) Issue 9, 2018

TABLE 9.1 Subsystem parameters.

Subsystems	Parameter	Value	Parameter	Value
Engine	Moment of inertia of crank J_{bi}	0.004 kg·m²	Moment of inertia of flywheel J_f	0.18 kg·m²
	Mass of reciprocating motion part m_j	0.88 kg	Shaft stiffness between cranks k_i	4.5e+5 N/m
	Shaft stiffness between crank 4 and flywheel k_4	4e+6 N/m	Shaft stiffness between flywheel and planet carrier k_5	5e+5 N/m
Motor/generator	Moment of inertia of engine J_M	0.023 kg·m⁵	Moment of inertia of motor J_{M2}	0.023 kg·m⁵
	Shaft stiffness between generator and sun gear k_6	2e+5 N/m	Shaft stiffness between motor and ring gear k_7	8e+5 N/m
Planetary gear system	Moment of inertia of ring gear J_r	0.08 kg·m²	Moment of inertia of sun gear J_s	0.002 kg·m⁵
	Moment of inertia of planet gear J_i	0.0004 kg·m²	Mass of planet gear m_p	0.72 kg
	Moment of inertia of planet carrier J_c	0.04 kg·m²	Module of planet gear Set	3
Reducer/ differential	Equivalent moment of inertia J_r'	0.0852 kg·m⁵	Transmission ratio i	4.113
Axle	Torsional stiffness of axle k_8	6600 N·m/rad		
Tire/body	Moment of inertia of wheels J_{lw}, J_{rw}	0.9 kg·m²	Moment of inertia of hub J_a	0.5 kg·m²
	Moment of inertia of wheel ring J_b	0.4 kg·m²	Torsional stiffness of tire k	4750 N/m
	Torsional damping of tire C_θ	100 N·m·s/rad	Mass of whole vehicle M_b	1361 kg

under different working modes. The boundary conditions of the planetary gear system under different working modes are given as follows.

1. In pure electric mode, the planet carrier is fixed and the engine shuts down. Under this condition, the vibration dynamics model of the transmission system satisfies the following constraints:

$$\begin{cases} \theta_{b1} = \theta_{b2} = \theta_{b3} = \theta_{b4} = \theta_f = 0 \\ \theta_c = 0 \end{cases}$$

(9.12)

2. In hybrid power mode, basic components of the planetary gear system are not fixed. Under this condition, there are no constraints in the vibration dynamics model of the transmission system.

3. In parking charging mode, the vehicle is stationary. Under this condition, the ring gear is fixed and the motor shuts down. The vibration dynamics model of the transmission system satisfies the following constraints:

$$\begin{cases} \theta_r = 0 \\ \theta_{M2} = 0 \end{cases}$$

(9.13)

in which θ_{M2} is the torsion displacements of the motor rotor.

To understand the transient torsional vibration characteristics of the hybrid powertrain system under different conditions, the typical working conditions and power control strategies in various working modes are presented as follows [12].

The typical analysis of a hybrid powertrain system must include the three power sources, the changes in their working conditions and mode switching. Different statuses

FIGURE 9.6   Typical working conditions.

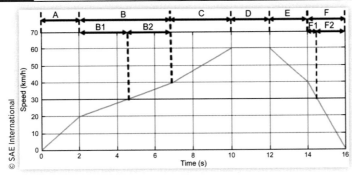

are set, including acceleration, uniform speed, and deceleration processes. By introducing the engine starting and stopping transient conditions, mode switching is achieved. The typical working conditions proposed in [1] are shown in **Figure 9.6**. Segment A, B, and C are the uniform acceleration process with different accelerations. Segment D is a uniform speed process. Segment E and F are uniform deceleration processes with different decelerations. The uniform speed is set to be 60 km/h. The working condition of segment A and B1 is a pure electric mode. After starting the engine, as shown in segments B2, C and D, the system switches into hybrid mode. Segment E and F are regenerative braking mode and segment F2 represents the engine stopping condition.

For the analysis of the power control strategy, forward simulation is used and the flow chart is shown in **Figure 9.7**. To simplify the analysis, the influence of the battery state of charge (SOC) is not considered in the simulation. The dynamic control strategy model includes the driver model and the controller model.

The driver model is actually a speed controller, as is shown in **Figure 9.8**. Based on the difference between input expected speed and actual speed that can be obtained from the vehicle longitudinal dynamic model, PID control is used to get required torque.

FIGURE 9.7   Forward simulation flow chart.

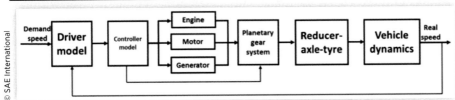

FIGURE 9.8   Driver model.

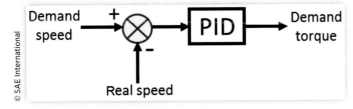

The Controller Model is briefed as follows,

(1) When $0 \leq \dot{v} \leq a$ and $v \leq 30$ (a is the maximum acceleration that can be reached in pure electric mode. A speed of 30 km/h is corresponding to the engine ignition time in B2 stage). The system works in pure electric mode. The engine shuts down. The generator idles. The system is driven only by the motor.

(2) When rapid acceleration is required, the required torque will be larger than the maximum torque that the motor alone can provide. It means that when the expected acceleration $\dot{v} > a$ or the vehicle speed reaches 30 km/h ($\dot{v} \geq a$ or $v \geq 30$), the engine will start and work on the optimum fuel consumption curve. The generator also starts and generates electricity. The system works in hybrid mode.

(3) When the vehicle decelerates ($\dot{v} < 0$), the system is in regenerative braking mode. Both the motor and the generator stop working and idle. When the vehicle speed goes down to slower than 30 km/h, the engine will shut down. If the brake deceleration is small, the motor torque that is used to decelerate is determined by the required braking force. If the brake deceleration is large, the motor will work on the maximum output power curve and the braking force will be also supplied by the brake.

Under the acceleration condition, the calculation formula of the torque that the motor needs to provide is as follows.

$$T_{demand} = \left(0.72 T_{engine} + T_{motor}\right) \times i \qquad (9.14)$$

T_{demand} is the required torque to the drive wheel, which can be obtained from the driver model. $0.72 T_{engine}$ is the torque to the ring gear that is provided by the engine. The value of this torque can be obtained from tables. i is the gear ratio of the reducer/differential assembly. Changes in torques while switching working modes are applied as smooth transitioning ramp functions.

Under deceleration condition, the calculation method of the single front wheel braking torque T_{bf} is as follows.

$$T_{demand} = i \times T_{motor} + 2\left(T_{bf} + T_{br}\right) = i \times T_{motor} + \frac{2}{\beta_0} T_{bf} \qquad (9.15)$$

T_{br} is the braking torque of the single rear wheel. β_0 is the braking force distribution coefficient. When the deceleration request is small, the friction braking torque is ignored. The motor generates electricity according to the demand condition. T_{motor} is calculated according to the acceleration condition calculation method, as is shown in the formula (9.15). When the deceleration request is large, the motor works on the maximum output power curve. At this time, T_{motor} can be obtained from tables. The parameters in the power control strategy are shown in **Table 9.2**.

9.4 **NVH Study of HEV Powertrain**

When the engine and motor excitation source input are set to be zero, the inherent characteristics of the system can be obtained. The transmission path and vibration model

TABLE 9.2 Parameters in the control model.

Sub-models	Parameter	Value
PID model	Proportional cycle K_P	5300
	Integral cycle K_I	20
	Derivative cycle K_D	80
Start and stop initial parameters	Start torque of generator	25 Nm
	Generation torque of generator	−20 Nm
	Locking torque of planet carrier	−200 Nm
Braking parameter	Force distribution coefficient of brake β_0	0.6

© SAE International

of the system are different because the system works in different modes including pure electric mode, hybrid mode, and parking charging mode. The natural frequencies of the system in different modes are shown in **Table 9.3**.

The following observations can be seen from Table 9.3:

1. In pure electric mode, the system has 11 degrees of freedom. There are 8 non-repetitive frequencies and 3 repetitive frequencies. The non-repetitive frequencies correspond to the overall torsional vibration mode of the system, which contains first-order rigid body mode. The 296.7 Hz vibration mode is basically only the planet gear in vibration, but it is different from the vibration mode of the planetary gear system that corresponds to the repetitive frequencies. The vibration modes of 24.0, 1058.8, and 1673.9 Hz are mainly corresponding to the left and right wheels, motor rotor, and sun gear in vibration. The body amplitude is almost zero in the overall torsional vibration mode of the system. The tires only at low frequencies of the 5.20, 24.0, and 24.7 Hz will be in obvious vibration. The vibration modes of 24.7 and 75.9 Hz are more complex. The repetitive frequency is 287.8 Hz, corresponding to the planet gear vibration mode. In this mode, only the planet gear is in vibration.

2. In hybrid mode, the system has 17 degrees of freedom. There are 14 non-repetitive frequencies and 3 repetitive frequencies. The non-repetitive frequencies correspond to overall torsional vibration mode of the system, which contains second-order rigid body mode. The vibration modes corresponding to the frequencies of 24.0, 296.7, 1058.8, and 1673.9 Hz are the same as those in pure electric mode. The vibration mode corresponding to the frequencies of 756.5, 2066.6, 3020.4, and 5388.4 Hz are mainly due to the engine vibrations. The vibration modes corresponding to the frequencies of 25.2, 90.3, and 530.6 Hz are complex and unidentified. The vibration of the body is very small. The wheels only have vibration (The vibration of third to fifth order) of low

TABLE 9.3 Natural frequencies under different modes.

Mode	Natural frequencies (/Hz)													
Electric	—	0	5.2	24.0	24.7	75.9	287.8	296.7	—	—	1058.8	1673.9	—	—	—
Hybrid	0	0	6.3	24.0	25.2	90.3	287.8	296.7	530.6	756.5	1058.8	1673.9	2066.6	3020.4	5388.4
Charging	—	0	—	—	—	68.8	287.8	289.9	530.6	756.5	—	1673.9	2066.6	3020.4	5388.4

© SAE International

frequency. The repetitive frequency is 287.8 Hz, corresponding to the planet gear vibration mode. In this mode, only the planet gear has vibration.

3. In parking charging mode, the system has 12 degrees of freedom. There are 9 non-repetitive frequencies and 3 repetitive frequencies. The non-repetitive frequencies correspond to overall torsional vibration mode of the system, which contains first-order rigid body mode. Most of the natural frequencies and their corresponding vibration modes are the same as those in hybrid mode, except for the frequency of 68.8 and 289.9 Hz. The natural frequency of 1058.8 Hz disappears in parking charging mode because the motor does not work. The repetitive frequency is 287.8 Hz, corresponding to the planet gear vibration mode. In this mode, only the planet gear has vibration.

4. Comparing the inherent characteristics of different working modes, it can be found that the 4 planet gears have the same vibration state in the vibration model corresponding to the non-repetitive frequencies. Under this situation, the system is in overall torsional vibration mode. Only the planet gears have vibration of the repetitive frequency. The other components are not vibrating. Under this situation, the system is in the planet gear vibration mode. According to the analysis of the inherent characteristics in different modes, the system has planet gear vibration in all working modes. The repetitive frequency is the same in different modes. This means that the natural frequency of 287.8 Hz is independent of the state of the planet carrier and the ring gear.

Assuming that the vibration mode vector corresponding to the planet gear vibration mode is as follows.

$$A_i = \begin{bmatrix} 0 & \cdots & 0 & a_1 & a_2 & a_3 & a_4 & 0 & \cdots & 0 \end{bmatrix} (i = 1,\, 2,\, 3) \tag{9.16}$$

Substituting the vector into the main mode equation in any working mode [13]:

$$\left(K - \omega_i^2 M \right) A_i = 0 \tag{9.17}$$

Then getting the equations:

$$\left(k_{rn} + k_{sn} - \omega_i^2 \frac{J_n}{r_n^2} \right) a_n = 0 \tag{9.18}$$

$$\omega_i = \sqrt{\frac{(k_{rn} + k_{sn}) r_n^2}{J_n}} \, (n = 1,\, 2,\, 3,\, 4) \tag{9.19}$$

It can be seen that the natural frequency (287.8 Hz as discussed above) depends on the moment of inertia of the planet gear J_n, the radius of the base circle r_n, the meshing stiffness between the planet gear and the sun wheel k_{sn} and the meshing stiffness between the planet gear and the inner ring gear k_{rn}.

Under the typical working conditions of the hybrid system, the errors between actual speed and expected speed are shown in **Figure 9.9**. It can be found that the errors are small. It means that the model can match the expected speed well and the system is

FIGURE 9.9   Simulation results of vehicle speed.

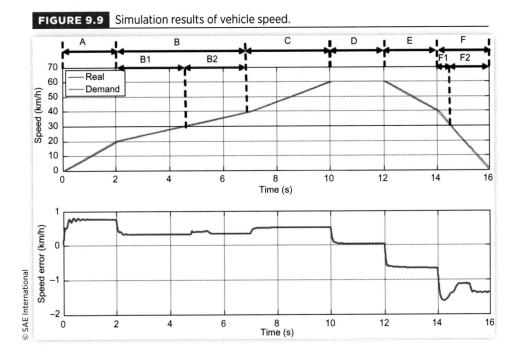

responsive. The simulation results of the engine, the electric motor, and the generator are consistent with the expected value of the dynamics control strategy, which proves that the control model is effective.

The speed and torque curves of engine, generator, and electric motor are shown in **Figures 9.10** and **9.11**. From 0 to 2 s, the system works in stage A of pure electric mode. When the electric motor is started, there is an obvious fluctuation of its rotate speeds and torque. It mainly results from the harmonic torque, which is formed by the specific

FIGURE 9.10   Speed and torque curve of motor/generator.

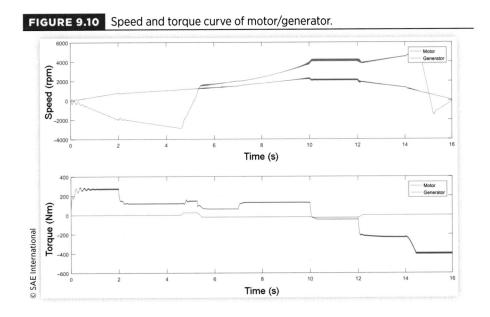

FIGURE 9.11　Speed and torque curve of the engine.

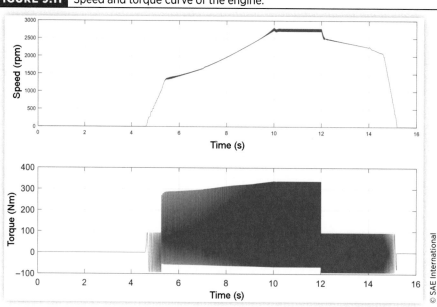

© SAE International

characteristics of the motor's structure. The speed of the generator is opposite to that of the electric motor. It also has obvious fluctuation and the torque is zero. It results from the relative motion characteristics of the components in the planetary gear system. The generator is affected by the torque ripple of the electric motor and idles. In this stage, both the speed and torque of the engine are zero. Because the planet carrier is fixed, and the engine does not work in pure electric mode. At 2 s, the vehicle's acceleration speed suddenly decreases, and the torque of the electric motor becomes smaller and fluctuates in a small range. It is because that the power demand of the system decreases and causes a transient impact.

From 2 to 4.5 s, the speed of the vehicle keeps growing up steadily and the system works in stage B1 of pure electric mode. The rotate speed of the generator keeps growing up steadily while the torque remains unchanged. The generator and the engine still do not work, and the rotate speed of the generator still keeps increasing reversely. At 4.5 s, the vehicle speed reaches 30 km/h and the system works in stage B2 of hybrid mode when the basic components of the planetary gear system are all unfixed. From 4.5 to 5.5 s, the rotate speed and torque of the electric motor rise slightly and fluctuate obviously. Driven by the generator's constant torque, the rotate speed and torque of the engine grow up gradually. At 5.5 s, the speed of the engine reaches 1000 rpm and the engine starts to work, while its rotate speed and torque fluctuate obviously. Meanwhile, as the engine starts to work, the electric motor's torque goes down and the rotate speed fluctuates obviously. The rotate speed of the generator goes up and the generator begins to generate electricity through constant torque.

From 5.5 to 7 s, the system works in stage B2 of hybrid mode. The rotate speed and torque of the engine go up rapidly. After that, both rotate speed and torque reach specific values corresponding to the best fuel consumption curve. The rotate speed of the generator increases continuously and fluctuates obviously, which is mainly resulted from the engine torque's fluctuation, At 7 s, the acceleration of the vehicle goes up suddenly and the system works in stage C of hybrid mode. Because of the system's increased torque demand, the torque of the electric motor increases suddenly, while the rotate speed and torque of the engine go up gradually. The rotate speed of the engine finally keeps at 2600 rpm and the torque keeps at 320 Nm. The state of the generator remains unchanged.

From 10 to 12 s, the acceleration of the vehicle goes down to zero and keeps working at a maximum speed of 60 km/h. The system works in hybrid mode. Because of the torque demand of the system, the torque of the electric motor decreases obviously. The rotate speeds of the electric motor, the generator and the engine all fluctuate obviously. From 12 to 14.5 s, the system works in stage E, F1 of regenerative braking mode. With the ring gear fixed, the vehicle brakes at a constant deceleration. The deceleration is high in stage F1 and the vehicle speed decreases gradually. Rotate speeds and torques of the engine and the generator go down obviously. The generator stops working at 12 s. After 14.5 s, the engine stops working. Rotate speeds and torques of the generator and the electric motor go down gradually and the rotate speed of the electric motor fluctuates obviously.

To analyze torque fluctuation frequency components of the electric motor and the generator, short-time Fourier transform (STFT) is used to conduct time-frequency analysis. The result is shown in **Figure 9.12**. It can be found that torque fluctuation frequencies of the electric motor and the generator are about six times the size of the current fundamental frequency. However, there is obvious torque fluctuation of the generator in low frequency, which is associated with the nonlinearity of the inverter and magnetic distortion of the motor.

Figure 9.13 is the time-frequency spectrum of engine torque. It can be found that there is an obvious order characteristic in the engine's torque excitation frequency. The second and fourth orders are the main ones.

FIGURE 9.12 Three-dimensional torque spectrum of motor and generator.

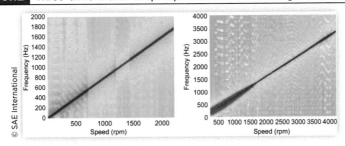

© SAE International

FIGURE 9.13 Engine torque time-frequency spectrum.

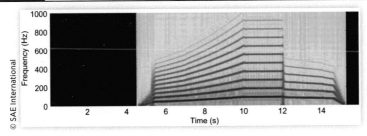

© SAE International

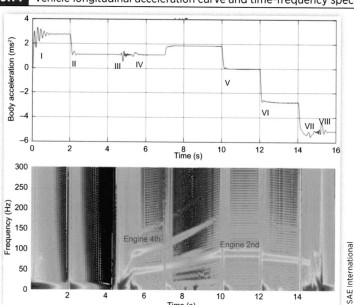

FIGURE 9.14 Vehicle longitudinal acceleration curve and time-frequency spectrum.

© SAE International

Figure 9.14 shows the longitudinal acceleration curve and time-frequency spectrum of the vehicle and marks the time when bigger longitudinal vibration happens. As the system works in pure electric mode and the electric motor starts working, the longitudinal vibration is mainly caused by the system's rotate speed and torque fluctuation. At 2 s, the system works in stage B1 in which acceleration is smaller and the torque of the electric motor decreases suddenly. Then, it contributes to the longitudinal vibration at II. Because the engine starts working, the longitudinal vibration at III and IV happens. According to the time-frequency spectrum, it can be found that frequency components mainly consist of second-order and fourth-order frequencies. It strongly suggests that the simple harmonic interference torque of the engine mainly causes the vehicle longitudinal vibration. Because the acceleration in stage C is bigger, the acceleration of the vehicle increases. Then the vehicle works in stage D, E, constant speed, and braking conditions. The acceleration goes down suddenly, so the acceleration of the vehicle decreases, which causes the vibration at V and VI. There is obvious fluctuation at VII and VIII when the engine shuts down and the locking torque of the planet carrier changes suddenly.

To sum up, there are obvious fluctuations of the vehicle longitudinal acceleration at the moments when the vehicle starts as well as the engine starts and stops. The fluctuations decrease slowly, as shown in stage I, III, and VIII in Figure 9.14. These are mainly caused by the transient impact, which is resulted from sudden changes of the electric motor's torque, the engine's torque, and the planet carrier's locking torque.

Figure 9.15 shows the angular acceleration curve of the planet carrier, the planet gear, the sun gear, and the ring gear. From the curve, it can be found that from 0 to 2 s, the system works in stage A of pure electric mode. Except for the locked planet carrier and the locked engine, acceleration fluctuations of other components are big and decrease slowly. It is mainly caused by transient impact, which results from sudden changes in

FIGURE 9.15   Angular acceleration curve of the planetary gear system.

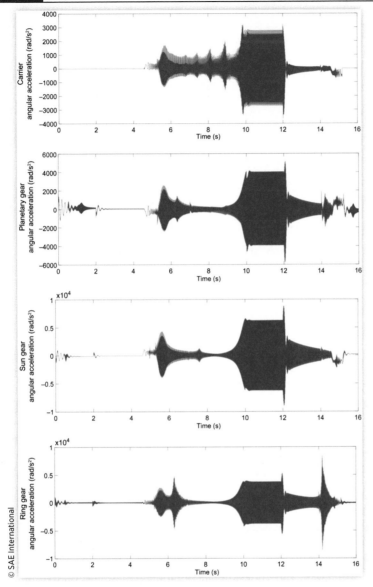

the electric motor's torque. At 5.5 s, the engine starts, and the system works in stage B2 of hybrid mode. Each component of the planetary gear system is not fixed, and the fluctuation of each component's angular acceleration increases obviously. From 6 to 10 s, the system works in acceleration conditions C, D of hybrid mode and each component is not fixed. Different from other components, there are many obvious fluctuations in the angular acceleration response of the planet carrier. When the system works in stage E of regenerative braking mode at 12 s, both the engine and the generator stop working, remain idle and cause impact, which contributes to the obvious transient impact in each component. At 15 s, the engine shuts down and it causes that the fluctuation of each component's angular acceleration increases obviously and decreases slowly.

FIGURE 9.16 Angular acceleration curve and time-frequency spectrum of the planet gear.

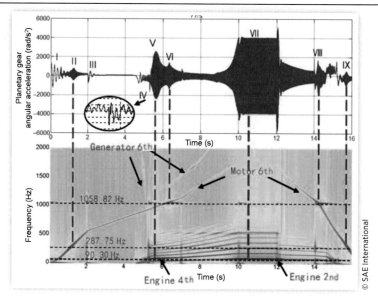

© SAE International

Comparing the angular acceleration curves of the planet carrier, the planet gear, the sun gear, and the ring gear, it can be found that the curve of the planet gear is the most typical one. Therefore, we take the planet gear as an example and conduct a time-frequency analysis of its angular acceleration response.

Figure 9.16 shows the angular acceleration curve and time-frequency spectrum of planet gear. From 0 to 2 s, the angular acceleration fluctuations at I and II are mainly electric motor torque's sixth-order fluctuation frequency. The reason is that the electric motor torque's sixth-order fluctuation frequency is almost the same as the natural frequency of 287.8 Hz in pure electric mode, and the planet gear resonates. At 6 s, the angular acceleration at V fluctuates greatly. The main reason is that the engine is at fourth-order harmonic frequency, which is almost the same as the natural frequency of 90.3 Hz in hybrid mode. Then the planet gear resonates. The main reason for the angular acceleration's great fluctuation at III is electric motor torque's sixth-order fluctuation frequency is almost the same as the natural frequency 1058.8 Hz in hybrid mode. Then the planet gear resonates. From 10 to 12 s, the system reaches a stable level of maximum speed and the angular acceleration at VII fluctuates greatly. The engine is mainly at fourth-order harmonic frequency, which is almost the same as the natural frequency of 90.3 Hz in hybrid mode. Then the planet gear resonates. From 14 to 16 s, the system works in regenerative braking mode. The engine and the generator stop working. The angular acceleration fluctuations at VIII and IX are mainly caused by the fluctuation of motor's torque. According to the angular acceleration curve and time-frequency diagram of the planet gear, it can be found that corresponding frequency components are existing between the angular acceleration response of the planet gear and the torque incentives of the engine, the electric motor, and the generator. Moreover, the torsional vibration of the planet gear mainly demonstrates in the system resonance, which is resulted from the power device torque fluctuation.

Based on the analysis of planetary gear system torsional vibration, it can be found that the system vibration is caused by the fluctuation excitation source due to the motor torque's fluctuation and the engine's starting. The active control method should be used to improve excitation source fluctuation and impact and to prevent system vibration. When the excitation source frequency is close to system natural frequency, the system will resonate. The torsional shock absorber at outputs of the electric motor and the engine should be used to improve planet gear's torsional vibration characteristics.

This study is mainly focused on a typical working condition and different working modes in a series-parallel hybrid powertrain system. It investigated the torsional vibration characteristics of a series-parallel hybrid powertrain system. The following conclusions can be drawn.

First, aiming at a series-parallel hybrid powertrain system that contains engine, motor, and planetary gear subsystems, this chapter considered a typical working condition that is based on the power control strategy and established the torsional vibration mechanical model of the hybrid powertrain system.

Second, the system has torsional vibration mode and planetary gear vibration mode under the electric mode, hybrid mode, and parking charging mode. The vibration state of the planetary gear remains the same in the torsional vibration mode. The natural frequency under the planetary gear vibration mode are the same and only related to the inertia and base circle radius of the planetary gear, meshing stiffness between the planetary gear, sun gear, and the ring gear. The nonrepetitive frequency under the hybrid mode and parking charging mode are close.

Third, the transient response under the typical working condition mainly includes torque and speed fluctuations of motor, generator, and engine, longitudinal vibration of vehicle body under low-frequency range and the vibration of the planetary gear system. The impact and torque fluctuations of the vehicle body are obvious during the engine ignition process, which can easily lead to longitudinal vibration under low-frequency range. The torsional vibration response of planetary gear has a similar frequency component to the excitation source of engine, motor, and generator torque, which in turn leads to the resonance of the system.

References

1. Zhang, L., Chen, W., Meng, D., Gu, P. et al., "Vibration Analysis of Series-parallel Hybrid Powertrain System under Typical Working Condition and Modes," SAE Technical Paper 2018-01-1291, 2018, doi:https://doi.org/10.4271/2018-01-1291.

2. Liang, R., "Research on Torsional Vibration of Electric Vehicle's Powertrain System," Ph.D. thesis, Tongji University, Shanghai, 2008.

3. Xiong, J., "Analysis and Control of Noise and Vibration of Hybrid Electric Vehicles," *Noise and Vibration Control* 29, no. 05: 96-100, 2009.

4. Zhao, T. and Lu, B., "NVH Control Technology of Hybrid Electric Vehicle," *Journal of Jilin University* 42, no. 06:1373-1377, 2012.

5. Tang, X. and Zhang, J., "Study on the Torsional Vibration of a Hybrid Electric Vehicle Powertrain with Compound Planetary Power-Split Electronic Continuous Variable

Transmission," *Proceedings of the Institution of Mechanical Engineers Part C: Journal of Mechanical Engineering Science* 228, no. 17: 203-210, 2014.

6. Ma, X. and Luo, K., "Modeling and Analysis of Torsional Vibration on Engine-generator System of Hybrid Electric Vehicle," *Proceedings of SAE-China Congress 2014: Selected Papers* (Berlin: Springer, 2015), 59-70.

7. Zhang, D., Analysis and Control of Torsional Vibration in HEV," *Drive System Technique* 28, no. 4: 3-8, 2014.

8. Yu, H. and Zhang, T., "Torsional Vibration Analysis of Planetary Hybrid Electric Vehicle Driveline," *Transactions of the Chinese Society of Agricultural Engineering* 29, no. 15: 57-64, 2013.

9. Tang, X. and Yang, W., "A Novel Simplified Model for Torsional Vibration Analysis of a Series-Parallel Hybrid Electric Vehicle," *Mechanical Systems & Signal Processing* 85: 329-338, 2017.

10. Zhang, B., *Internal Combustion Engine Dynamics* (Beijing; National Defense Industry Press, 2009), ISBN:9787118063219.

11. Yu, Z., *Internal Combustion Engine Dynamics* (Beijing, China Machine Press, 2009), ISBN:978-7-111-02076-9.

12. Wang, L., "Study on Planetary Gear Unit for Parallel Hybrid Electric Vehicle," Ph.D. thesis, Xi'an University of Technology, Shaanxi, 2008.

13. Wang, S. and Zhang, C., "Natural Mode Analysis of Planetary Gear Trains," *China Mechanical Engineering* 16, no. 16: 1461-1465, 2005.

14. Jin, X. and Zhang, L., *Analysis of Vehicle Vibration* (Shanghai: Tongji University Press, 2002), ISBN:978-7-5608-2406-2.